Artificial Mathematical Intelligence

Danny A. J. Gómez Ramírez

Artificial Mathematical Intelligence

Cognitive, (Meta)mathematical, Physical and Philosophical Foundations

 Springer

Danny A. J. Gómez Ramírez
Research's Labs Center Parque i
Instituto Tecnológico Metropolitano (ITM)
Medellín, Antioquia, Colombia

ISBN 978-3-030-50275-1   ISBN 978-3-030-50273-7   (eBook)
https://doi.org/10.1007/978-3-030-50273-7

Mathematics Subject Classification: 03-02; 08-02; 13-02; 14-02; 18-02; 68-02

This Springer imprint is published by the registered company Springer Nature Switzerland AG.
The registered company address is: Gewerbestrasse 11, 6330 Cham, Switzerland

To Maria Batjacob, Luz Stella Ramírez Correa, José Omar Gómez Torres, and Jesús Benjosé

Preface

Imagine an outstanding researcher who has a great ability to mathematically model particular phenomena in their specific field of research. However, at the end of this modeling process, the researcher often has to solve purely formal (mathematical) problems which require a huge amount of time and effort. Sometimes, this part of the research takes even more time and creative energy than the work required to produce the mathematical data. Then, one day, the researcher discovers there is a technological device that can assist them in the second (and often exhausting) part of the research. In other words, this interactive artificial agent is effectively able to assist the researcher during the conceptual process of generating either human-style proofs or counterexamples of the corresponding (mathematical) conjectures in an understandable way. The researcher is astonished because in at least 80% of previous cases, such a device offers the right conceptual answers. Thus, after months of active usage of this co-creative device, the researcher is quite satisfied because (s)he has been able to mathematically model more instances in their particular field of study. Further, (s)he finds that their general insight about their subject of inquiry has become more universal and accurate than ever.

This story immediately leads to questions such as: How close are we to developing such a technological co-creative device? What kind of scientific discipline(s) should support the development of such an artifact? How should our fundamental understanding regarding mathematical entities be refined to enable the construction of such a technology? And finally, how far are we from understanding, at a global level, the way in which our minds solve mathematical inquiries and create mathematical theories at any level of abstraction?

This book aims to provide initial answers to these questions from a global perspective using an inter- and multidisciplinary approach. An alternative name for it which also encompasses its global structure is *Artificial Mathematical Intelligence: On the Multidisciplinary Foundations of Cognitive Metamathematics*. Based on a multidisciplinary spectrum of modern and classic results belonging to cognitive science, (classic) metamathematics, logic, mathematics, (cognitive) linguistics, (cognitive) psychology, physics, philosophy (of mind), and computer science fields, I present the (initial) foundational cornerstones of a new integrative research program aiming to resurrect a modern version of one of the original central "dreams" of artificial intelligence: the creation of a global artificial mathematical agent that is capable of simulating human-style intelligence regarding the mathematical conceptual generation and problem-solving at arbitrary levels of complexity.

Due to the wide scope of this vision, the main focus of this book is on the theoretical foundations. So, the algorithmic pillars of this program will be developed (as indicated throughout the book) after having carried out a meta-analysis (along the lines of the methods described here) of dozens (and dozens and dozens) of mathematical structures. Before this, it would be methodologically "too early" to grasp the computational substratum of artificial mathematical intelligence.

In the first part of the book, I establish a grounding program called New Cognitive Foundations for Mathematics, where I redescribe some of the most foundational notions of mathematics with the additional and enhancing perspective of cognitive sciences (and related disciplines). In particular, I study the cognitive reality of mathematics and the mathematics of cognitive reality, and the ontological scopes and limitations of mathematics in nature. Also, the degree of coherence between continuous models of the universe and "singular" extrapolations of artificial intelligence, the development of a refinement of the notion of natural number given through "the physical numbers," and the cognitive "singularities" that the classic set-theoretical (foundational) semantic frameworks possess along with the most important (future) challenges of this program.

In the second part of the book, I provide an explicit global taxonomy of the most fundamental cognitive metamathematical mechanisms used by the (human) mind during the intellectual task of mathematical research. I introduce and formalize the cognitive abilities of conceptual blending, analogical reasoning, and conceptual substratum in three different chapters. For the first mechanism, an initial conceptual generation (and implementation) of seminal concepts in Fields (and Galois) theory is presented based primarily on this seminal ability. Finally, I present and formalize a global list of more than 20 further cognitive mechanisms based on a specific formalization of the notion of mathematical concept, and, more generally, of mathematical structure.

In the last part of the book, I describe the cognitive and conceptual power of the previous global taxonomy by presenting cognitive meta-generations of two classic mathematical proofs in number theory and Euclidean geometry as well as dozens of fundamental mathematical concepts in several mathematical sub-disciplines (e.g., set theory, topology, abstract algebra, category theory, sheaf theory, commutative algebra, and algebraic geometry). The degree of conceptual complexity of such notions varies significantly, beginning with quite elementary notions like mathematical function and advancing to more sophisticated ones like the concept of (mathematical) scheme in modern algebraic geometry.[1] Finally,

---

[1] These cognitive meta-generations can be seen as instances of conceptual constructions of an ideal (non-necessarily embodied) mathematical artificial agent.

I describe the most fundamental future challenges of this new interdisciplinary program as well as plausible extensions of it for similar scientific disciplines, such as physics, chemistry, biology, and (mathematical) finances.

In the introductory chapter, I provide the initial global motivation and inspiration of this cognitive metamathematical program, together with a more detailed description (chapter by chapter) of the whole book.[2]

Due to the wide methodological spectrum of this book, I describe its audience in terms of a similarly broad spectrum of students and professionals. In particular, this work is suitable and desired to be read by cognitive scientists, computer scientists, mathematicians, philosophers, physicists, AI researchers, as well as researchers in related areas. In fact, the second chapter is particularly devoted to the non-specialist (mathematician or/and logician) and describes in a very compact way the minimal (meta)mathematical and logical background that is needed in order to understand better the whole AMI program, and more than 80% of the meta-generations are presented in Chap. 11.[3] In fact, the reader can use this chapter as needed and read it simultaneously when (s)he required some notions to understand a special section of the book. Each chapter—with the exception of Chaps. 6 and 11—is to a large extent self-contained. Chapter 6 requires minimal knowledge of the von Neumann–Bernays–Gödel (NBG) class theory. For a better understanding of Chap. 11, basic knowledge in modern mathematics is desired but not needed, since the interested reader can learn (simultaneously) all the (cognitively meta-generated) notions from classic references, which will be enough to obtain an initial understanding of the corresponding cognitive meta-generations.[4] For the entire second part of the book, minimal knowledge in mathematical logic is highly desired. With exception of Chap. 12, each chapter can be read in an independent way. However, the content of Chap. 10 can be better understood if one already knows the content of Chaps. 7, 8, and 9. In order to maintain each chapter as self-contained as possible, I (re-)describe quite a few (seminal) facts in different forms in at most two chapters, for instance. Notwithstanding the former thematic clarifications, I will always recommend to read/study the whole book for obtaining a (more) solid and genuinely global understanding of what artificial mathematical intelligence really means.

---

[2]So, the reader can consider the introductory chapter as a suitable complement to this preface.

[3]I will recommend to the non-specialist (mathematician/logician) reader to go gradually and methodically with the most technical parts of the book. After some time, (s)he will benefit enormously from the holistic and global methodological view offered in the AMI multidisciplinary program.

[4]Along the book, I use the name "cognitive meta-generation" (or sometimes cognitive generation) for denoting the way in which mathematical structures are cognitively produced in terms of the global taxonomy of cognitive mechanisms.

I use the first-person plural pronoun "we" during the whole book mainly for implicitly acknowledging all the researchers who have inspired me in this unique quest and whom I eventually cite throughout each of the chapters. Nonetheless, the results and ideas described in this work are of my own sole authorship, and when co-authors were involved I explicitly reference/acknowledge their contribution.

Medellín, Colombia   Danny A. J. Gómez-Ramírez
October 2020

Acknowledgements

I thank my parents José Omar Gómez Torres and Luz Stella Ramírez Correa for their incomparable love and support even before my conception. They gave me a part of their essence, blood, and flesh to be able to exist here, thanks for all even beyond the words.

I would like to express my sincere acknowledgments to each person who inspired me in some way or another. Especially, I sincerely thank my friends Edisson Gallego Gonzalez, Rafael Betancur, Juan Pablo Hernandez, Yoe Herrera, Jose Manuel Jimenez, Ismael Rivera, Edisson Mauricio Rivera, Sergio Daladier Molina, Jose Gregorio Rodriguez; and to Ahmed M. H. Abdel-Fattah, Christoph Benzmüller, Bruno Buchberger, Razvan Diaconescu, Marlon Fulla, Juan Manuel Palacio, Juan Carlos Gómez, Hernan Giraldo, Ulf Krumnack, Kai-Uwe Kühnberger, Peter König, Sergio Daladier Molina, Diego Alejandro Mejia, Jutta Müller, Alison Pease, Omar Dario Saldarriaga, Pedros Stefaneas, Sergio Tobón, Alvaro A. Velásquez T., Stefan Hetzl, Ramiro Grisales, Gianny Rosso, Diony González, Alexandra Agudelo, Sabina Kieninger, and Alfons Rodlauer for the (cooperative) support. Also, thanks go to William Restrepo and children, Johannes Farmer, Sebastian Wilde, Julia Schatzer, Gloria Rodriguez, Tünde Fülöp, Juan Guillermo Pérez R., Eduard Alberto Garcia, and Valeria María. Many special thanks go to Mark Turner for believing in the potential of this work and for all his additional support; to Elizabeth Loew, Dahlia Fisch, and Marcel Danesi for their kindness and help.

Some parts of this work were supported by the Vienna Science and Technology Fund (WWTF), Vienna Research Group 12-400. Sincere thanks to the Instituto Tecnológico Metropolitano (ITM) and to the Institución Universitaria Pascual Bravo for all the kindness and support and to Vision Real Cognitiva s.a.s. In particular, sincere acknowledgments to Juan Guillermo Rivera, Juan Pablo Arboleda, Carmen Elena Úsuga, and Heber Lopez for their constant support and for believing in this multidisciplinary program. Huge thanks go to Judith Mia Kieninger for the unconditional friendship, love, and support.

Contents

Part III   Towards a Universal Meta-Modeling of Mathematical Creation/Invention: Meta-Analysis of Several Classic and Modern Proofs and Concepts in Pure Mathematics

11   Meta-Modeling of Classic and Modern Mathematical Proofs and Concepts 201

Acronyms

m.-s. Morpho-syntactic: It denotes the purely symbolic part of a conceptual entity together with the domain-specific syntactic rules that configure its atomic parts. For instance, the morpho-syntactic part of a mathematical entity is understood as the specific notation used for describing it symbolically.

Chapter 1
Global Introduction to the Artificial Mathematical Intelligence General Program

1.1 A Quite Revolutionary "Artificial Mathematical" Vision

More than eight decades ago, a brilliant scientist astonished the mathematical community with his simple and, at the same time, powerful formal notion of what an (autonomous) machine should be. With his new precise concept, he was able, on the one hand, to set initial bounds to the deductive scope that such machines possess regarding decision paradigms in formal mathematical thinking. On the other hand, he was able implicitly to show how powerful, useful, and universal his new devices could be with regards to enlightening and manipulating not only numerical, but also conceptual issues in mathematics.

Fourteen years, he once again surprised a broader community with an even simpler, more suggestive idea: could it be possible to verify in later, a pragmatic way if a physical realization of his formal machines can imitate "human intelligence" in an indistinguishable manner? Perhaps without knowing it, he indirectly also established a global interest in determining more precisely the real bounds for the general pragmatic scope that (his conceptual) machines have regarding the wide spectrum of intellectual activities that human beings can do.

More specifically, the former question is a generalization of the following deep request: Can we construct a specific instance of the former kind of machines with the ability to imitate an ideal universal mathematician,[1] when we restrict the conversation to the search of a (human-style understandable) solution of a specific

---

[1] By "an ideal universal mathematician," we mean a human being with extraordinary mathematical abilities, who has a universal mathematical ability and a basic knowledge of all mathematics done until the present time. In particular, (s)he should be able to provide solutions to (meta-)mathematical problems, (resp. problems with a mathematical nature) in the style of Arquimedes, Diophantus, Gauss, Euler, Riemann, Cauchy, Hilbert, Klein, Gödel, or/and von Neumann, among others.

© Springer Nature Switzerland AG 2020

D. A. J. Gómez Ramírez, *Artificial Mathematical Intelligence*,
https://doi.org/10.1007/978-3-030-50273-7_1

mathematical inquiry?[2] If one takes a closer look into the intellectual and 'secret' work of our inspiring thinker,[3] one can perceive that he was inspired by questions similar to the former one for inventing most of his astonishing concepts and devices. So, as some readers may already suspect, our mysterious figure is one of the leading founders of modern computer science and artificial intelligence—Alan Mathison Turing.[4]

If we compare the legacy of Alan Turing, as a mathematician and logician, with the work of his (contemporary and predecessor) colleagues, we can affirm that he was essentially the first one who quite seriously and in a pragmatic way offered structural insights to the question concerning the possibility that the intellectual human activity of doing (abstract) mathematics could be simulated concretely by an artificial engine.

Inspired by the most outstanding scientific and practical achievements of Alan Turing, we would suggest that the construction of a real artificial device able to perform mathematical intelligence at a global level, with a human-style manner and simulating, and even improving the (mathematical) minds of the most outstanding mathematicians (as individuals as well as a group) could be considered as one of the most important scientific programs and implicit visions that Alan Turing indirectly gave us. This vision is heuristically supported by the fact that mathematical generation, in general, obeys quite clear deductive and methodological rules, which are closely related to mechanical and systematic (artificial) processes.

Let us call the former global vision *Artificial Mathematical Intelligence* (AMI), in honor of one of the seminal founders of Artificial Intelligence: Alan Turing.

We will now systematically explain the central foundational principles of this vision.

First, let us note that this (AMI) vision possesses a broad multi- and interdisciplinary methodological nature. In other words, the fact that this kind of updated Turing's vision involves concepts like mathematics, intelligence, simulation and the (human) mind implies that it requires the integration and subsequent combination of results belonging not only to mathematics, metamathematics, logic, computer science and artificial intelligence, but also (cognitive) linguistics, (cognitive) psychology, neuroscience, cognitive science, physics and philosophy (of mind), among others. This methodological requirement is highly desirable and virtually necessary due to two facts: First, on the one hand, we want to get a "surgical" knowledge of how the human mind proceeds by solving a mathematical inquiry. On the other hand, the scientific knowledge that we have about the mind (together with its physical

---

[2]By mathematical inquiry, we mean any kind of mathematical formal computation, exercise, and conjecture at an arbitrary degree of sophistication.

[3]It is worth mentioning that during the time between the initial invention of his (conceptual) machines and the publication of his seminal ideas about intelligence and machines, he was applying all the pragmatic power of his technological instruments in a special team that was able to shorten the second world war by at least 2 years and save millions of lives worldwide.

[4]For the references supporting the former paragraphs see [28, 48, 49] and www.turingarchive.org.

"mirrors" like the brain and more generally the body) is not centralized and, in fact, it is spread among several disciplines like the ones mentioned above.

Second, due to the fact that some aspects of the AMI vision have already been intensively studied since the beginning of AI, most of the partial solutions offered have a mono-disciplinary nature (e.g., each of them possesses a purely metamathematical, linguistic, or philosophical character). For instance, Turing himself, as well as Kurt Gödel and Alonso Church, offered local negative answers, mainly from a logical point of view, to a purely metamathematical reading of the AMI plan [9, 17, 48]. More specifically, by doing a closer and detailed reading of the former classic results, one observes the following:

On the one hand, the concepts and arguments used have a structural mathematical character when the central demonstrations are presented. Nonetheless, the initial and final conclusions are extended at a meta-level and therefore lastly they have a metamathematical scope (e.g. the notion of Turing machine can be completely formalized in mathematical terms and simultaneously be used to infer metamathematical results like the insolubility of the Entscheidungsproblem). Contrastingly, a purely logical approach to the AMI vision turns out to be essentially blind regarding concept formation in pure mathematics (see, for instance, [31]), which is one of the most fundamental aspects underlining mathematical generation.

Now, if we wear a "methodological glasses" coming from cognitive science, then we can affirm based on the former fact that those results were essentially a cognitive product of three (quite brilliant) minds described with an intrinsic mathematical style. So, inspired by the most outstanding results in cognitive science regarding formal reasoning (e.g., assuming the veracity of the computational nature of the mind, at least in relation with mathematical creation [29]), there is no methodological obstacle for the construction of a kind of universal mathematical and logic artificial agent that is able to give answers similar to those given by Turing (himself), Gödel and Church, to the same kind of questions.[5,6]

Therefore, the former results are valuable more from a mono-disciplinary point of view. However, if we assume a more pragmatic perspective regarding the way in which mathematics is concretely produced at a global scale, we see (after enough time of search) that roughly speaking more than 90% of the mathematical results

---

[5]It is worth mentioning at this point that although the former three classic results possess a high level of brilliance in their central ideas, none of them develop an explicit explanation of how human-made mathematics are done from a cognitive perspective. Perhaps the only one who was able to achieving that was Turing with the development of his seminal concept of Turing machine.

[6]It is also important to mention that a lot of methods used by the former authors for obtaining their limiting results were based on some kind of meta-physical assumptions about the nature of space, time, and spacetime, like the existence of infinite collections of numbers and temporal processes. Now, these assumptions are, strictly speaking, not based on standard physical laws, therefore can be classified from a physical and cognitive perspective, as subjective mental assumptions from the corresponding authors, with a wide level of (subjective) acceptance by the general community. For a deeper discussion about these quantitative issues please see Chap. 5.

produced each year involve solvable and decidable mathematical inquiries.[7] So, our AMI program has a fundamental pragmatic importance in spite of the former limiting results.

A further ontological revision of the former three classic meta-theorems reveals that each of them use as implicit foundational principle—the existence of the natural numbers at a classic indefinite basis, i.e., the existence of infinite collection of objects (e.g., numbers) generated sequentially; or an equivalent version of this fact. Enhancing our cognitive glasses with additional "formal lenses" coming from seminal results in modern physics, one can see that the former hypothesis possesses more a mental nature than a practical and authentic physical substratum (for more details see Chap. 5). Hence, this shows that since those meta-results were based implicitly on a mental construct (e.g., a classic understanding of the natural number as a sequential structure without final quantity), the deductive scope should be preponderantly mental and less pragmatic as argued previously.

Third, after the groundbreaking work of Alan Turing regarding the foundations of AI at the beginning of the twentieth century, we have seen a tremendous number of theoretical and practical advances in our global understanding of how the mind works and how we can simulate, and sometimes improve, specific aspects and abilities of it through artificial devices (see, for instance, [4, 12, 35] and Sect. 3.1, Chap. 3). This gives the necessary inspiration to think about the AMI vision in terms of the creation of a *universal mathematical artificial agent* (UMAA) whose percepts and actions (in a classic AI sense [43, Ch.2]) consist exactly of mathematical structures (e.g., mathematical concepts, theorems, facts, conjectures) described in a clear, syntactic, and human-style way; its environment is delimited by the mathematical information that a user would like to provide it; and finally its sensors, actuators, and internal engine are formed in terms of specific and computationally-feasible formalizations of the most fundamental cognitive abilities used by the mind during mathematical research, among others (see, for instance, part II).

Finally, its performance measure is structured in direct relation to the conceptual solutions that it provides to the problems asked, for example, in the form of (mathematical) total or partial proofs or counterexamples.

So, arguing from a more philosophical perspective, an UMAA would be capable of reminding us an ideal universal mathematician-logician whose formal thinking is rapid, effective and not so constrained by physical, sensorial, or cultural influences. An abstract thinker who is able to create sophisticated mathematical concepts and (counter-)examples based on the questions[8] and (formal) evidence[9] provided by the interlocutor (e.g., user). In particular, this universal mathematician-logician should

---

[7]Here, the reader can conduct a self-tour through the articles and books published at renowned mathematical journals and by publishers, to obtain a stronger conviction of this global qualitative claim. For example, one can verify that more than 90% of the last 100 papers (until the first semester of 2019) published in renowned journals like Annals of Mathematics involve solvable problems dealing with the usage of classic concepts and with the constructions of new mathematical concepts, theories, and some physical applications.

[8]For example, a mathematical conjecture.

[9]For instance, successful cases where the conjecture was verified.

be able to solve formal mathematical inquiries and generate mathematical structures at a new level of sophistication.[10]

In other words, we conceive the fulfillment of the AMI program more from an interactive point of view than from an automatic one. Explicitly, let us imagine that the user provides not only the formal version of a mathematical conjecture, but also additional information describing special cases where the conjecture turns out to be truth (e.g., "formal specific evidence"). Such "special" interactive conditions and extra information can cut back several additional challenges that a purely automatic approach could potentially possess.

At this point, it is worth mentioning that the notions of algorithmic complexity and efficiency should be conceived with slightly different "eyes" within the AMI program. Effectively, one of the main goals of our vision will be to produce detailed and gradually explainable solutions to formal (domain-specific) mathematical problems, which would take less time to be found than the time required by a professional researcher (in mathematics or related areas). For example, our vision goes in the direction of imaging that a robust version of a UMAA which should require some months of (interactive) work (e.g., 6 months) for solving a Ph.D.-level mathematical problem, which turns out to be fine in comparison with the standard time that a Ph.D. thesis takes to be done (e.g., 3–4 years). Additionally, suppose that the rate of success of it is around 80%, this would represent a huge step towards an universal and cognitively-inspired (interactive) mechanization of mathematics.[11]

In summary, we want to "resurrect" through the AMI program a new form of what we could call, in modern terms, one of the biggest dreams of Alan Turing: the fulfillment of artificial intelligence within the special domain of (pragmatic) mathematical creation/invention.

> "Those who can imagine anything, can create the impossible."
>
> Alan Turing

1.2 Towards Conceptual Computation

Let us make a simple comparison between the brain (and in an extended manner the (embodied) mind) and the computer: First, regarding working speed, the human mind processes information (roughly speaking) around six million eight hundred thousand times slower than an (average) computer and it is around one billion (10^9)

---

[10]See, for example, Chap. 11 for initial formal evidence regarding the universal way in which such an (artificial) researcher should be able to generate several mathematical structures from different mathematical domains.

[11]A quite simpler version of the AMI vision was described initially by the author in [20] at a very elementary and condense way.

times less accurate (regarding the error's rate per number of operations performed) [1]. On the other hand, the human mind is able to do very simple (and at the same time very powerful) conceptual inferences like if y =cymbal and Y =tambour, then $yYyYyYy$ =drum set; or, if A =house and B =boat, then AB =houseboat;[12] while this ability is essentially non-existence in (modern) computers. Furthermore, more than 95% of the scientific results generated in history are product of the (research of the) human mind, with all the former limitations and strengthens. Thus, why not to simulate the deductive-pragmatic functioning of the mind (taking into account the former spectrum of features) with all the strengthens of modern computation?

This kind of simple methodological approach is not so common in auto-matic deduction or computational logic. In fact, one of the main methodological approaches is to reduce the conceptual complexity of the problem to solve until it can be fully verified or refused computationally, instead of modeling com-putationally the manner in which the mind approaches the problem (without performing necessarily an ontological reduction on the way). The latter form of obtaining computationally-feasible solutions should be explored more deeply, because we have a sufficiently robust constellation of results in cognitive sciences (and related fields) describing a wide spectrum of (deductive) features of the mind. This motivates the quest for a new form of *conceptual computation* paradigm in computer science and artificial intelligence. In fact, the initial motivation of Alan Turing to create its famous Turing machines was to find a concrete and pragmatic formalization of the way in which a mathematician's mind perform quantitative tasks (involving, for example, the calculation (by hand) of a function on the natural numbers) [48]. Therefore, we can think in extending the classic (Church-)Turing Thesis (or Turing Theorem (TT)) to a general metamathematical conceptual framework:

Thesis 1.1 (Towards a Conceptual Extension of the Church–Turing Thesis) *A Mathematical structure (e.g., a concept, a proof, a counterexample, a theory) is effectively calculable (i.e., generated) by a human being('s mind) if and only if it can be computed by a "conceptual" (Turing) Machine (e.g., UMAA (Universal Mathematical Artificial Agent)).*

One of the main purposes of this book is to offer a general idea of how such a conceptual machine should look likes in terms of initial formalizations of a global taxonomy of fundamental cognitive mechanisms (see Chap. 10). In fact, we will prove in Chap. 9 that just with the cognitive ability of conceptual substratum it is possible to recover and to reinterpret cognitively the classic Church–Turing Thesis. Thus, this represents a starting evidence of the fact that conceptual machines should be at least as powerful as classic Turing machines. On the other hand, in Chaps. 7 and 11 it will be extensively shown that conceptual machines are a more appropriate formal device for producing artificial conceptual generations of

---

[12]Here, we assume the standard and relatively sophisticated meaning of the word "houseboat" in English.

dozens of mathematical structures from the most simple until the most sophisticate ones, which currently is not the case for the contemporary literature in automated deduction (and related fields).

1.3 Former and Current (Local) Advances Towards the AMI Vision

A lot of (mechanical and computational) aspects of our AMI vision have captured the attention of a considerable number of researchers during the last decades. Most of them have done amazing, valuable work which can be seen in our context as local evidence and support in favor of its ("near") fulfillment. In this section, we will mention some of the most outstanding results together with further remarks concerning their main original goals.[13]

At its very beginning, the research field of automated deduction had, as part of its central motivations, the construction of software able to generate (and implicitly solve) concrete mathematical work (e.g., outstanding mathematical theories/books). For example, Whitehead and Russell's Principia Mathematica [38, 51]; elementary plane Euclidean geometry [16], (some parts of) Newton's Principia [14], and, of course, many instances of propositional calculus [3], among (a few) others. On the other hand, some specific mathematical challenges as the Robbins problem, the four color's problem, the Kepler's theorem, and the Feit–Thompson theorem have given (significant) additional inspiration for developing more sophisticated (automated) theorem provers [22, 23, 25, 36].

Furthermore, nowadays there are many kinds of (free and paid) computer programs which can assist the researcher in mathematics (and related areas) on different tasks. Now, these have always involved a relatively small collection of mathematical areas, for instance, numerical and symbolic computation, the drawing of technical graphics, solving particular classes of systems of equations, inequalities, Diophantine and differential equations and quantifier elimination, among others [7, 34, 46, 52] (for a more general list see "The Guide to Available Mathematical Software"[14]).

Other kinds of outstanding software are used for finding proofs in several classes of propositional calculi and for proof verification and proof generation in some specific logics which, in principle, do not cover completely the scope of the mathematics done every day, not only by professional mathematicians, but also for researchers working in related fields [3, 42]. Furthermore, some instances of the later kind of software mentioned possess, in general, such a highly technical syntax that for the non-specialized mathematician (or related researcher) it is not

---

[13]It is not the purpose of the present section to give an exhaustive list of all of them, due to the fact that the literature in this direction is considerably vast.

[14]https://gams.nist.gov.

straightforward to begin to use it in his/her daily work, mainly because it would require several weeks (or even months) of regular and quite technical study to understand and manipulate practically their main semantic and syntactic features.

There are also a third kind of valuable programs aiming to produce human-style proofs by integrating the more robust account of the linguistic dimension involved in mathematical generation. However, its scope involves only very particular kinds of problems within quite specific theories, e.g., metric space theory (see [15] and the references there).

Furthermore, there are new proposals for setting general foundational frameworks for mathematics that aim to facilitate the implementations of mathematical proofs in computers, for instance, the univalent foundations project [50].

Most of the former works belong to what can be roughly called "The (classic) Mechanization of Mathematics" (see, for instance, [2]). They are "classic" in the sense that their methodologies and goals possess essentially a more purely logical, metamathematical, and algorithmic nature, and, on the other hand, they provide a reduced (and often nonexistent) formal account of the cognitive causes that underlines the origin and structure of the corresponding explanatory frameworks. In other words, an (implicit) external ontological point of view of mathematics is perceived in those works, i.e., the corresponding mathematical phenomena is analyzed as external entities that may or may not have a cognitive origin.

For instance, one (classic) trend in this direction, and one not so intimately related with a cognitively-inspired model of mathematical invention, is formed by the main techniques coming from resolution theorem proving, whose slightly different motivation and orientation emerges more from the need of finding efficient methods in proof verification and proof generation [42, Ch. 2].

So, keeping in mind the great value and brilliance some of those results can have, most of them possess the limitation that they cannot be generalized in a straightforward way to other mathematical domains, because their explanatory ontology depends structurally on the particular mathematical entities in consideration.

However, there is a complementary trend of formal and computational frameworks based more on a cognitive understanding of mathematical generation and starting with the identification and formalization of fundamental cognitive abilities used by the mind during creative thinking (see, for instance, [5, 8, 13, 18, 19, 21, 32, 33, 37, 44, 45, 47]).

Other quite outstanding works explore and exploit domain-specific mathematical heuristic at a computational basis for (automated) concept and conjecture generation and verification [10, 11, 39].

Moreover, there are more traditional treatises with a more philosophical touch, and simultaneously with deep insights regarding the identification of fundamental (mathematical) heuristics and cognitive strategies used in mathematical creation/invention like the classic works of G. Polya and I. Lakatos [31, 40, 41]. In fact, [39] presents a creative computational account of Lakatos' work where an initial formalization of the social dimension of mathematical generation plays a central role.

The former second type of results represent strong formal evidence in favor not only of the thesis that specific mathematical thinking can be gradually understood and subsequently simulated in a computational way, but also bringing an additional and enlightening new perspective into the AMI vision that classic purely logical approaches have only been barely able to suggest. Nonetheless, these more cognitively- and heuristically-based works present a clear constrain regarding the level of sophistication of the mathematical structures meta-analyzed and simulated. These works deal essentially with elementary problems belonging to, for example, geometry, real and complex analysis, algebra and number theory, among others. So, more abstract and general mathematical sub-disciplines are virtually not meta-studied, for instance, modern algebraic geometry [24, 27, 30], which represents an integrative, illuminating and fascinating case of study for the fulfillment of the AMI program. This is due to the high level of technical sophistication and elegance of its concepts and methods.

1.4 A New Foundational and Integrative Program

After exploring in the former section the strengths and weaknesses of some of the existent (local) results towards the fulfillment of the AMI vision, we will now describe the precise way in which we aim to used and integrate classic and modern techniques and perspectives and to create new ones for filling some important foundational and pragmatic gaps existent in the literature, as well as for setting a stronger inter- and multidisciplinary basis.

Virtually all the former (local) results towards a positive solution of the AMI program, with exception of the univalent foundations project and the classic works of Polya and Lakatos, propose almost immediate algorithmic formalizations of the particular classes of mathematical inquiries to be solved. This is usually done without a previous solid and deeper exploration and search into the foundational properties of the mathematical structures to be computationally modeled.

Methodologically speaking, this can be done without problem, however, such straightforward approaches have the limitation that they should create highly technical representations for the semantic content of the corresponding mathematical structures involved, which has the cost of sacrificing the "cognitive naturalness" of the whole framework. Furthermore, an ontological gap used to remain implicit "in the air" between the intrinsic nature and (cognitive) meaning of the mathematical entities involved, and the corresponding "artificial" juxtapositions of symbols used for representing them syntactically as well as semantically in artificial devices.

Contrastingly, aiming to go directly to the development of algorithmic frameworks after having analyzing (only) local mathematical data has the clear limitation of implicitly ignoring further heuristic, syntactic, morphological, and semantic principles that ground other mathematical areas and that should be mandatory for any kind of global explanatory (cognitive) metamathematical framework. In fact, we will see in further chapters that the meta-analysis of several kinds of conceptual substrata (Chap. 9) is necessary for the subsequent fulfillment of the AMI vision.

In particular, it involves an integrative symbolic, semantic, and cognitive meta-study of prototypical substrata belonging to a not-small collection of mathematical sub-disciplines, which remains (as far as we know) a non-accomplished task in (classic) metamathematics. Therefore, the present work focuses essentially on the theoretical foundations of the AMI program, together with some relevant remarks for the algorithmic aspects.

Explicitly, we propose in the first part of this book the establishment of a research program aim to set new cognitive foundations for mathematics, which includes implicitly a computational component.

We describe in Chap. 3 the main reasons supporting the necessity of a new foundational program for mathematics and its most fundamental future challenges. In addition, we discuss seminal issues involving the cognitive substratum of a mathematical proof in a wide generality. Further, we enlighten a central fact of the most successful natural machine producing mathematics, i.e., we describe the methodological implications for the AMI program from the fact that the whole mathematical product that we know today is basically the systematic accumulation of billions of conscious outputs of the human mind, considered also collectively.[15]

Inspired by "cosmological" and "synthetic" considerations, we do a deeper philosophical exploration (in Chap. 4) into the (cognitive) reality of mathematics and into the mathematics of the (cognitive) reality. Moreover, in our methodological framework we update the notions of observer's perspective at the macro, mecro, [16] and micro level. We argue in favor of the thesis that nature at any level of observation possess a kind of mathematical precision, and that, in fact, entities in nature possess a real mathematical substratum which structure them. So, mathematics understood in the widest sense of the word constitutes an existing dimension of the universe that structures any part of it. Subsequently, we state and support the existence of a kind of unpredictability principle at the mecro level. In other words, we argue that natural human will represent an example of a concrete entity in nature that qualitatively bounds the predictive scope of the kind of phenomena that any form of UMAA could model. Further, taking inspiration from some thought experiments and the development of a "continuous notation" for real numbers, we show the incompatibility of the following: the fulfillment of the "singularity" as an extreme form of artificial intelligence [6], and the fact that space and time can be modeled with a continuous framework, e.g. using the set of real numbers.[17] In summary, we conduct a concise intellectual exploration using philosophical, physical, mathematical, and logic tools for estimating more precisely the ontological status of the explanatory scope that a UMAA can have, not only concerning mathematical questions, but also (mathematical reformulations of) questions belonging to other scientific disciplines.

---

[15]This simple fact has important consequences on the way in which some cognitive abilities required in mathematical research are identified and formalized (see part II).

[16]For a more concrete description of this notion see Chap. 4, Sect. 4.3.

[17]Strictly speaking, we use a third fact for doing that, namely the unsolvability of the halting problem [48].

Chapter 2 serves as a compact and quite concise (meta)mathematical preparation for the non-specialist (mathematician/logician) reader. It briefly revises the notions of propositional and predicative logic, the most outstanding logical frameworks for modern mathematics (e.g., ZFC and NBG set theory, Peano arithmetic), and the notion of category and some of its derived notions. Moreover, a short description of fundamental algebraic, topological, and geometric notions are presented that are mostly required in Chaps. 7 and 11.

In Chap. 5, we start with the development of a seminal (cornerstone) topic within the new cognitive foundations' program. In other words, inspired by a formal and multifaceted analysis of our basic understanding of (mental) counting processes, we propose a concrete cognitive refinement of one of the most well-known structures in mathematics, namely the natural numbers. In fact, we present the *physical numbers* as a more precise quantitative notion which includes, and at the same time, refines classic perceptions that we use on a daily basis when we estimate the "number" of elements of collections of objects.

More explicitly, we enhance the standard notion of "counting" by the new notion of partitioning, and we show that the former can be considered a particular form of the latter where our minds can potentially gain a more global and precise perception of numerical (and subsequently mathematical) entities.

Additionally, we state that the physical numbers have an initial as well as a final entity, which is bound by the number of physical quanta in our universe. This allows us to make a finer (cognitive) taxonomy of the natural numbers, (i.e., natural number n "smaller" or equal that such a bound, which we denote by ω, are considered "physical natural numbers" since they count on a physical support represented by collections of (external) entities (e.g., elementary particles) having exactly n elements). On the other hand, a natural number m strictly surpassing ω will be considered simply as a "mental natural number," because it can be (cognitively) produced recursively as a concrete conceptual blend of physical natural numbers. It is a well-known fact that conceptual blending (see Chap. 7) can produce purely mental objects by combining two input concepts which have (or have not) physical realizations in the external realm [13].

Moreover, we establish a quite significant distinction between the (mental) notion of "infinity" (in its several variants) and the (more physical) notion of "immensity." In particular, we propose notions of small and immense numbers based on the specific conscious and unconscious patterns required for the mind in order to understand them.

In addition, we explore in a global way the fact that the explanatory range of mathematical frameworks based on numerical structures being finite as a whole, is mature enough to allow us to develop a lot of our most fundamental mathematical and physical theories.

Finally, we offer an initial formal framework for the physical numbers with (physical) division as main operation. We also propose a new kind of research heuristic in (classic) number theory which can be informally called "physical number theory." This essentially consists of doing an initial verification of the (non-)validity of an arithmetic conjecture for the physical (natural) numbers in order to test firstly its "physical (or external) veracity," and after that using (eventually) additional methods for the "proof" of a more mental component of it.

It is important to note here that the development of coherent refinements on the way in which we understand and manipulate pragmatically as well as theoretically (the notion of the) natural numbers and the particular "counting" (cognitive) processes underlying them would possess a huge influence not only on the foundations of mathematics, but also on the foundations of (theoretical) computer science and physics, among many others. So, this topic represents a central pillar of the AMI program with consequences beyond the AMI vision.

Delving deeper into the quantitative dimension of the new foundations' program, we show explicitly in Chap. 6 a "singular" phenomenon happening into "foundational bricks" of mathematics, (i.e., Zermelo–Fraenkel set theory with Choice (ZFC)). In other words, we prove formally that we can construct an identical (i.e., meta-isomorphic) version of (standard) mathematics (i.e., mathematics classically constructed from ZFC), called "Dathematics" (or Dual Mathematics), where instead of sets, one uses a special kind of proper classes as foundational bricks. This fact turns out to be surprising not only from a purely metamathematical perspective, but even more from a cognitive point of view. Effectively, proper classes are, strictly speaking, mental constructions without any kind of physical realization at any quantitative level. So, the fact that we can support in a semantic way a meta-isomorphic copy of our (in some sense daily life's) mathematics strictly based on objects that do not have any kind of physical counterpart in nature (by definition) implies that our current basic logic-deductive frameworks are grounding the semantic content of mathematical structures more in a purely formal and syntactic way and much less in a physical and (more) "tangible" manner. This implies, among other things, that deeper intuitions about mathematical structures could be highly limited by the mono-thematic formalizations that have been developed for them classically.[18]

In the second part, we focus our attention on the specific cognitive mechanisms used by the mind during mathematical creation/invention. Here, we take inspiration from a wide spectrum of classic and new results in cognitive science, cognitive linguistics, psychology, and from the classic works on the philosophy of mathematics of G. Polya and I. Lakatos.

First, we dedicate an entire chapter to the study of one of the most fundamental of these processes–conceptual blending (Chap. 7), or, informally, the ability of the mind to create genuine conceptual fusions of two (or more) input concepts. Explicitly, we use a classic formalization of conceptual blending in terms of colimits embedded in a categorical many-sorted framework for mathematical concepts. For such a formalization one can generate implementations of concrete blends in the Heterogeneous Tool Set (HETS). Moreover, we show how to generate fundamental notions of Fields and Galois theory recursively only in terms of conceptual blends starting from five elementary concepts coming from different mathematical sub-disciplines like group theory, fixed point's theory, and abstract algebra. This is

---

[18]This issue will be illuminated in Sect. 3.1, Chap. 3.

an initial case study regarding concept generation from a cognitive as well as a logic perspective that aims to fill the gap existing in the automated deduction's literature.[19]

In Chap. 8, we present an initial cognitively-based formalization of (atomic and best) analogy and analogical space of two formulas, starting with a classic Hilbert's style calculus for propositional logic. Additionally, we illustrate the explanatory power of the former notions for offering meta-descriptions of the generation of classic (elementary but non-trivial) proofs of some tautologies. Moreover, a new formalization of conceptual blending is described in terms of the former notions. Finally, some notions are extended to a first-order setting.

In Chap. 9, the new cognitive (metamathematical) mechanism of conceptual substratum is introduced (i.e., the ability of producing specific morpho-syntactic configuration of symbols with intrinsic meaning, which allow our minds to manipulate essential formal features of (mathematical) concepts in sophisticated deductive tasks). We present two formalizations of this notion in different deductive contexts and their relations with classic tools in automated deduction and logic like Skolemization and Diophantiveness. Furthermore, based on a first-order formalization of (functional) conceptual substratum, we state an explicit cognitive characterization of the Church–Turing Thesis, which can be seen as a modern (and more cognitively-supported) description of this classic and foundational principle. Moreover, we show how to construct equivalent versions of the sequent calculus for first-order logic with equality over a language L, including deductive rules codifying (functional) conceptual substratum inside. Such deductive systems can be seen as slightly improved versions of the classic (Gentzel) sequent calculus from the point of view of their (increased) cognitive soundness. Lastly, we introduce conceptual lining as the dual cognitive ability of conceptual substratum.

In Chap. 10, we present an initial global taxonomy of the most fundamental cognitive mechanisms used in mathematical research, together with the corresponding formalizations in terms of a more global notion of mathematical concept (and mathematical structure) than the one initially presented in Chap. 7. In that chapter, one can see in a more concrete way the multi- and interdisciplinary nature of the whole AMI program from a methodological point of view. The formalizations are presented assuming a minimal robustness of the logical frameworks underlining the (local) mathematical theories that can be used as the object of meta-study. In particular, the meta-notions are presented at a level of generality that includes the possibility of meta-analyzing (local) concepts described over a wide spectrum of logics. In addition, more general versions of the three initial cognitive abilities of conceptual blending, analogical reasoning, and conceptual substratum are presented to match the level of abstractness needed subsequently in further chapters.

This chapter has an additional central relevance from the point of view of cognitive science because it offers (as far as we know, for the first time) a global and formal classification of all the essential mechanisms that the mind uses during

---

[19]We will tackle this issue in quite more detail in Chap. 11.

mathematical research. In particular, the intellectual activity of producing abstract mathematics is broad enough to represent an outstanding case study towards the development of more general formal frameworks explaining the general functioning of the mind which is a central research goal by cognitive scientists.

Furthermore, Chap. 10 has also a seminal relevance for the foundations of computer science because the mechanisms described there are more plausible to be modeled symbolically as well as algorithmically. And, in some sense, the concrete formalizations developed there, which possess a more finite nature, begin to "knock down" methodological "walls" that could emerge from "over-extrapolations" of classic (unsolvability) results.

In the last part, we present in an explicit and extended manner the concrete evidence for the universality of all the formal meta-tools developed so far.

Explicitly, in Chap. 11 we offer global formal support to fill the gap that the majority of the classic methods used in standard automated deduction have regarding the development of meta-explanations of conceptual generation [2, 26, 42]. In other words, we show explicit cognitive meta-generations (i.e., meta-explanations) not only of the proofs of two classic theorems in elementary geometry and number theory, but also of dozens of fundamental mathematical concepts belonging to several mathematical disciplines like topology, set theory, abstract algebra, category theory, sheaf theory, commutative algebra, and (classic and modern) algebraic geometry. These cognitive meta-constructions can be seen as explicit evidence of the creative power of an ideal (non-necessarily embodied) mathematical artificial agent (which is one of the main goals of the AMI program).

In particular, we exhibit a recursive and explicit cognitive meta-generation of one of the conceptual cornerstones of modern algebraic geometry (i.e., the notion of (mathematical) scheme). This notion was chosen in advance due to its technical sophistication to show that the multifaceted tools developed in the previous chapters are strong enough to generate higher abstract mathematics. This fact starts to fill simultaneously another gap existing in the literature involving the elementary scope that more cognitively-inspired accounts of conceptual creation possess.[20] It is worth mentioning at this point that we could have potentially chosen any other sophisticated mathematical notion instead of the one of scheme, however, we choose it due to the central role that it plays in modern mathematics and even beyond algebraic geometry.

In addition, all these cognitive meta-generations are presented more from the point of view of a global version of a UMII. So, some of them have more qualitative commonalities with the original historical reconstructions of the corresponding concepts (e.g., (mathematical) categories), some possess less and can be seen as new ways of generating those concepts (following the integrative guidelines of the AMI vision) (e.g., sheaves).

In Chap. 12, we present the most outstanding (future) challenges of the AMI program not only from a theoretical and foundational perspective, but also from a

---

[20]See, for example, the references presented at the end of Sect. 1.3.

more pragmatic and algorithmic point of view. In addition, we describe plausible extensions of the AMI vision to others scientific disciplines close to mathematics in some foundational aspect.

Along the lines of such an extension is exactly where one perceives the importance that a mature version of the AMI vision can have regarding the way in which we currently do scientific research at essentially a purely human level.

Finally, due to the multifaceted methodological dimension and the cognitive nature of our new metamathematical (AMI) program, we can also use the more classic name of *Cognitive Metamathematics* for it. In fact, this more neutral name has the advantage that, on the one hand, it stresses deeply the integrative scientific discipline grounding the AMI vision, and, on the other hand, it emphasizes with a new clarity the theoretical aspect of the AMI vision in a concise way, extricating along the way the AMI program from being understood only in terms of the computational challenge behind it.

1.5 Ethical Considerations

From the very beginning, this new inter- and multidisciplinary AMI program was conceived for improving and enhancing our theoretical, constructive, and practical understanding of mathematics in the widest sense of the word, and, subsequently our understanding of nature at several levels of observation. So, from a middle- and long-term perspective, any new product, technology, invention, and community emerging and largely (in-)directly based on (applications coming from) the AMI program should pursue respectful, deserving, peaceful, and integrative purposes regarding a pacific living with our fellows and with nature. So, the AMI program and all its future applications are strictly envisioned to increase and to protect (our quality of) life inside and outside earth at any stage of development. More generally, the Alisomar principles of the Future of Life Institute represent a valuable source for the global ethical principles that should be observed on any materialization of Artificial Mathematical Intelligence.[21]

References

1. Beck, H.: Scatterbrain: How the Mind's Mistakes makes Human Creative, Innovative and Successful. Greystone Books Ltd (2019)
2. Beeson, M.J.: The mechanization of mathematics. In: Alan Turing: Life and legacy of a great thinker, pp. 77–134. Springer (2004)
3. Biere, A., Heule, M., van Maaren, H.: Handbook of satisfiability, vol. 185. IOS press (2009)
4. Boden, M.A.: Mind as machine: A history of cognitive science. Oxford University Press (2008)

---

[21] For more details, please consult the website https://futureoflife.org/ai-principles/.

5. Bou, F., Corneli, J., Gomez-Ramirez, D., Maclean, E., Peace, A., Schorlemmer, M., Smaill, A.: The role of blending in mathematical invention. Proceedings of the Sixth International Conference on Computational Creativity (ICCC). S. Colton et. al., eds. Park City, Utah, June 29-July 2, 2015. Publisher: Brigham Young University, Provo, Utah. pp. 55–62 (2015)
6. Chalmers, D.: The singularity: A philosophical analysis. Journal of Consciousness Studies **17**(9–10), 7–65 (2010)
7. Char, B.W., Geddes, K.O., Gonnet, G.H., Leong, B.L., Monagan, M.B., Watt, S.: Maple V library reference manual. Springer Science & Business Media (2013)
8. Chiu, M.M.: Metaphorical reasoning in mathematics: Experts and novices solving negative number problems. (1994)
9. Church, A.: An unsolvable problem of elementary number theory. American journal of mathematics **58**(2), 345–363 (1936)
10. Colton, S.: Automated theory formation in pure mathematics. Ph.D. thesis, University of Edinburgh (2001)
11. Colton, S., Bundy, A., Walsh, T.: Automatic concept formation in pure mathematics. In: Proceedings of the 16th international joint conference on Artificial intelligence-Volume 2, pp. 786–791. Morgan Kaufmann Publishers Inc. (1999)
12. Colton, S., Wiggins, G.A., et al.: Computational creativity: The final frontier? In: Ecai, vol. 2012, pp. 21–16. Montpelier (2012)
13. Fauconnier, G., Turner, M.: The Way We Think. Basic Books (2003)
14. Fleuriot, J.D., Paulson, L.C.: A combination of nonstandard analysis and geometry theorem proving, with application to newton's principia. In: International Conference on Automated Deduction, pp. 3–16. Springer (1998)
15. Ganesalingam, M., Gowers, W.T.: A fully automatic theorem prover with human-style output. Journal of Automated Reasoning pp. 1–39 (2016). https://doi.org/10.1007/s10817-016-9377-1
16. Gelernter, H.: Realization of a geometry theorem proving machine. In: IFIP Congress, pp. 273–281 (1959)
17. Gödel, K.: Über formal unentscheidbare sätze der principia mathematica und verwandter systeme i. Monatshefte für mathematik und physik **38**(1), 173–198 (1931)
18. Goguen, J.: An introduction to algebraic semiotic with application to user interface design. *In* Computation for metaphors, analogy and agents. C. L. Nehaniv, Ed. Vol. 1562 pp. 242–291 (1999)
19. Goguen, J.: Mathematical models of cognitive space and time. Proceedings of the Interdisciplinary Conference on Reasoning and Cognition **123**, 125–148 (Keio University Press, 2001)
20. Gomez-Ramirez, D., Smaill, A.: Formal conceptual blending in the (co-)invention of (pure) mathematics. In: R. Confalonieri, A. Pease, M. Schorlemmer, T. Besold, O. Kutz, E. Maclean, M. Kaliakatsos-Papakostas (eds.) Concept Invention: Foundations, Implementation, Social Aspects and Applications, pp. 221–239. Springer International Publishing, Cham (2018)
21. Gomez-Ramirez, D.A.J., Hetzl, S.: Functional conceptual substratum as a new cognitive mechanism for mathematical creation. arXiv preprint arXiv:1710.04022 URL https://arxiv.org/pdf/1710.04022.pdf
22. Gonthier, G.: Formal proof–the four-color theorem. Notices of the AMS **55**(11), 1382–1393 (2008)
23. Gonthier, G., Asperti, A., Avigad, J., Bertot, Y., Cohen, C., Garillot, F., Le Roux, S., Mahboubi, A., O'Connor, R., Biha, S.O., et al.: A machine-checked proof of the odd order theorem. In: International Conference on Interactive Theorem Proving, pp. 163–179. Springer (2013)
24. Grothendieck, A., Dieudonné, J.: Eléments de Géométrie Algébrique I. Springer (1971)
25. Hales, T.C.: A proof of the Kepler conjecture. Annals of mathematics **162**(3), 1065–1185 (2005)
26. Harrison, J.: Handbook of practical logic and automated reasoning. Cambridge University Press (2009)
27. Hartshorne, R.: Algebraic Geometry. Springer-Verlag, New York (1977)
28. Hodges, A.: Alan Turing: The Enigma. Random House (2012)

29. Horst, S.: The computational theory of mind. Stanford Encyclopedia of Philosophy (2011)
30. Kobayashi, S., Nomizu, K.: Foundations of differential geometry, vol. 2. Interscience publishers New York (1969)
31. Lakatos, I.: Proofs and refutations: The logic of mathematical discovery (Cambridge Philosophy Classics). Cambridge university press (2015)
32. Lakoff, G., Núñez, R.E.: Where mathematics comes from: How the embodied mind brings mathematics into being. AMC **10**, 12 (2000)
33. Martinez, M., Abdel-Fattah, A., Krumnack, U., Gómez-Ramírez, D., Smail, A., Besold, T., Pease, A., Schmidt, M., Guhe, M., Kühnberger, K.U.: Theory blending: Extended algorithmic aspects and examples. Annals of Mathematics and Artificial Intelligence pp. 1–25 (2016)
34. MatLab, M.: The language of technical computing. The MathWorks, Inc. http://www.mathworks.com (2012)
35. McCorduck, P.: Machines who think: A personal inquiry into the history and prospects of artificial intelligence. AK Peters/CRC Press (2009)
36. McCune, W.: Solution of the Robbins problem. Journal of Automated Reasoning **19**(3), 263–276 (1997)
37. Moreno, R., Mayer, R.E.: Multimedia-supported metaphors for meaning making in mathematics. Cognition and instruction **17**(3), 215–248 (1999)
38. Newell, A., Shaw, J., Simon, H.: Empirical explorations with the logic theory machine: A case study in heuristics. Automation of reasoning **1**, 1957–1966 (1957)
39. Pease, A.: A computational model of Lakatos-style reasoning (2007)
40. Pólya, G.: Mathematics and plausible reasoning: Induction and analogy in mathematics, vol. 1. Princeton University Press (1990)
41. Pólya, G.: Mathematics and plausible reasoning: Patterns of plausible inference, vol. 2. Princeton University Press (1990)
42. Robinson, A.J., Voronkov, A.: Handbook of automated reasoning, vol. 1. Elsevier (2001)
43. Russell, S.J., Norvig, P.: Artificial intelligence: a modern approach. Malaysia; Pearson Education Limited, (2016)
44. Schorlemmer, M., Smaill, A., Kuehnberger, K.U., Kutz, O., Colton, S., Cambouropoulos, E., Pease, A.: COINVENT: Towards a computational concept invention theory. In: 5th International Conference on Computational Creativity (ICCC)
45. Schwering, A., Krumnack, U., Kuehnberger, K.U., Gust, H.: Syntactic principles of heuristic driven theory projection. Cognitive Systems Research **10**(3), 251–269 (2009)
46. Stein, W., Joyner, D.: Sage: System for algebra and geometry experimentation. ACM SIGSAM Bulletin **39**(2), 61–64 (2005)
47. Boy de la Tour, T., Peltier, N.: Computational Approaches to Analogical Reasoning: Current Trends, chap. Analogy in Automated Deduction: A Survey, pp. 103–130. Springer-Verlag, Berlin, Heidelberg (2014)
48. Turing, A.M.: On computable numbers, with an application to the entscheidungsproblem. Proceedings of the London mathematical society **2**(1), 230–265 (1937)
49. Turing, A.M.: Computing machinery and intelligence. In: Parsing the Turing Test, pp. 23–65. Springer (2009)
50. Voevodsky, V., et al.: Homotopy type theory: Univalent foundations of mathematics. Institute for Advanced Study (Princeton), The Univalent Foundations Program pp. 2007–2009 (2013)
51. Wang, H.: Toward mechanical mathematics. IBM Journal of research and development **4**(1), 2–22 (1960)
52. Wolfram, S.: The Mathematica book, wolfram media, 2003. Received: November **2** (2015)

Chapter 2
Some Basic Technical (Meta-)Mathematical Preliminaries for Cognitive Metamathematics

2.1 Introduction

In this chapter, we will introduce some classic logic and (meta-)mathematical terminology needed in several chapters of this book. So, the present chapter is mainly devoted to non-mathematicians (e.g., cognitive scientists, AI specialists) who want to acquire a minimal technical knowledge of some of the fundamental theories used implicitly along the AMI meta-program.[1] In this presentation, we will describe essentially foundational notions and results without proofs. It is worth to clarify that we offer in this chapter the minimal syntactic descriptions of most of the notions needed to get a better technical understanding of the initial applications of the AMI formal framework to the meta-generation of a wide spectrum of concepts in pure mathematics.[2]

2.2 Propositional and First-Order Logic

The main reference for this section is the classic treatise of E. Mendelson [11]. Propositional logic deals with one of the most simple ways of articulate deductive (semantic and syntactical) procedures among propositions. In other words, one generates recursively more complex propositions starting with atomic ones and using a suitable collection of logic connectives with some resemblance with natural language (e.g., "and" ($\wedge$), "if $\cdots$ then $\cdots$" ($\rightarrow$)). The possible truth value of

---

[1]Therefore, the working mathematician or logician can easily skip most of this chapter without any substantial lack of preparation for understanding the rest of this book.

[2]Hence, we recommend that the (non-specialist) reader consult also the references (better simultaneously) for a more complete and detailed presentation of the corresponding topics.

© Springer Nature Switzerland AG 2020
D. A. J. Gómez Ramírez, *Artificial Mathematical Intelligence*,
https://doi.org/10.1007/978-3-030-50273-7_2

a proposition is either "true" or "false" and the truth value of a compounded proposition depends completely on the truth value of these atomic components and of the (truth tables of the) corresponding logical connectives. Additionally, one possesses fixed deductive rules and (logic and proper) axioms. We will use the following concrete system for propositional logic in this book.

2.2.1 A Formal System for Propositional Logic

Explicitly, the basic symbols of our language are $\neg, \rightarrow, (,)$, (primitive connectives and parenthesis) and upper-case letters $A, B, \ldots$ (statement letters). As usual, a well-formed formula (wf) is defined recursively as follows: all statement letters are wfs and, if $\mathscr{A}$ and $\mathscr{B}$ are wfs, then $(\neg \mathscr{A})$ and $(\mathscr{A} \rightarrow \mathscr{B})$ are wfs.

Let us use the special symbols $\#_1, \#_2,$ and $\#_3$ to denote propositional variables ranging over all wfs. So, for any assignation of particular wfs on the former variables the following wfs are (logic) axioms of our system:

(A1) $(\#_1 \rightarrow (\#_2 \rightarrow \#_1))$
(A2) $((\#_1 \rightarrow (\#_2 \rightarrow \#_3)) \rightarrow ((\#_1 \rightarrow \#_2) \rightarrow (\#_1 \rightarrow \#_3)))$
(A3) $(((\neg \#_2 \rightarrow \neg \#_1) \rightarrow (((\neg \#_2) \rightarrow \#_1) \rightarrow \#_2))$

We have three axiom schemes in our formal system. The only inference rule that we use is modus ponens (MP), namely, we can deduce directly a wf $\#_2$ from the wfs $\#_1$ and $\#_1 \rightarrow \#_2$.

By a (formal) proof (in propositional logic) we will understand a constructive proof in the following sense: $\mathscr{W}$ is a syntactic consequence of $\Gamma = \{\mathscr{H}_1, \cdots, \mathscr{H}_n\}$ (i.e., $\Gamma \vdash \mathscr{W}$) if and only if there exists a collection of wfs $\mathscr{A}_1, \cdots, \mathscr{A}_m$; such that $\mathscr{A}_m = \mathscr{W}$, and for any j, $\mathscr{A}_j$ is either an exemplification of an axiom scheme, or one of the $\mathscr{H}_i$, or it is a direct consequence (by MP) of two of the formers $\mathscr{A}_k, (k < j)$.

We can speak also of an (indirect) "proof" of the fact that $\mathscr{T}$ is a consequence of the axioms of our system (i.e., $\mathscr{T}$ is a theorem, $\vdash \mathscr{T}$), in the sense of being able to verify that the wf $\mathscr{T}$ is a tautology (i.e., for any assignation of truth values of its atomic components, the resulting truth value is always true). Clearly, this is equivalent to the fact that there exists a formal proof of $\mathscr{T}$ in the former sense due to the completeness theorem for propositional logic, in other words, the notions of theorem (syntactic consequence) and tautology (semantic consequence) coincide [11, Ch. 1].

2.2.2 First-Order Logic

We want to extend the former (zeroth-order) logic system by including the possibility of expressing universal and existential quantifications of formal variables. For example, we wish to formalize sentences like "For all natural numbers x and y,

$x(x + 1) + y(y + 3)$ is an even number," which cannot be expressed non-trivially in propositional terms, due to the limited expressiveness of the language.

First, let us define the m.-s. part of a first-order theory. Explicitly, a *first-order language* L contains propositional connectives $\neg$, $\rightarrow$[3] and the universal quantifier symbol $\forall$;[4] (optionally) punctuation marks ":", parenthesis "(" and ")", comma ","; countably many (single) variables $y_1, y_2, y_3, \cdots$; a (possibly empty) finite or countable set of function letters (e.g., f_s^m) and individual constants (e.g., a, b); and a nonempty set of predicate letters (e.g., A_r^n).

The notion of *(atomic) term* is defined recursively starting with variables and constants; and applying these to function letters, i.e., $f_s^m(t_1, \cdots, t_m)$ is a higher level terms, where $t_1, \cdots, t_m$ are terms of lower (syntactic) level.

The notion of (well-formed (wf) *(atomic) formula* is defined also recursively starting with atomic formulas of the form $A_r^m(t_1, \cdots, t_m)$, where A_r^m is a predicate letter and $t_1, \cdots, t_m$ are atoms, i.e., if $\mathscr{C}$ and $\mathscr{D}$ are (lower level) formulas, then $(\neg\mathscr{C})$, $(\mathscr{C} \rightarrow \mathscr{D})$, and $((\forall y)\mathscr{C})$ are (the corresponding next higher level) formulas.

Let L be a first-order language, on the semantic level we define *an interpretation* N of L as the following collection: a set D (the domain of the interpretation); for any relation symbol A_s^m, one fixes $m-$ary relation $(A_s^m)^N$ of D (i.e., $(A_s^m)^N \subseteq D^m$); for each function letter f_r^n, one chooses a function $(f_s^n)^N : D^n \rightarrow D$; and for each single constant a_r, one fixes an element $(a_r)^N$ belonging to D. There exists a straightforward and unique way of extending the interpretation recursively to terms which only involve constant and functional symbols.

In this context, variables on L are thought of as ranging over D. Quantifiers and logical connectives possess the natural (contextual) meaning. A wf formula ϕ whose variables are all under the scope of a quantifier (i.e., ϕ has no free variables) is called a *sentence*. This kind of formulas represents either true or false statements with respect to N. For example, if ϕ is an atomic sentence of the form $A_s^m(t_1, \cdots, t_m)$, where the terms $t_1, \cdots, t_m$ do not involve any variable, but only constant and functional symbols, then ϕ is true for (the interpretation) N, if and only if the $m-$tuple $((t_1)^N, \cdots, (t_m)^N)$ belongs to $(A_s^m)^N \subseteq D^m$. In the case that the terms of ϕ have variables contained in $\{y_1, \cdots, y_q\}$ (more generally in $\{y_1, \cdots, y_q, \cdots\}$), we say that ϕ is true for N, if and only if for any countable sequence $s = (s_1, \cdots, s_m, \cdots)$ of elements of D, the corresponding sentence $A_s^m(s * (t_1), \cdots, s * (t_m))$ is true for N (or that s *satisfies* ϕ), where $s * (-)$ is the natural extension of the assignation s for arbitrary terms in L.

In the next step, one extends also recursively the former notion of satisfiability for arbitrary wf formulas. Using this notion, one expands the notion of truth for any wf formula, exactly as in the case of atomic ones. The standard notation for saying that ϕ is true for the interpretation N is $\models_N \phi$. One says that ϕ is false for N, if no

---

[3]One can additionally include $\vee$ and $\leftrightarrow$, or these connectives can be defined in terms of the former ones in terms of suitable meta-expressions.

[4]Again, the existential quantifier symbol $\exists$ can be included explicitly, or it can be defined in terms of the universal one, i.e., $(\exists x)P(x) :\cong \neg(\forall x)(\neg P(x))$.

(countable) sequence s (over D) satisfies ϕ. Additionally, if Ω denotes a collection of wf formulas over L, N is a *model* for Ω if and only if for any formula ψ in Ω, $\models_N \psi$. It can be proved that the former notions of satisfiability, truth, and falsehood fulfilled all the standard intuitive properties expected classically [11, Ch 2. §2.2.]. A wf formula ϕ in a language L is *logically valid* if it is true for any interpretation of L.

Let L be a first-order language. A *first-order theory* T is a formal theory with symbols and wf formulas taken from L (defined as before) and with (logical and proper) axioms and inference rules defined as follows:

Let $\#_1, \#_2$, and $\#_3$ denote (meta-)variables (i.e., they take wf formulas as specific possible values). Then, the *logical axioms* (which are, strictly speaking, meta-axioms since they involved meta-variables) are

(A1) $(\#_1 \rightarrow (\#_2 \rightarrow \#_1))$

(A2) $((\#_1 \rightarrow (\#_2 \rightarrow \#_3)) \rightarrow ((\#_1 \rightarrow \#_2) \rightarrow (\#_1 \rightarrow \#_3)))$

(A3) $(((\neg \#_2 \rightarrow \neg \#_1) \rightarrow (((\neg \#_2) \rightarrow \#_1) \rightarrow \#_2)))$

(A4) $(\forall x_i)\#_1(x_i) \rightarrow \#_1(t)$, where t is a term in L that is free for x_i in $\#_1(x_i)$ (i.e., no free occurrence of x_i in $\#_i$ lies within the scope of an universal quantifier involving a variable of t).[5]

(A5) $(\forall x_i)(\#_1 \rightarrow \#_2) \rightarrow (\#_1 \rightarrow (\forall x_i)\#_2)$, where all the (possible) occurrences of x_i in $\#_1$ are bounded.

Typically, any instantiation of a logical axiom is logically valid.

The *proper* axioms are the specific axioms of the particular theory which structure the semantic essence of the fundamental objects specified, e.g., sets, groups, fields.[6]

The *rules of inference* used are modus ponens ($\#_2$ follows from $\#_1$ and $\#_1 \rightarrow \#_2$ and generalization ($(\forall x_i)\#_1$ follows from $\#_1$).

One defines the notion of (formal) proof in a first-order theory similarly as in the propositional case, but using more (logic and (potentially) proper) axioms and adding an extra rule of inference, i.e., generalization. A wf sentence ϕ is called a *theorem* of a theory T if ϕ is a consequence of the axioms of T, in other words, there exists a (formal) proof in T for ϕ starting exclusively with proper axioms and/or (instantiations of) logical axioms.

Now, essentially all the standard deductive rules and meta-facts used by proving mathematical facts can be (meta-theoretically) derived from the former meta-facts (see for example [11, Ch. 2]).

The completeness theorem (Goedel) in this context extends the former version (in the context of propositional logic) as follows: let T be a first-order theory in a (well-orderable) language L, let Γ be a collection of wf formulas in T, and let ψ be another wf formula. Then, ψ is a semantic consequence of Γ, $\Gamma \models \psi$ (i.e., any

---

[5]Intuitively, one avoids degenerated cases where the replacement of the corresponding variable t would generate an intrinsically different wf formula, e.g., if the former condition is not fulfilled.

[6]In the next sections we will present several enlightening examples.

model M of Γ is also a model of ψ) if and only if ψ is a syntactic consequence of Γ, $\Gamma \vdash \psi$ (i.e., there exists a formal proof of ψ starting from wf formulas of Γ (and possibly instantiations of axioms)). In other words, in the former kind of theories the notions of semantic and syntactic consequence coincide [11, Ch. 2]. A theory T is *consistent* if one cannot derive syntactically a formal contradiction, there is no wf formula ϕ such that $\vdash_T \phi$ and $\vdash_T \neg\phi$. It is equivalent to the existence of a model for the theory T [11, Ch. 4].

A *many-sorted first-order theory* is an important variant of a first-order theory, where one has an additional collection of *sorts* in the language which is meant to be used in order to create a taxonomy on the particular range of each of the variables (and indirectly on the domain of definition of the relation and function symbols). This technical trick can be reconstruct by a classic first-order theory by defining basically an (explicit) unary relation symbol in the language for any sort. Thus, both notions are meta-equivalent. We will see an enlightening example of this kind of theories in Chap. 7.

2.3 Foundational Instantiations of First-Order Theories in Mathematics

2.3.1 Zermelo–Fraenkel Set Theory with the Axiom of Choice (ZFC)

One of the most used and famous foundational (first-order) theory for (a large part of) modern mathematics is the theory of sets, originally developed in a primitive form by Georg Cantor, and based on the proper axioms of Ernst Zermelo and Abraham Fraenkel, including the axiom of choice [8]. From a cognitive perspective, one of the biggest reasons why ZFC is so widely used is its wide easy-going appealing use of mental and intuitive images as part of the phenomenological way of understanding its main objects and the relations between them.

ZFC set theory is a first-order theory with a canonical membership (binary) relation ($\in$), so $a \in b$ is expressed as "a is an element of b" or "a belongs to b."[7] Let us describe the proper axioms of ZFC in a compact and intuitive way. In later chapters one can gain a deeper idea of the way in which these axioms can be written more formally. However, it is more enlightening for an initial presentation if we mainly appeal to intuition.

---

[7]One can also add the equality relation as a primitive relation (assuming the fulfillment of the standard properties as proper axioms) or one can define it in terms of the membership relation. Here, we assume the first variant to stress prominently the main intuition behind it and to avoid excessive technicalities.

1. **Axiom of Extensionality.** Two sets are equal when (and only when) both have exactly the same elements.
2. **Axiom of Pairing.** Given two sets a and b there exists a (unique) set $\{a, b\}$ having exactly a and b as its elements.
3. **Axiom of Union** For any set a there exists a set b containing exactly all the elements of a. This set is called the union of a and is denoted as $\cup a$.
4. **Axiom of Power Set** For any set a there exists a set containing as elements all the subsets of a. This set receives the name of power set of a and is denoted as $P(a)$.
5. **Axiom of Infinity** There exists a set with infinitely many elements.
6. **Axiom of Regularity** Any nonempty set a has an element e, such that a and e has no common elements.
7. **Axiom Schema of Separation** Let $Q(a, b)$ be a formula (describing an (unary) property where the free variable b is fixed, i.e., b is a parameter). Then, for any (fixed) set w and (fixed) parameter b, there exists a set v containing precisely the elements of w fulfilling Q, in other words, $v = \{a \in w : Q(a, b)\}$.
8. **Axiom Schema of Replacement** Let $\psi(x, y, p)$ be a formula that describes a function F with parameter p,[8] (i.e., for any sets x, y, and z, if $\psi(x, y, p)$ and $\psi(x, z, p)$ hold, then $y = z$. This unique element y is also denoted as $F(x)$). Thus, the image of any set u under ψ is a set. In other words, for any u there exists a set v such that $v = \{F(d) : d \in u\}$.
9. **Axiom of Choice** Any family of nonempty sets (which can be expressed as $\cup a$ for some set a) possesses a choice function f, i.e., $f : a \mapsto \cup a$ and for any $x \in a$, $f(x) \in x$.

2.3.2 Von Newmann–Bernays–Gödel (Class and) Set Theory (NBG)

In some mathematical (modern) theories (like category theory) one usually needs to construct notions involving the collection of all sets and (very large) sub-collections of it. Now, due to the fact that such collections do not represent sets anymore,[9] one needs a kind of suitable extension of ZFC that do not increase the potential for the existence of inconsistencies that ZFC already possesses.

In this sense, Von Newmann–Bernay–Gödel set theory (NBG) is a coherent candidate for a broader theory maintaining the "working feeling" very similar to the one in ZFC and allowing to talk about classes as natural (semantic) extensions of sets.

---

[8]In this axiom p can be replaced by several parameters $p_1, \cdots, p_n$. However, for simplicity we describe the version with only one.

[9]This is based on the existence of Paradoxes emerging from the assumption that such collections are sets, e.g., Russell's paradox [13].

Here we adopt essentially the approach presented in [11, Ch. 4]. NBG set theory is a first-order theory with a binary (enlarged) membership relation ($\in$) and a primitive notion of class. We define equality between classes in terms of extensionality: $A = B$ stands for $(\forall C)(C \in A \leftrightarrow C \in B)$.

A *set* is defined as a class that belongs to some other class. A class that is not a set is called a *proper class*. As a matter of terminology, we denote classes by upper-case letters (e.g., X, Y, and Z) and sets by lower-case letters. Let us describe the proper axioms of NBG set theory in a more formal way:

1. **Axiom T**

$$A = B \rightarrow (\forall C)(A \in C \leftrightarrow B \in C)$$

2. **Axiom P (Axiom of Pairing)**

$$(\forall a)(\forall b)(\exists c)(\forall d)(d \in c \leftrightarrow d = a \lor d = b)$$

3. **Axiom N (Empty Set)**

$$(\exists a)(\forall b)(b \notin a)$$

4. **Axiom F (Axiom of Regularity)**

$$(\forall a)(a \neq \emptyset \rightarrow (\exists b)(b \in a \land a \cap b = \emptyset)$$

5. **Axiom E1 (Set-theoretical membership Relation-Class)**

$$(\exists A)(\forall b)(\forall c)(\langle b, c \rangle \in A \leftrightarrow b \in c)$$

6. **Axiom E2 (Intersection of Classes (Conjunction))**

$$(\forall A)(\forall B)(\exists N)(\forall c)(c \in N \leftrightarrow c \in A \land c \in B)$$

7. **Axiom E3 (Complement of a Class (Negation))**

$$(\forall A)(\exists C)(\forall b)(b \in C \leftrightarrow b \notin A)$$

8. **Axiom E4 (Domain (Existential Quantifier))**

$$(\forall A)(\exists B)(\forall c)(c \in B \leftrightarrow (\exists d)(\langle c, d \rangle \in A))$$

9. **Axiom E5 (Product by the Universal Class)**

$$(\forall A)(\exists B)(\forall c)(\forall d)(\langle c, d \rangle \in B \leftrightarrow c \in A)$$

10. **Axiom E6 (Circular Permutation)**

$$(\forall A)(\exists B)(\forall c)(\forall d)(\forall e)(\langle c, d, e\rangle \in A \leftrightarrow \langle d, e, c\rangle \in B)$$

11. **Axiom E7 (Transposition)**

$$(\forall A)(\exists B)(\forall c)(\forall d)(\forall e)(\langle c, d, e\rangle \in A \leftrightarrow \langle c, e, d\rangle \in B)$$

12. **Axiom U (Union (Sum) Set)**

$$(\forall a)(\exists b)(\forall c)(c \in b \leftrightarrow (\exists d)(c \in d \wedge d \in a)$$

13. **Axiom W (Power Set)**

$$(\forall a)(\exists b)(\forall c)(c \in b \leftrightarrow (\forall e)(e \in c \rightarrow e \in a))$$

14. **Axiom S (Subsets)**

$$(\forall a)(\forall B)(\exists c)(\forall d)(d \in c \leftrightarrow (d \in a \wedge d \in B))$$

15. **Axiom R (Replacement)**
    Let V be the universal class containing all the sets. If A is a class, let $Fnc(A)$ be the formal statement saying that A is a function, i.e.,

$$A \subseteq V^2 \wedge (\forall a)(\forall b)(\forall c)(\langle a, b\rangle \in A \wedge \langle a, c\rangle \in A \rightarrow b = c).$$

Then

$$(\forall A)(Fnc(A) \rightarrow (\forall b)(\exists c)(\forall d)(d \in c \leftrightarrow (\exists e)(\langle e, d\rangle \in A \wedge e \in b))$$

16. **Axiom I (Axiom of Infinity)**

$$(\exists a)(\emptyset \in a \wedge (\forall b)(b \in a \rightarrow b \cup \{b\} \in a))$$

17. **Axiom G (Axiom of Global Choice)**

$$(\exists G)(Fnc(G) \wedge (\forall a)(a \neq \emptyset \rightarrow (\exists b)(b \in a \wedge \langle a, b\rangle \in G)))$$

We describe explicitly each of the former axioms for its particular foundational importance. However, the former list is, strictly speaking, non-minimal, i.e., some of the axioms can be deduced from some of the remaining ones. Nonetheless, we would not discuss this kind of technical issues in this very short presentation, mainly for pragmatic reasons.[10]

NBG set theory is a finitely axiomatizable theory, due to the fact that one can codify the membership and the equality relation for sets, together with the notions of intersection, complement, domain, and product (at the level of sets) as particular (binary) classes.

NBG set theory turns out to be an extension of ZFC set theory, without strictly bigger chances of generating formal contradictions. In other words, NBG is a conservative extension of ZFC, i.e., NBG extends formally a ZFC and one of them is consistent if and only if the other one so is [11, Ch,. 4].

A second fundamental working meta-principle assumed by a large portion of working mathematicians and logicians regarding ZFC is that one can simulate and ground almost any mathematical structure (e.g., concept and notion) used for example in mathematical analysis, abstract algebra, and (differential and algebraic) geometry (among many others) with set-theoretical structures. Notwithstanding, this theoretical meta-fact possesses more a platonic importance, because in the concrete mathematics done at a daily basis in research centers, it is pragmatically impossible to do the concrete and real grounding explicitly.

2.3.3 Peano Arithmetic

In this section, we will define the first-order theory needed for obtaining a modern syntactic formalization of the natural numbers in order to be able to establish the grounding framework for formal number theory.[11]

The *language of arithmetic* (in this formal context) includes a single predicate letter for equality ($=$), and individual constant (0) and three function letters f_1^1, f_1^2, and f_2^2, with the following conventions on notation $f_1^1(a) = a'$, $f_1^2(a, b) = a + b$, and $f_2^2(a, b) = a \cdot b$. Finally, the proper axioms are the following:

1. $(\forall y_1, y_2, y_3)(y_1 = y_2 \rightarrow (y_1 = y_3 \rightarrow y_2 = y_3))$
2. $(\forall y_1, y_2)(y_1 = y_2 \rightarrow y_1' = y_2')$
3. $(\forall y_1)(y_1 \neq 0)$
4. $(\forall y_1, y_2)(y_1' = y_2' \rightarrow y_1 = y_2)$
5. $(\forall y_1)(y_1 + 0 = y_1)$
6. $(\forall y_1, y_2)(y_1 + y_2' = (y_1 + y_2)')$
7. $(\forall y_1)(y_1 \cdot 0 = 0)$
8. $(\forall y_1, y_2)(y_1 \cdot y_2' = (y_1 \cdot y_2) + y_1)$
9. For any wf formula $\mathscr{D}$, $\mathscr{D}(0) \rightarrow ((\forall y)(\mathscr{D}(y) \rightarrow \mathscr{D}(y')) \rightarrow (\forall y)\mathscr{D}(y))$

---

[10]Our main goal here is to give a global view of the most foundational axiomatic aspects for mathematics, reviewing only some of the most seminal notions, axioms, and results.

[11]Here we adopt the terminology given, for instance, in [11, Ch. 3].

The last axiom is also known as the *principle of mathematical induction* (in one of its several forms). We explore the ontological, cognitive, and physical limitations of the former formalization implicitly in more detail in Chap. 5.

2.3.4 Categories

In the setting of ZFC set theory and from an intuitive point of view, the main objects of this theory (i.e., sets) are characterized in terms of their intrinsic features, e.g., their elements (axiom of extensionality). Even more, the working mathematician in order to be able to understand (geometrical and/or algebraic) properties of specific sets (e.g., manifolds, rings) used to consider internal properties involving their elements and their inner structure. Now, category theory emerged ontologically from subtle observations done in algebraic topology which motivated the quest for constructing a new formal framework based basically on a dual methodological position with respect to sets, i.e., extrinsicality [10]. Explicitly, in category theory a single object exists and is completely defined in terms of its relations with the other objects of the category (e.g., universal property). Moreover, from the explicit way in which categories are defined there are no concrete reference to an internal feature of its objects, in other words, "everything happens from a relational (external) point of view."

More formally, a *category* $\mathscr{C}$ consists of the following data:

1. A class $\mathrm{ob}(\mathscr{C})$ (in a lot of cases a proper class), whose elements are called *objects* of the category.
2. A class of morphisms $\mathrm{Hom}(\mathscr{C})$ defined as follows: for any objects $a, b \in \mathrm{ob}(\mathscr{C})$, there exists a set $\mathrm{Hom}_{\mathscr{C}}(a, b)$ of *morphisms from a to b*. An element $f \in \mathrm{Hom}_{\mathscr{C}}(a, b)$ is also denoted as $f : a \mapsto b$, and a is called the *source object* and b the *target object*. For every triple of objects $a, b, c \in \mathscr{C}$, there exists a *composition* function

$$\circ_{a,b} : \mathrm{Hom}_{\mathscr{C}}(a, b) \times \mathrm{Hom}_{\mathscr{C}}(b, c) \to \mathrm{Hom}_{\mathscr{C}}(a, c),$$

fulfilling the following properties (here the image of the pair (f, g) is denoted simply as $g \circ f$): for each object a there exists an *identity morphism* $id_a : a \mapsto b$, such that for any object b and any morphism $f : a \mapsto a$ and $g : b \mapsto a$, $f \circ id_a = g$ and $id_a \circ g = g$. And, the composition is associative, in other words, for any objects $a, b, c, d \in \mathscr{C}$, and for any morphisms $f \in \mathrm{Hom}_{\mathscr{C}}(a, b); g \in \mathrm{Hom}_{\mathscr{C}}(b, c)$ and $h \in \mathrm{Hom}_{\mathscr{C}}(c, d), h \circ (g \circ f) = (h \circ g) \circ f$.

One of the simplest and most fundamental examples of a category consists is the (universal) class, having sets as the object, functions between them, as the morphisms and set-theoretical composition of functions as the (categorical) composition. Usually, this category is denoted as **Set**.

One can compare categories by means of a special type of formal correspondence. Explicitly, let $\mathscr{C}_1$ and $\mathscr{C}_2$ be categories. Then a *(covariant) functor F* from $\mathscr{C}_1$ until $\mathscr{C}_2$ consists of a function (at the level of (proper) classes[12] from $\mathrm{ob}(\mathscr{C}_1)$ to $\mathrm{ob}(\mathscr{C}_2)$; and from $\mathrm{Hom}(\mathscr{C}_1)$ to $\mathrm{Hom}(\mathscr{C}_2)$; such that for any objects $a, b, c \in \mathrm{ob}(\mathscr{C}_1)$ and any morphisms $f : a \mapsto b$ and $g : b \mapsto c$, it holds $F(id_a) = id_{F(a)}$, $F(f) : F(a) \mapsto F(b)$, $F(g) : F(b) \mapsto F(c)$ and $F(g \circ f) = F(g) \circ F(f)$. F is a *(contravariant) functor* if under the same terminology, it holds $F(id_a) = id_{F(a)}$, $F(f) : F(b) \mapsto F(a)$, $F(g) : F(c) \mapsto F(b)$, and $F(g \circ f) = F(f) \circ F(g)$.

There is a secondary notion that will have a considerable importance in Chap. 7. Explicitly, let $\mathscr{C}$ be a category and let $f_1 : g \mapsto a_1$ and $f_2 : g \mapsto a_2$ be a pair of morphisms with common domain (such a pair is usually called a V−diagram (V_D)). Then, V_D has a *pushout* if there exits an object $c \in \mathscr{C}$ and two morphisms $h_1 : a_1 \mapsto c$ and $h_2 : a_2 \mapsto c$ such that for any object w and morphisms $j_1 : a_1 \mapsto w$ and $i_2 : a_2 \mapsto w$ such that $j_1 \circ f_1 = j_2 \circ f_2$, then exists a unique $h : c \mapsto w$ such that $j_1 = h \circ h_1$ and $j_2 = h \circ h_2$ (see Fig. 2.1). For the kind of categories that we will consider in Chap. 7, the former notion is a particular form of the more general notion called *colimit* (for the particular case of a (categorical) diagram of the form $\bullet \leftarrow \bullet \rightarrow \bullet$) which is a more technical notion [10, Ch. III].[13]

In the category **Set** the notion of pushout coincides with the operation of taking disjoint union identifying along the way all the elements with a common preimage (regarding the corresponding morphisms).

Fig. 2.1 Diagrammatic representation of the universal property of a pushout

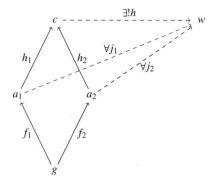

---

[12]In NBG set theory, one can prove the existence of general functions from classes to classes, even proper ones (see for instance [11, Ch. 4]).

[13]The purpose of this presentation is to give to the (non-specialist) reader the minimal intuitive and technical theoretical elements for a better understanding of the rest of the book, without saturating the chapter with an excessive amount of technicalities out of the context of this foundational book.

2.4 Further Seminal (Categorical and Set-Theoretical) Mathematical Notions

In this section, we will present additional mathematical concepts that will be used as test samples of the validity of our meta-formal taxonomy of fundamental cognitive mechanisms employed during abstract mathematical creation (see Chap. 10). Most of the following concepts can be seen in two ways: firstly, as instances of first-order theories, where the explicit conditions defining the structures corresponds to the proper axioms and, simultaneously, they give the essential elements of the (first-order) language in consideration; and, secondly, as categories, where the (proper) class corresponds to the collection of all models of the corresponding first-order theory, and the morphisms are given by the corresponding class of functions preserving the fundamental algebraic properties characterizing the corresponding mathematical structures (e.g., (group, ring) homomorphisms). So, the next conceptual examples have a double purpose, enlightening the former logic and (meta)mathematical notions, and preparing the way for a deeper understanding of the whole AMI program. For a more detailed description of the notions described in this section the reader can consult [4–6, 12] and [9, Ch5–6].[14] Note that the following notions are only required in Chaps. 7 and 11, the rest of the book can be read without an explicit knowledge of them.

Definition 2.1 A *relation r* (between a set a and a set b) is simply a subset of the Cartesian product between them, i.e., $r \subseteq a \times b$. A *(mathematical) function f* with domain a and codomain b is defined by a relation $f \subseteq a \times b$ such that for any $x \in A$ and $y, z \in b$, if $(x, y) \in f$ and $(x, z) \in f$, then $y = z$. If g is another function from b to c, then one defines the *composition function* $g \circ f$ from a to c, by the rule $g \circ f(x) := g(f(x))$, for any $x \in a$. A function $f : a \mapsto b$ is *injective* if for any $x, y \in a$, if $a \neq b$ then $f(a) \neq f(b)$; it is *surjective* if for any $z \in b$, there exists a $x \in a$ such that $f(x) = z$; it is *bijective* if is both injective and surjective.

Let a be a set. A relation $r \subseteq a \times a$ is an *equivalence relation* if it is *reflexive* (i.e., for all $x \in a$, $(x, x) \in r$), *symmetric* (i.e., for all $x, y \in a$, $(x, y) \in r$ if and only if $(y, x) \in r$, and *transitive* (i.e., for all $x, y, z \in a$, if $(x, y) \in r$ and $(y, z) \in r$, then $(x, z) \in r$). An equivalent relation r generates a partition of a into *equivalent classes* which are the subsets $c_w \subseteq a$ (where $w \in a$) such that for all $x \in a$, $a \in c_w$ if and only if $(w, a) \in r$.[15] The collection of equivalence classes is denoted as a/r and, sometimes, it is referred as the *quotient set* (where the relation should be clear from the context).

---

[14]Most of the notions are well-known structures in contemporary mathematics, others are less-known robust concepts which have an intermediate usage within the book, and, therefore, possess a local naming.

[15]Note that the element w is not canonical. In fact each element in c_w can also represent its equivalent class.

Definition 2.2 An abelian group is a set A with a binary operation $+$ and a special (neutral) element $0 \in A$ such that the following axioms hold:

1. $(\forall a \in A)(a + 0 = 0 + a = a)$.
2. $(\forall a \in A)(\exists b \in A)(a + b = b + a = 0)$.
3. $(\forall a, b, c \in A)((a + b) + c = a + (b + c)))$.
4. $(\forall a, b \in A)(a + b = b + a)$.

A is a group if it fulfills conditions 1–3.

The most elementary example of an (abelian) group are the integers $\mathbb{Z}$ with the addition operation and the zero element.

Definition 2.3 A pointed (abelian) group is a set B with a binary operation $*$ and a distinguished element $b \in B$ such that $(B \setminus \{b\}, *_{|B \setminus \{b\} \times B \setminus \{b\}})$ is an (abelian) group and, $b * c = c * b = b$ for all $c \in B$.

Well-known examples of pointed abelian groups are the rational, real, and complex numbers with the zero element and the product operation, respectively. Moreover, for any nonempty set with a distinguished element there exists at least one structure of pointed group for it. In fact, it could be shown that this statement is equivalent to the axiom of choice [7].

Definition 2.4 A distributive space consists of two sets D y K with two operations $\oplus : D \times D \to D$ and $\otimes : K \times D \to D$ such that

$$(\forall x \in K)(\forall y, z \in D)(x \otimes (y \oplus z) = (x \otimes y) \oplus (x \otimes z)).$$

Instances of distributive spaces are Boolean algebras, the space of square matrices with entries over a field (e.g., the real or complex numbers) with the standard sum and product operations, and clearly the natural, integer, rational, real, and complex numbers with addition and multiplication, respectively. In all these cases, $D = K$.

Contrastingly, we also obtain an example of a distributive space if $(D, \oplus)$ is a vector space over a field K and $\otimes$ denotes the corresponding scalar product. If $\dim D > 1$, then clearly $D \neq K$.

Definition 2.5 An action of a group $(G, +, 0)$ on a set X is simply a function $* : G \times X \to X$ such that the following two conditions hold:

1. $(\forall a, b \in G)(\forall x \in X)((a + b) * x = a * (b * x))$.
2. $(\forall x \in X)(0 * x = x)$.

Definition 2.6 An algebraic substructure $\mathbb{S} = ((A, +_A, 0_A), (B, +_B, 0_B), i : A \to B)$, consists with two sets, two binary operations defined over each of them, two special constants, and an (structural) embedding i fulfilling the following properties:

1. i is an homomorphism: $i(0_A) = 0_B$ and $\forall x, y \in A(i(x +_A y) = i(x) +_B i(y))$.
2. i is injective: $(\forall x, y \in A)(i(x) = i(y) \Rightarrow x = y)$.

3. $(\forall x \in B)(\forall y \in A)((x +_B i(y) = 0_B) \Rightarrow (\exists z \in A)(i(z) = x))$.

The last condition can be rephrased as follows: the "potential inverses" of elements of A, considered as elements in B, belong as well to A.

Usual examples of algebraic substructures are given by the natural injections $i_1 : \mathbb{Z} \to \mathbb{Q}$, $i_2 : \mathbb{Q} \to \mathbb{R}$, and $i_3 : \mathbb{R} \to \mathbb{C}$ (as well as the remaining meaningful combinations) with the addition operation and the zero element, respectively.

　　The main intuition of this definition is that when $(B, +_B, 0_B)$ has additionally an algebraic structure, as the one of a monoid, a semi-group, or a group, then $(A, +_A, 0_A)$ would automatically inherit the same structure.

　　This definition is a stronger notion than the one of embedding (i.e., an injective morphism) commonly used in the mathematical literature, since, in principle, sets A and B have a basic algebraic structure; e.g., we do not even require associativity for the corresponding operations. However, we impose the typical conditions for an embedding in (1) and (2) and additionally, we request condition (3) for including potential inverses of the smaller structure into itself. If we restrict ourselves to the category of monoids, semi-group, and groups, these two notions coincide, because we can prove that under these hypothesis, (3) would follow from (1) and (2).

Definition 2.7 If X denotes a set and F is a collection of functions from X to X, then a subset Y of X is called the space of fixed points of F, if

$$(\forall x \in X)((\forall f \in F)(f(x) = x) \leftrightarrow x \in Y).$$

Typical examples of spaces of fixed points appear in topology and in the setting of retractions between topological spaces [12].

Definition 2.8 A field is a set $(F, +, 0, *, 1)$, such that $(F, +, 0)$ and $(F \setminus \{0\}, *, 1)$ are abelian groups and the operation $*$ distributes with respect to $+$.

　　Canonical examples of fields are the rational, the real, and the complex numbers with the corresponding operations of addition and multiplication and the distinguished constants zero and one.

Definition 2.9 A bigroup is a set Q with two binary operations $+$ and $*$ such that $(Q, +, 0)$ is an abelian group and $(Q, *)$ is a pointed abelian group with distinguished element 0.

　　Examples of bigroups are the rational, real, and complex numbers with the standard operations and the zero element as distinguished constant in any case. In fact, let us show a concrete example of a bigroup which is not a field. Let us define in the group $(R = \mathbb{Z}/4\mathbb{Z}, +)$, the following second binary operation $*$: $a * b = 0$, if either $a = 0$ or $b = 0$. For the subset $R' = R \setminus \{0\}$, we define $*$, in terms of the following bijection $\phi : \mathbb{Z}/3\mathbb{Z} \to R'$, defined by $\phi(0) = 3, \phi(1) = 1$, and $\phi(2) = 2$. Here, we translate the addition in $\mathbb{Z}/3\mathbb{Z}$ to R' by means of ϕ. It is straightforward to show that $(R, +, *)$ is a bigroup. However, $1*(1+1) = 1*2 = 3$ and $1 * 1 + 1 * 1 = 2 + 2 = 4 = 0$, thus $1 * (1 + 1) \neq 1 * 1 + 1 * 1$. So, R is not a field. In conclusion, this notion represents additional mathematical concepts with new models, going beyond the ones of field theory.

Definition 2.10 A field extension E/F is just a pair of fields F and E, such that F is contained in E not only as a set, but also as a field; i.e., one can compute the binary operations in F by restricting the respective binary operations in E to the subset F.

Example of a field extension is $\mathbb{R}/\mathbb{Q}$, and $\mathbb{R}/\mathbb{C}$.

Definition 2.11 The group of automorphisms of a field E, denoted by $\mathrm{Aut}E$, is the collection of functions $\alpha : E \rightarrow E$, such that (1) α is a bijection; and (2) α is compatible with the binary operations of E; i.e., for any elements $a, b \in E$, $\alpha(a \oslash b) = \alpha(a) \oslash \alpha(b)$, where $\oslash$ denotes $+$ or $*$.

Definition 2.12 The group of automorphisms of a field extension E/F fixing the base field F, denoted by $\mathrm{Aut}_F(E)$, consists of the elements β of $\mathrm{Aut}E$, i.e., automorphisms of E, such that for all $a \in F$, $\beta(a) = a$, it means that β is the identity function when it is restricted to the base field F.[16]

Additionally, let us remember one of the "conceptual cornerstones" in commutative algebra and algebraic geometry [3], namely, the notion of commutative rings with unity, together with some of its most outstanding derived notions:

Definition 2.13 A set R with two binary operations $+$ and $*$, and two constants 0 and 1 is a commutative ring with unity if the following conditions hold:

1. $(\forall a \in R)(a + 0 = 0 + a = a)$
2. $(\forall a \in R)(\exists b \in R)(a + b = b + a = 0)$
3. $(\forall a, b, c \in R)((a + b) + c = a + (b + c)))$
4. $(\forall a, b \in R)(a + b = b + a)$
5. $(\forall a \in R)(a * 1 = 1 * a = a)$
6. $(\forall a, b, c \in R)((a * b) * c = a * (b * c)))$
7. $(\forall a, b \in R)(a * b = b * a)$
8. $(\forall a, b, c \in R)(a * (b + c) = a * b + a * c)$

In other words, $(R, +)$ is an abelian group, $(R, 1)$ is a *commutative monoid*, and $*$ distributes with respect to $+$.

A subset $I \subseteq R$ is called an *ideal* of R if it satisfies the following axiom:

$$(\forall i, j \in I)(\forall a, b \in R)((a + j = 0 \rightarrow i + a \in I) \wedge b * i \in I).$$

We also define

$$\mathrm{Id}(R) = \{I \subseteq R : I \text{ is an ideal of } R\}.$$

---

[16]This last concept is one of the most fundamental in Galois theory, since $\mathrm{Aut}_F(E)$ is exactly the Galois Group of E/F, when this extension is Galois [9, Ch. 6,§1].

Ideals can be multiplied together using the following definition:

$$I \cdot_\iota J = \left\{ \sum_{k=1}^{n} i_k \cdot j_k : n \in \mathbb{N} \wedge i_1, \ldots, i_n \in I \wedge j_1, \ldots, j_n \in J \right\}.$$

Besides, an ideal P is a *prime ideal* if for any element $a, b \in R$ if $a * b \in P$, then $a \in P$ and $b \in P$.

The collection of all prime ideals of R is called the *prime spectra of* R and is denoted as $\mathrm{Spec}(R)$.

The prime spectra of R can be seen as a topological space where the open sets are simply the complements of collections of ideals containing a fixed ideal I_r (this set is denoted for example as $V(I)^c$). This is called the *Zariski topology for* $\mathrm{Spec}(R)$.

A pair (R, m) is called a *local ring* if R is a commutative ring with unity and m is the only ideal that is maximal, i.e., $m \neq R$ and m is maximal among all the other ideals of R fulfilling this property (i.e., being proper ideals).

Let R_1 and R_2 be commutative rings (with unity), then an *homomorphism* $f : R_1 \mapsto R_2$ is a function such that $f(0_1) = 0_2$, $f(1_1) = 1_2$ and for all $x, y \in R_1$, $f(x +_1 y) = f(x) +_2 f(y)$ and $f(x *_1 y) = f(x) *_2 f(y)$. If f is bijective it is called an *isomorphism* and the rings are *isomorphic*. Let S be a subset of R_1. Then S is called a *sub-ring of* R_1 if $0_1, 1_1 \in S$ and $(S, 0_1, 1_1, (+_1)', (*_1)')$ is a commutative ring with unity, where $(+_1)'$ and $(*_1)'$ are the restrictions of the operations to $S \times S$, respectively.

A commutative ring with unity is a *multiplicative ring* (or *containment-division ring*) if for any ideals I, J of R, $I \subseteq J$ if and only if J divides I, i.e., there exists another ideal H of R such that $I = J \iota H$. A commutative ring with unity R is a *Dedekind domain* if R is an integral domain (i.e., $a * b = 0$ implies that $a = 0$ or $b = 0$) and any ideal of R can be written as a finite product of prime ideals.[17]

A subset M of a commutative ring with unity R is a *multiplicative system* if $1 \in M$ and for all $x, y \in M$, $x * y \in M$. In this case, one defines the *localization of* R *at* M is the quotient set (denoted as) $M^{-1}R$ defined through the equivalence relation $\subseteq R \times M$, defined by $(r_1, m_1) (r_2, m_2)$ if there exists a $r \in M$ such that $r(r_1 * m_2 - r_2 * m_1) = 0$. One can see that the localization inherits the structure of a commutative ring with unity with the induced operations. In the case of choosing an ideal I of R instead of a multiplicative system M, one defined the quotient ring R/I by means of the following equivalence relation on R: for all $x, y \in R$, $x \ y$ if and only if $x - y \in I$. Again $R\iota$ possesses the derived structure of commutative ring with unity in the natural way.

Let $\{x_1, \cdots, x_n\}$ be a collection for formal symbolic units (also known as variables) and let R be a commutative ring with unity. The *ring of polynomials over* R *with finitely many variables* $R[x_1, \cdots, x_n]$ is the unique commutative ring

---

[17]This is one of the many equivalences that the reader can find of this notion in the mathematical literature.

with unity having R as sub-ring and extending the addition and multiplication to the variables symbolically.

A *finitely generated algebra over a field k* is a commutative ring with unity that is isomorphic to a quotient of a polynomial ring over k with finitely many variables.

Let W be a subset of $k[x_1, \cdots, x_n]$, let $\mathbb{A}^n(k) := \{(a_1, \cdots, a_n) : a_1, \cdots, a_n \in k\}$ be the *affine space* (of dimension n over k). Then, an *algebraic set* (over k) V is a set of the form

$$V = V(W) := \{(a_1, \cdots, a_n) \in \mathbb{A}^n(k) : g(a_1, \cdots, a_n) = 0, \text{ for all } g \in W\}.$$

Dually, if $V \in \mathbb{A}^n(k)$ is an algebraic set, then the *ideal of polynomials associated to V* is defined as

$$I(V) := \{f \in k[x_1, \cdots, x_n] : f(b_1, \cdots, b_n) = 0 \text{ for every } (b_1, \cdots, b_n) \in V\}.$$

Moreover, the *ring of coordinates (of the algebraic set)* V is defined as

$$k(V) := k[x_1, \cdots, x_n]/I(V).$$

Let us define the seminal conceptual "brick" of general (point-wise) topology; i.e., the notion of topological space [12].

Definition 2.14 A *topological space* is an ordered pair (X, T), where X is a set and T is a collection of subsets of X (called *open sets*) satisfying the following axioms:

1. $\emptyset \in T$.
2. $X \in T$.
3. For all $A, B \in T, A \cap B \in T$.
4. For all family of sets $\{U_\alpha : \alpha \in I\}$, if $U_\alpha \in T$ for all $\alpha \in I$, then $\bigcup_{\alpha \in I} U_\alpha \in T$.

One of the most important instances of a topological space is the real line with the collection consisting of arbitrary unions of open intervals.

A topological space (X, T) can be seen as a category whose objects are the open sets and the arrows are giving by the natural inclusions if the source open is contained in the target open, and by the empty function otherwise.

Definition 2.15 A subset B of open sets of a topological space (X, T) is called a *base for T*, if the following conditions hold:

1. For all $U \in T$ and for all $u \in U$, exists a $W \in B$ such that $u \in W$ and $W \subseteq U$.
2. For all $U, V \in T$ and for all $u \in U \cap V$, exists $W \in B$ such that $u \in W$ and $W \subseteq U \cap V$.

A canonical base for the real line with the (standard) former topology consists of all the open (bounded) intervals.

Definition 2.16 Let (X, T) be a topological space and let $\mathscr{C}$ be a category. A ($\mathscr{C}$-valued) *pre-sheaf* F on X is a contravariant functor from X (viewed as a category) and $\mathscr{C}$.

Definition 2.17 A *sheaf with values on the category of sets* is a (**Set**-valued) pre-sheaf F fulfilling the following condition: Let $\{U_i\}_{i \in I}$ be an open covering of an open set $U \in X$. Assume that for each $i \in I$ there exists a (section) $f_i \in F(U_i)$ such that for any $i, j \in I$,

$$f_i \mid_{U_i \cap U_j} := F_{i_1 : U_i \cap U_j \mapsto U_i}(f_i) = F_{i_2 : U_i \cap U_j \mapsto U_j}(f_j) = f_j \mid_{U_i \cap U_j}.$$

Then, there exists a unique global section $f \in F(U)$ that reconstructs all the (local) sections f_i, for all $i \in I$, i.e.,

$$f \mid_{U_i} = F_{i : U_i \mapsto U}(f) = f_i \text{ for all } i \in I.$$

A quite enlightening example of a sheaf is the collection of real-valued continuous functions from open real sets, together with the restriction maps induced by the inclusions.

For the last and more sophisticated notions used in Chap. 11 like ringed spaces, stacks of a sheaf and (affine) schemes the reader may consult, for instance, [6], due to constraints of space in this presentation.

Remark 2.1 Implicitly, we assume throughout the book also a standard and pragmatic understanding (by the reader) of the former numerical systems (e.g., $\mathbb{N}, \mathbb{Z}, \mathbb{Q}$ and $\mathbb{R}$) as well as of the elementary notion of set, the membership relation, and the corresponding basic operations (e.g., union, intersection, power set and complement, among others). Additionally, the formal understanding of a unique set possessing no elements at all will be also assumed; i.e., the empty set $\emptyset$.

More explicitly, the natural numbers ($\mathbb{N}$) will be understood as the unique "algebraic structure" satisfying the (higher-order) Peano axioms (see Sect. 2.3.3). In other words, $\mathbb{N}$ is essentially the only countable well-ordered commutative monoid used for "counting" concrete as well as purely abstract entities, among others [11, Ch.3].[18]

The integers ($\mathbb{Z}$) is the uniquely defined algebraic structure emerging by adding formal additive inverses of natural numbers called the negative numbers. In particular, $\mathbb{Z}$ fulfills the axioms of a commutative ring with unity. Moreover, the elementary algebraic substructures of $\mathbb{Z}$ will be assumed, e.g., the set of even numbers, and more generally the structure(s) consisting of multiples of a fixed integer m.[19] Elementary finite structures derived from the former ones are also assumed, e.g., the integers "modulo" 27, (i.e., $\mathbb{Z}/(27 \cdot \mathbb{Z})$ is a quotient ring), (resp. modulo m). In addition,

---

[18]Moreover, concerning the natural numbers, almost all the sections in Chap. 5 can be seen as a multidisciplinary "enlightenment" of our standard perceptions of this ancient numerical system.

[19]This property describes, in fact, all the ideals within the integers.

we also assume a working understanding of the notion of prime number over $\mathbb{Z}$ (or even in a more general structure possessing a suitable binary operation with a neutral element and a divisibility relation).

The rational numbers ($\mathbb{Q}$) are understood as the symbolic fractions constructed by adding the multiplicative inverses of non-zero integers.

The real numbers ($\mathbb{R}$) are understood as the unique complete ordered field containing "a (standard) copy" of the rational numbers. Additionally, $\mathbb{R}$ has associated a fixed collection of subsets of it consisting of (arbitrary unions of) open balls of the form

$$B_r(a) := \{x \in \mathbb{R} : |x - a| < r\},$$

for any $r, a \in \mathbb{R}$, with $r > 0$. This is one of the most fundamental and basic constructions done with the real numbers in order to set the foundational setting of standard calculus [2].[20]

We assume a pragmatic "familiarity" with quite elementary spaces of real-valued functions defined over open real sets, e.g., constant and locally constant functions, and with the real unitary circle.

Finally, a basic knowledge of the simple classic structure of the real projective plane is assumed (see for instance [1] for simple models of this classic structure).[21]

As mentioned at the beginning, we encourage the (non-specialist) reader to take the time to really obtain a working understanding of the former notions, also with the help of the references and additional material for being able to appreciate better the meta-generations (of most of the notions) presented in Chap. 12.

References

1. Apéry, F.: Models of the real projective plane. Friedr. Vieweg & Sohn (1987)
2. Bloch, E.D.: The real numbers and real analysis. Springer-Verlag New York (2011)
3. Eisenbud, D.: Commutative Algebra with a View Toward Algebraic Geometry. GTM. Springer-Verlag (1995)
4. Fraleigh, J.B.: A first course in abstract algebra. Pearson Education India (2003)
5. Görtz, U., Wedhorn, T.: Algebraic Geometry: Part I: Schemes. With Examples and Exercises. Springer (2010)
6. Hartshorne, R.: Algebraic Geometry. Springer-Verlag, New York (1977)
7. Horward, P., Rubin, J.E.: Consequences of the Axiom of Choice, vol. 59. Mathematical surveys and monographs, American Mathematical Society, USA (1998)
8. Jech, T.: Set theory. Springer Science & Business Media (2013)

---

[20] In some of the sections of Chap. 11, we will employ both structures: one consisting only of open balls and one consisting of arbitrary unions of open balls.

[21] The former mathematical structures are elementary and intuitive at the level of freshers subscribed in math-related programs.

9. Lang, S.: Algebra (revised third edition). Graduate Texts in Mathematics 211, Springer-Verlag, New York (2002)

10. Mac Lane, S.: Categories for the working mathematician, *GMT*, vol. 5. Springer Science+Business Media, New York (2013)

11. Mendelson, E.: Introduction to Mathematical Logic (Fifth Edition). Chapman & Hall/CRC (2010)

12. Munkres, J.: Topology. Second Edition. Prentice Hall, Inc (2000)

13. Tait, W.W.: Cantor's grundlagen and the paradoxes of set theory. Between Logic and Intuition: Essays in Honor of Charles Parsons (ed. G. Sher and R. Tieszen) pp. 269–290 (2000)

Part I
New Cognitive Foundations for Mathematics

Chapter 3
General Considerations for the New Cognitive Foundations' Program

3.1 General Introduction

All the wonderful and inspiring mysteries living inside the mathematical world, together with all its astonishing (structural) relations with the foundational mechanisms of different phenomena in nature, are one of the most beautiful and precise products of thousands of human minds throughout thousands of years on intensive and methodical research. During the last century, the main discoveries in psychology, psychiatry, neurobiology, artificial intelligence, and cognitive science have shown us that, although each human being is unique in their anthropological and biological dimension, there exist universal laws and principles governing and structuring any human mind independent of its origin and time of existence on earth.

In particular, from the former two facts we can conclude that there exist universal principles shaping and structuring mathematical creation/invention, and implicitly mathematics as a whole. Effectively, if we see mathematical generation (and therefore important aspects of what mathematics represents) simply as a concrete product of human cognition, then we can affirm that the cognitive engine called the human mind is a central materializer of mathematical structures in nature, independently if such structures exist in additional parts of the universe. In fact, for the purposes of this work, we can see human minds as highly effective processing engines that take input in the form of sensory-motor and phenomenological information and produce (descriptions of) mathematical structures, among many other things.

Thus, the intellectual enterprise of grounding mathematics as a whole should be structurally related to the nature (e.g., meaning and form) that the entities produced by the mind possess. This unique kind of phenomenological nature is formed through the fundamental cognitive mechanisms and principles used by the mind, together with all the physical mechanisms and laws that influence them from a (neuro)biological perspective. Therefore, it is necessary that any modern foundational program for mathematics includes seminal inputs not only from

© Springer Nature Switzerland AG 2020
D. A. J. Gómez Ramírez, *Artificial Mathematical Intelligence*,
https://doi.org/10.1007/978-3-030-50273-7_3

logic, mathematics, and metamathematics but also from disciplines like cognitive sciences, psychology, physics, (neuro)biology, and Artificial Intelligence, among others.[1]

The *New Cognitive Foundations for Mathematics* is a program for grounding mathematics from a inter- and multidisciplinary perspective. The need for a new kind of "colorful" methodological and global framework for mathematical creation/invention and for mathematics as a whole is based on the fact that virtually all the foundational programs for mathematics that have been proposed throughout centuries of (meta)mathematical inquiry tend to be more mono-disciplinary than multidisciplinary (i.e., they involve essentially at most two scientific disciplines, and in most of the cases one, as main grounding focus). Explicitly, some of the most outstanding programs and views with this kind of restricted disciplinary spectrum are Platonism [3], constructivism (e.g., intuitionist) [27], (neo)logicism [26, 31], formalism [30], purely cognitive approaches [1, 8, 14, 21],[2] synthetic views [35]; mathematical cognition [6]; mathematical universe [25]; Quine's new foundations [19]; reverse mathematics [23]; (strict) finitism [34]; and (some parts of) automated reasoning [20], proof theory [5], and model theory [15], among many others [22]. The general vision of the present foundational program is to be able to combine and subsequently fuse seminal aspects and methods of most of the former single approaches in order to get a more robust and solid grounding meta-theory of mathematical generation and mathematics in general.

Summarizing our initial motivation, we can affirm that one of the central principles for developing these new foundations emerges from the observation that, independently of the degree of geniality that some mathematical theories can possess, they is always a product of the finite and bounded human cognition, which at the same time is immersed in our immense, but finite, universe.

The foundational relation between the physical realm and the nature of mathematical structures (e.g., concepts and proofs) is deeper than one can perceive at first glance. In fact, the history of science has seen a lot of inspirational breakthroughs in mathematics coming originally from research done in theoretical and experimental physics (see, for example, [2, 32, 33] and [24]).

On contrary, how strong is the influence that formal physical (and computational) frameworks and principles have when explaining and grounding mathematics? This classic question starts to gain importance in current research because the most essential results in cognitive sciences are supporting (more and more) the fact that

---

[1]Here it is important to clarify that, implicitly, cognitive science includes, in its original and modern approaches, results and methods from psychology, (neuro)biology, computer science, anthropology, and AI, among others. However, it is always valuable to explore new connections that cognitive science, as a well-established discipline, can have with the latest results of the former disciplines for understanding (the fundamental principles of) the mind, and therefore of mathematical research. Therefore, all the other scientific disciples are explicitly mentioned, although they are partially included in cognitive science.

[2]This specific approach can be included as a part of the classic cognitive foundations for mathematics, introduced at the same time in [14].

a lot of aspects of human mathematical creation/invention are susceptible to be modeled computationally [11, 12].

3.2   Essential Aspects of the New Cognitive Foundations Program

This program encompasses the following essential parts:

First, the discovery and subsequent formalization of improved deductive systems (e.g., logics) that can explain in a more pragmatic way how our minds make inferences involving mathematical content. In particular, a carefully multifaceted revision of Zermelo–Fraenke set theory with Choice (ZFC) [16] is highly desired, due to the fact that ZFC is considered one of the most fundamental (classic) theories for grounding (most parts of) modern mathematics. For instance, what are the strengths and weaknesses that ZFC, as a first-order theory, possesses from a cognitive and physical point of view? To what extent are the logical operators of negation and disjunction understandable by the mind in comparison with conjunction and affirmation? In what way does the method of proof by contradiction substantially limit the deductive scope of the mind regarding mathematical invention? How could a sound framework be developed for defining and effectively proving the existence of concrete mathematical models without relying on the consistency of any underlying (FOL) syntactic theory?

For the last question, it is necessary to take a precise and detailed study of the cognitive plausibility of Gödel's Completeness and Incompleteness theorems [9]. In fact, regarding the completeness theorem we could ask the following questions: should the semantic soundness of our (new) foundational (metamathematical) frameworks not depend entirely on the existence of more fundamental (physical) entities of our universe, instead of depending structurally on the deductive consistency of them? Dually, should the deductive consistency of our (new) foundational frameworks for mathematics not essentially rely on the underlying pragmatic consistency of the mind, as inferential agent, and beyond the existence of abstract (and again cognitively created) mathematical models?

Note that here we are assuming the well-known validity of these classic metafacts and, simultaneously, we are focusing on its limitations from a cognitive and a physical point of view. Similarly, regarding the incompleteness theorems, we could ask the question: to what extent are the existence of statements that are simultaneously truth and unprovable in ("consistent") systems including the arithmetic, as well as the proof of their own consistency, are directly based on a nonphysical fact (e.g., a meta-physical/mental statement), i.e. the (formal) "existence" of the natural numbers? In other words, should the whole deductive coherence and soundness of our current seminal deductive frameworks (e.g., first-order logic and

ZFC) rely on the existence of an entity that possesses no physical or cognitive counterpart at all (i.e., the natural numbers?).[3] [4]

By contrast, we note that although the importance and technical beauty of the completeness and incompleteness theorems is evident, from a purely pragmatic point of view, we estimate (as we mentioned in the introductory chapter) that at least 90% of the mathematics generated by working mathematicians and related researchers involves solvable and decidable questions.[5] So, from the perspective of AMI, such meta-limitations turn out to have a smaller practical effect, mainly because we pursue to effectively simulate the mind as a mathematical engine from a cognitive metamathematical point of view. This means, in particular, that future versions of a UMAA would be able to simulate the way in which our minds generate even (meta-)proofs of fact, for example, the former classic Gödelian results, which have a quite mathematical intrinsic nature in their methods.

Second, the development of a more physical and cognitive grounded notion of mathematical model, whose existence should be supported more by fundamental interdisciplinary laws of nature (e.g., physical, cognitive, biological laws), and less by syntactic constructions and purely mental objects [15]. For example, let us choose one of the most fundamental notions for mathematics: set in the context of ZFC, or more generally, in the context of Von Newmann–Bernays–Gödel set theory (NBG) [16, Ch.4]. Although this notion is quite familiar and seminal for any mathematician, the concrete existence of such objects turns out to be not so clear and generally based on syntactic and logic considerations (like the completeness theorem), which reduces the answer of a quite pragmatic ontological question to a more meta-physical, agnostic issue. Furthermore, the basic intuitions (and elementary manipulations) that we have (and make) about sets as collections of (existent physical) objects provide us a stronger feeling of certainty about their existence than the kind of unprovable dimension given by a purely logical approach. Now, let us explore in deeper detail the ontological limitations of the classic line of argumentation about the fact that one cannot prove inside ZFC (or even NBG) the existence of sets due to the fact that it is equivalent to the consistency of ZFC (resp. NBG).

Effectively, let us consider the notion of proper class (in the context of NGB), which represents "huge" objects that working mathematicians would usually avoid due to fact that they could possess paradoxes (for instance, versions of the classic Russell's paradox) and would provide not so clear intuitions about practical (e.g., small) mathematical objects. From a pragmatic, ontological (and even physical) point of view, the existence of sets is a more evident fact than the existence of

---

[3]The interested reader can take a look into Chap. 5 concerning the physical numbers, which can be seen as a formal refinement of the natural numbers with a stronger cognitive and physical basis.

[4]At this point, it is worth noting that the incompleteness theorems would need, as one of its minimal hypothesis, the existence of infinity structures (e.g., an infinity tree), which are again essentially non-physical statements [4].

[5]See the corresponding footnote on Sect. 1.1, Chap. 1.

proper classes. However, following the classic logic guidelines, we can prove that there is a meta-isomorphic version of standard mathematics[6] grounded on objects called "seds," which are a special kind of proper class (see Chap. 6). This dual mathematics, that we call "Dathematics," is from a syntactic and deductive point of view exactly the same as the daily mathematics that most mathematicians do. Nonetheless, from a semantic perspective, it is based on quite different objects that indirectly force our minds to do more meta-physical representations to gain some heuristic intuition about them, due to the fact that there are no physical entities characterizing any of them from a quantitative point of view.

The former meta-fact is just an example showing the clear restrictions that a kind of mono-disciplinary approach to metamathematics can have. It also reminds us that a lot of basic and central questions about the foundations of mathematics are still open and needed to be answered from a wider methodological perspective.

The former two components of the AMI program constitute, in more general terms, the development of a global, cognitive, and pragmatic new logic that can offer a general explanatory framework of mathematical creation/invention viewed, among others, as an innovative process of the mind. The main reason is that the classic logic frameworks used in mathematics like first-order (and higher-order) logic are too narrow to be able to explain human reasoning as embedded in a spatio-temporal context, which is essentially the case of mathematical generation.[7]

In other words, those well-known logic frameworks from their very origin did not take into consideration any solid formal account of human cognition in their initial development because, on the one hand, there were no such (cognitive) general accounts in the literature of that time, and, on the other hand, they aimed to give a purely metamathematical answer based on (mono-disciplinary) (classic) philosophical and mathematical considerations [10, 13, 16].

Third, the development of improved versions of the most fundamental numerical systems like the natural, integer, rational, real, and complex numbers. In this case, such refinements should fulfill, as before, stronger ontological, pragmatic, physical, and cognitive requirements. At the same time, these new numerical systems would allow us to simulate and re-represent coherently the essential and practical features that our current numerical constructions possess. In Chap. 5, we initiate this quest with the development of the "physical numbers," which can be seen as a formal refinement of the natural numbers in the multidisciplinary direction of our new cognitive foundational program. In particular, we classify the fundamental formal features and intuitions about the (so called) natural numbers into physically supported parts (i.e., the physical numbers), and meta-physical ones (i.e., purely mental entities).

---

[6]Standard mathematics in this context means all the standard theories that we can ground based on ZFC set theory, which would represent, let us say, more than 95% of modern mathematics.

[7]For more details about the strong methodological limitations of classic logic frameworks for the construction of more general artificial intelligent agents, the reader may consult [28, 29].

Fourth, an important aspect worth analyzing for these new cognitive foundations is a carefully multifaceted revision of the way in which we represent morphological-syntactically mathematical structures, proofs and sketches about them, and how much more effective such representations could be for the sake of enlightening the creative process of the researcher's mind. A relevant related question is the following: to what extent does the current mathematical notation[8] fulfill minimal properties of uniqueness and contextual freedom with regard to being suitable to be integrated into a bigger artificial framework simulating human-style deductive mechanisms involving mathematical creation/invention? If we take a small historical tour through the way in which the notation used in mathematics has evolved since Euclid until the present time, we immediately see that the gradual accumulation of tiny improvements only in the graphic configurations used for denoting mathematical structures and deductive inferences has empowered considerably our effectiveness for understanding, manipulating, and proving quite sophisticated facts in virtually any mathematical discipline (see, for example, [17, 18]).

Fifth, the generation of a complete list of the central cognitive mechanisms used by our minds during mathematical creation, which, in principle, form a finite collection of mechanisms. This is due to the bound nature of our conscious and unconscious mind. These mechanisms underline and shape in many aspects basic features of the way in which we make logical inferences during mathematical research. In fact, in the second part of the book, we present an initial taxonomy for such a collection of mechanisms together with the respective formalizations. In Chap. 11, we show explicitly with plenty of examples how they are used for grounding conceptual and deductive creation in several mathematical areas.

Sixth, to be able to obtain global heuristic principles and meta-results that allow us to ground (virtually) any mathematical result and structure in terms of the new formal ontology developed before.[9]

Seventh, there is a collection of (classic) concepts and methods, which deserve to be independently reviewed and eventually refined from this multidisciplinary "optic" due to their outstanding importance not only in mathematics but (in a lot of cases) in physics and related fields. Instances of such entities are the notions of continuity, differentiability, infinite set, large and inaccessible cardinal, power set, infinitesimal, geometrical figure, real and complex number, formal negation, and the method of proof by the sake of contradiction, among others.

As mentioned before, it would be useful to find refinements of the former notions and methods; not only more physically-accessible and computationally-feasible but also possessing minimal structural requirements for modeling (at least) as much (physical and) mathematical phenomena as their classic counterparts.

---

[8]Here we mean the formal as well as the informal notation that we use for getting intuitions and for subsequently formalizing (mathematical) arguments.

[9]For more about this, see Chap. 12, where we describe the most fundamental future challenges of the AMI (meta-)program.

Eighth, the development, refinement, and subsequent integration of classic and new (interactive) computer programs that can codify, process, and simulate in a human-style way this new (cognitive) syntactic and semantic framework.[10]

Ninth, the generation of a new and enhanced notion of mathematical proof possessing a deeper cognitive soundness (see Sect. 3.3 in this chapter for more details).

3.3   The Cognitive Substratum of a Mathematical Proof

Throughout hundreds of years, our minds have looked for general arguments, proofs, and (abstract and concrete) evidence to prove formal statements of a wide range of mathematical disciplines (e.g., number theory, calculus, topology, and geometry). Within that time, the concept of what "an argument," "a proof," and "(abstract and concrete) evidence" (e.g., a counterexample) are has evolved considerably, and currently we count on precise meta-notions to helps us distinguish between informal intuitions and formal and valid arguments, at least from a purely theoretical perspective [5, 15, 16]. Nonetheless, it happens again and again that outstanding researchers (in mathematics and related fields), in their effort to solve important (abstract) problems, find new kinds of argumentative frameworks that push strongly the prevailing notions of proof and counterexample, subsequently forcing us to re-consider our methods of demonstration more deeply.[11] Therefore, if we really want to uncover the mystery behind the generation of mathematical proofs, we should focus our attention more closely on the invisible machine behind them (i.e., the human mind). Simultaneously, we need to be flexible and open enough to renew the "present" formalization of such a seminal notion (i.e., proof) with regard to new insights and results about the technical working processing of the mind.

In particular, let us do a brief cognitive meta-analysis of one of the most well-known notions of formal mathematical proof that we use today in mathematical research, i.e. a proof as a sequence of formulas obtained gradually starting with some fixed axioms, axiom schemes, and fixed hypothesis by applying valid inference rules (like modus ponens and generalization) for obtaining a desired thesis (described again as a formula) [16, Ch.2].

First, it is a pragmatic fact of mathematical practice that almost no mathematical proof written in a mathematical journal explicitly fulfills all the requirements presented before. The main reason for that is the working mathematician used to have a central focus in presenting the "key ideas," "the essence of the proof," instead of writing down all the (primary and secondary) details. What is more doing a complete description of all the inferences and instantiations of the axioms

---

[10] Again, the reader can find more information in Chap. 12.

[11] For a more detailed collection of such researchers, together with general considerations about the impact of their work, see [35].

starting from a (chosen) foundational framework (e.g., ZFC) would be basically an impossible, non-enlightening task.

Some of the logical operations that are implicitly used in such a notion like negation ("¬") and disjunction ("∨") are, from a cognitive perspective, more artificial to understand and manipulate in comparison with other operations like implication ("→") and conjunction ("∧"). For instance, there is a more natural and straightforward cognitive comprehension of the sentence "Let a be an even number" than of the sentence "Suppose that b is not an even number." Effectively, in the first case, one can immediately imagine a representation of the form $a = 2 * c$ (for some $c \in \mathbb{Z}$) (see Chap. 9), for getting a working understanding of the statement. However, in the second statement, one is forced to generate a similar description of b, but starting with the notion of evenness and finding a more "comprehensible" representation of the fact that a number is not even, namely, that it is odd, it can be written as the addition of an even number and one ($b = 2 * d + 1$). Another example is the statement, "Assume that m is not a prime number." If we find such an affirmation at the beginning of a conjecture, and we need to use the information inside it in a pragmatic way, then our (trained) minds look immediately for an equivalent formulation of it without any kind of external negation over it and in this way one can manipulate the hypothesis in a concrete manner (e.g., we re-interpret the statement in the pragmatic form "m can be written as the product of two numbers n and p, whose absolute value is strictly bigger than 1").[12]

More generally, our minds look for equivalent and more pragmatic versions of "negated statements" generating useful (mental) representations with an outstanding "affirmative" nature. So, although from a purely (classic) logical point of view the propositions $Q \equiv x$ is a prime number and $\neg(\neg(\neg(\neg Q)))$ are equivalent, from a cognitive perspective, the first one is a quite elementary and understandable statement whereas the latter is an almost intelligible affirmation.

Additionally, for statements involving a formal disjunction of clauses, let us say $P \vee Q$, we cognitively save each of the statements P and Q in an independent manner, together with the fact that if we want to deduce a clause R from such a disjunction, then we should deduce it either from each of them separately (disjunction of cases) or we should use other methods of demonstration along the way.

By contrast, logical conjunction (e.g., $G \wedge H$) is a more sound connective from a cognitive point of view, because it reflects on its own the natural fact of having two statements whose validity can be assumed mentally without any kind of additional cognitive "token" that can restrict any subsequent reasoning done with constituting clauses (e.g., G and H). Similarly, logical implication (e.g., $A \rightarrow B$) possesses a similar cognitive nature. Effectively, typical logical inferences done with it involves the fact that one can assume the hypothesis (A) in its fully spectrum of certainty and from that one should be able to prove the veracity of the thesis (Q). Finally,

---

[12]This statement has a more explicit content suitable to be combined with additional facts involved in the corresponding conjecture.

logical equivalence $(X \leftrightarrow Y)$ is quite closely related to implication, also from a pragmatic perspective. In addition, this fundamental cognitive operation mirrors the cognitive ability of conceptual identification (see Sect. 10.8, Chap. 10). Effectively, from a heuristic point of view, a statement of the form $X \leftrightarrow Y$ usually creates a conceptual identification of the mathematical structure(s) defining X with the corresponding structure defining Y. Although from a purely logic perspective we can identify certain logical connectives in terms of other ones (e.g., $P \rightarrow Q$ is equivalent to $\neg P \vee Q$), from a cognitive point of view this is not exactly the case.

The fact that the deductive structure of a proof, under the terms of the former version, is fundamentally delimited by the operational rules of the corresponding logical connectives (e.g., $P \wedge \neg P \rightarrow Q$, for any statement Q), implies that unique proof's methods can be used, for instance, reductio ad absurdum. The roots of these methods lie in the reduction of the underlying logic responsible for creating/inventing mathematical structures to formal symbolic frameworks with a binary quantification of truth (e.g., first- and higher-order logic). From a cognitive metamathematical perspective, many ontological aspects of such deductive methods possess pragmatic limitations. For instance, there are a lot of theorems in algebra, analysis, topology, and geometry, which affirm the existence of certain mathematical objects fulfilling fixed properties, but from the corresponding proofs one cannot identify concretely (even in elementary cases) how such special objects look or what they are concretely.[13]

In conclusion, a stronger, deeper and more precise meta-analysis (and updated version) of the notion of proof is needed that can enhance the sound aspects of the current and most used formalizations, and, at the same time, that fulfills minimal cognitive and ontological requirements in a more explicit way. The AMI program aims to fill a gap that exists between what we could call "abstract proof theory" (as a formal sub-discipline, part of automated deduction) and "actual proof theory" (the actual demonstrations produced by working mathematicians in daily research).

3.4 The Local Nature of the Conscious Mind

One of the most pragmatic aspects of human reasoning is its local nature, namely, all the scientific, technological, technical, and artistic (co-)creations we have achieved are largely the gradual, local, and particular effort of the human conscious (and unconscious) mind.[14] For instance, the most general and sophisticated mathematical constructs that we have today (e.g., modern [derived] algebraic geometry and

---

[13]This is, in fact, one of the common objections that constructive mathematics does to the "standard" proof methods [27].

[14]Based on the state of the art of neurobiology and psychology, it is clear that our unconscious mind also plays a primary role in conceptual creation. However, we want to emphasize here the primary features of the conscious part of the mind, which are fundamental during abstract creation as well.

algebraic topology, together with all kinds of the corresponding [co-]homology and homotopy theories) are the result of millions of completely specialized "unities" of conscious thought.[15]

In fact, behind each (public) invention/creation of the human mind, there are lots of (private) trials, informal graphic sketches, partial enlightening ideas, conceptual comparisons with similar results, detailed experiments with the corresponding recompilation of data, analysis of (tiny) examples, detailed syntactic computations (in most of the cases done in a sequential way), informal mental (3D) representations, and small conceptual operations, among others. A commonality of all the former heuristics is that they are executed through precise and directed processes of the conscious mind(s), which are essentially of a local nature. In other words, the "magic" wide generality of lots of mathematical statements (e.g., theorems) is, in most cases, the intellectual outcome of dozens of "real" (local) steps, characterized by being partial, sectional, restricted, and specific.

Explicit examples illustrating this kind of deductive phenomena are present in the great majority of the theorems proved in (Euclidean) Geometry, (Classic) real analysis, and graph theory (among others), where one of the strongest sources of creative inspiration is given by the concrete drawing of pictures providing a minimally coherent (cognitive) laboratory for gaining highly abstract intuitions and technical insights.

It is natural that our cognitive meta-theory for mathematical invention/creation can take this kind of "local" and "informal" dimension into account. Therefore, initially it will be common for us to formalize certain cognitive abilities in terms of partial functionals, selected substructures of the range of some (partial) functions, sub-collections of axioms, and local specifications of functions/functionals, among others.[16]

It is fundamental to clarify here that we are looking for a universal and (at the same time) pragmatic meta-theory of mathematical invention/creation and, therefore, in our AMI quest we are located one step above modern mathematics. In particular, this fact together with the multi- and interdisciplinary approach that we want to use implies that we are not necessarily looking for a "mathematical" meta-theory for explaining the origin of mathematical theories in a circular way. We are looking for much more than that; we are in the search of a universal formal framework for general mathematics, which takes into account the actual way in which mathematics are created/invented in the human mind(s). In particular, it should give more practical information of the way in which we comprehend, combine, and materialize cognitive information with mathematical content.

In particular, such meta-theory should be able to (meta-)formalize the essential aspects of the "trial and error" procedures that human mathematicians and related

---

[15]The usage of the term "unity" is inspired by the unified nature of conscious experience, as well as for the quest of finding a formal quantification for it [7].

[16]This machinery will be developed in much more detail in the second part of the book.

researchers use when they discover/create new mathematical results. Thus, the term *accuracy* in our context acquires a new and enhanced meaning transcending the classic usage of it in a purely mathematical setting.

References

1. Alexander, J.C.: Blending in mathematics. Semiotica **2011**(187), 1–48 (2011)
2. Atiyah, M.: Topological quantum field theories. Publications Mathématiques de l'Institut des Hautes Études Scientifiques **68**(1), 175–186 (1988)
3. Balaguer, M.: Platonism and anti-platonism in mathematics. Oxford University Press on Demand (2001)
4. Berto, F.: There's Something About Gdel: The Complete Guide to the Incompleteness Theorem. John Wiley & Sons (2011)
5. Buss, S.R.: Handbook of proof theory, vol. 137. Elsevier (1998)
6. Campbell, J.I.: Handbook of mathematical cognition. Psychology Press (2005)
7. Cleeremans, A.E.: The unity of consciousness: Binding, integration, and dissociation. Oxford University Press (2003)
8. Dehaene, S.: The number sense: How the mind creates mathematics. OUP USA (2011)
9. Feferman, S., Dawson, J.W., Kleene, S.C., Moore, G.H., Solovay, R.M.: Kurt gödel: Collected works, vol. i: Publications 1929–1936 (1998)
10. Ferreirós, J.: Labyrinth of thought: A history of set theory and its role in modern mathematics. Springer Science & Business Media (2008)
11. Horst, S.: The computational theory of mind. Stanford Encyclopedia of Philosophy (2011)
12. Jackendoff, R.: Consciousness and the computational mind. The MIT Press (1987)
13. Kunen, K.: Set theory an introduction to independence proofs, *Studies in Logic and the Foundations of Mathematics*, vol. 102. Elsevier (2014)
14. Lakoff, G., Núñez, R.: Where Mathematics Comes From: How the Embodied Mind Brings Mathematics into Being. Basic Books, New York (2000)
15. Marker, D.: Model theory: an introduction, vol. 217. Springer Science & Business Media (2006)
16. Mendelson, E.: Introduction to Mathematical Logic (Fifth Edition). Chapman & Hall/CRC (2010)
17. Novaes, C.D.: Formal languages in logic: A philosophical and cognitive analysis. Cambridge University Press (2012)
18. Novaes, C.D.: Mathematical reasoning and external symbolic systems. Logique et Analyse **56**(221) (2013)
19. Quine, W.V.: New foundations for mathematical logic. The American mathematical monthly **44**(2), 70–80 (1937)
20. Robinson, A.J., Voronkov, A.: Handbook of automated reasoning, vol. 1. Elsevier (2001)
21. Schwering, A., Krumnack, U., Kuehnberger, K.U., Gust, H.: Syntactic principles of heuristic driven theory projection. Cognitive Systems Research **10**(3), 251–269 (2009)
22. Shapiro, S., Wainwright, W.J., et al.: The Oxford handbook of philosophy of mathematics and logic. OUP USA (2005)
23. Simpson, S.G.: Reverse mathematics. In: Proc. Symposia Pure Math, vol. 42, pp. 461–471 (1985)
24. Tegmark, M.: The mathematical universe. Foundations of Physics **38**(2), 101–150 (2008)
25. Tegmark, M.: Our mathematical universe: My quest for the ultimate nature of reality. Penguin UK (2014)
26. Tennant, N.: Logicism and neologicism. Stanford Encyclopedia of Philosophy (2013)

27. Troelstra, A., Van Dalen, D.: Constructivism in mathematics, vol. 121 of studies in logic and the foundations of mathematics (1988)
28. Wang, P.: Non-axiomatic reasoning system: Exploring the essence of intelligence. Ph.D. thesis, University of Indiana (1995)
29. Wang, P.: Cognitive logic versus mathematical logic. In: Proceedings of the Third International Seminar on Logic and Cognition (2004)
30. Weir, A.: Formalism in the philosophy of mathematics. Stanford Encyclopedia of Philosophy (2011)
31. Whitehead, A.N., Russell, B.: Principia Mathematica. (3 vols). Cambridge University Press (1910,1911,1912)
32. Witten, E.: Topological quantum field theory. Communications in Mathematical Physics **117**(3), 353–386 (1988)
33. Witten, E.: Quantum field theory and the Jones polynomial. Communications in Mathematical Physics **121**(3), 351–399 (1989)
34. Ye, F.: Strict finitism and the logic of mathematical applications, vol. 355. Springer Science & Business Media (2011)
35. Zalamea, F.: Synthetic philosophy of contemporary mathematics. Urbanomic (2012)

Chapter 4
Towards the (Cognitive) Reality of Mathematics and the Mathematics of (Cognitive) Reality

4.1 Introduction

The ancient question regarding the ontological place occupied by "mathematics"[1] within the (meta-)physical realm has seen a new (minimal interdisciplinary) enlightenment during the last decades not only from a philosophical, historical, and logical perspective [25, 26] but also from a cosmological one [23]. However, most of the treatments concerning this kind of foundational question are considerably narrow in their multidisciplinarity. This has as a general consequence that most of the seminal conclusions obtained are either too meta-physical (i.e., almost unprovable or irrefutable), although significantly bright [25, 26]; too reductionist from a physical point of view [16], or too reductionist from a formal point of view [23] (i.e., they represent oversimplifications of the problem on several directions). Hence, the more multidisciplinary and bright our perspective is, the more it can supply both former shortages.

In the following sections, we will present an initial, unique, compact, and multidisciplinary approach to the (cognitive) realm of mathematics and the mathematics lying behind the physical (and cognitive) realm. Furthermore, our approach assumes substantially moderate versions of the thesis presented in [16, 25, 26] and [23]; i.e., we would agree with partial versions of the main theses presented in the former seminal works. This means, for example, that a conclusion of the form "In fact, all physical entities are mathematical objects,"[2] would mean for us a thesis of the form "In fact, all physical entities are, in part (and from a structural point of view) mathematical objects." In other words, each of the former perspectives possesses a basic and necessary degree of general (ontological) truth, however partial and incomplete.

---

[1] Let us understand the term "mathematics" in the widest sense of the word.

[2] This can be seen as one of the (paraphrased) main conclusions of Tegmark's work [23].

© Springer Nature Switzerland AG 2020
D. A. J. Gómez Ramírez, *Artificial Mathematical Intelligence*,
https://doi.org/10.1007/978-3-030-50273-7_4

4.2   Towards the Reality of Mathematics

4.2.1   *Qualitative Commonalities of Several Possible Physical Scenarios*

Imagine that a seller has 20 copies of a new book to sell. One day, three people come in simultaneously and ask him/her for 12, 7, and 1 copies of that book, respectively. After a couple of seconds, (s)he realizes that there are enough copies for all the customers and sells them the books.

Now, imagine that we replace the word "book" with the words TV-s, dinner, bicycle, auto, orange, banana, and house. Also, we can change the word seller with dealer (TV-s), chef (dinner), and seller machine. Question: What kind of (ontological) "nature" does the commonalities of the several possible situations have?

A compact answer to this question is that such commonalities are just mathematical structures (some natural numbers, e.g. 1, 7, 12, and 20; the addition operation) and mathematical properties (the associativity law of the addition).

When we think about the case of a seller machine doing the addition of the three numbers in order to determine if this sum is smaller than or equal to 20, we can imagine (assuming that we have any amount of time and resources at our disposition) that this machine is basically constructed of any kind of matter of the universe; let us think about the first computer constructed in history as Charles Babbage's Analytical Engine (1910), Konrad Suze's Z1–Z4 (1938), Tony Flower's Colossus (1936), ADC (1942) and ENIAC (first digital computers (vacuum tubes)) (1946), IBM's 701 (the first commercial scientific computer), Whirlwind machine (magnetic core RAM), TX-O (Transistorized Experimental Computer), IBM 5100 first portable computer (55 pounds) (1975), Dell's Turbo PC (1985), D-WAVE quantum computers (the processor environment has a temperature of -273.13 C and an extremely low magnetic environment); or more generally, let us consider any (potentially constructable) computer made of a plausible combination of physical materials. All these situations do not have any kind of macro-physical commonalities, although, of course, at a quantum level all of them have basically the same elementary particles, which can be configured in many ways.

The former technical considerations were described in order to show that the nature of the common features of this collection of "selling" examples needs ontological descriptions beyond the standard physical realm, namely, they exhibit a pronounced "mathematical" structure.

Besides, note that this collection of examples has, in principle, a concrete physical nature, namely, each of them can be, by definition, physically reproduced.

So, we arrive to mathematical structures like numbers, starting from purely physical constructs.[3]

Let us distinguish between the intrinsic existence of a natural number and either its syntactic representation and/or its logic formalization [4].[4] Thus, one moderate conclusion of the former thought experiment is that the whole ontological essence of a natural number cannot be strictly contained in a particular physical entity. Assuming a naturalistic dualism [8] as global ontological taxonomy for nature, we infer that a natural number is an entity (of nature) lying at least in a phenomenological (i.e., mental) dimension.

In this way, we encounter one of the most fundamental questions in the philosophy of mathematics: What (and "where") is the place that mathematics occupy in (the existing) nature?

During the last decades, there has been a lot of progress concerning a general answer to this question, which has arisen not only from a philosophical or logic-mathematical perspective but also from a cognitive and cosmological one.

More explicitly, the work of [16, 19] and [18] strongly emphasizes the role that cognitive abilities, like metaphorical reasoning, plays in the generation of mathematical entities. Hence, they give an outstanding importance to the human mind for inventing mathematics.

Contrastingly, there is a valuable and deep work coming from a cosmological perspective, which is trying to give a concrete answer to the question of what our physical reality can be. Explicitly, in [22] and [23], Max Tegmark employs a considerably large number of quantitative and qualitative aspects of the most outstanding formal theories used in modern physics (e.g., general and special relativity, inflation, quantum mechanics, string theory, among many others) in order to justify the fact that what we call ("external") reality,[5] has a seminal mathematical component. Moreover, he argues extensively for explaining the nature of reality in terms of the nature of mathematical structures.

Although during the last centuries we have developed sophisticated logical and mathematical syntactic theories studying a wide range of formal entities, from the semantic point of view, namely, from the perspective of finding explicit and "external" descriptions of the mathematical structures grounding our most outstanding mathematical theories, we are at the very beginning of our scientific inquiry. In fact, let us start considering one of the most well-accepted foundational frameworks for standard modern mathematics, i.e. Zermelo–Fraenkel set theory with Choice (ZFC) [10, 15, 17]. Regarding the (formal) existence of mathematical

---

[3]The former collection of thought experiments form an additional approach for emphasizing the structural relation between the mathematical and the physical world, which goes beyond the standard line of argumentation summarized by the fact that mathematics emerges naturally from our search for understanding physical phenomena at a more precise level. In fact, in our thought experiments we just "observe" the several possible scenarios without doing any active inference inside of them.

[4]Formalizations based on set theory [17], Lambda Calculus [3], among others.

[5]See, for example, the External Reality Hypothesis [22].

entities at a fundamental level, the completeness theorem for first-order logic states that the existence of a genuine seminal model for ZFC is equivalent to the consistency of ZFC itself.

On the other hand, assuming the second incompleteness theorem, we can infer that the task of producing a canonical existent model of ZFC goes beyond the ZFC formalism; i.e., we should look for additional (closely related) scientific disciplines in order to be able to construct an authentic model grounding ontologically modern mathematics.

Here, we should clearly differentiate between the "formal mathematical dogmata" (e.g., the existence of sets, numbers, geometrical structures, the implicit consistency of ZFC, among many others), which comes from our natural intuition for understanding mathematics and the syntactic consequences and sophisticated formal concepts that we have developed based on these dogmata. Thus, with respect to the first ones, we have, almost by definition, a very informal knowledge; we believe to have a deeper understanding of the second ones, mainly because we have gotten used to them (at least on a mental level).

However, what the mathematicians call "mathematical structures" is essentially a formal syntactic construct based on this kind of intuitive mathematical dogmata. Therefore, in order for such (mathematical) structures to be able to serve as an ontological source of explanation for our physical realm, they should be structurally better understood, not only from a logical (metamathematical) perspective but also from a (meta-)physical one.

By the same token, what we call (external) "reality" also includes all the other formal structures used in several scientific disciplines, like psychology, anthropology, medicine, chemistry, biology, sociology, computer sciences, philosophy, among others. So, should we not also include the new formal structures emerging from these types of sciences as part of reality? They are as important as the structures studied by the working mathematicians. However, they come from unique and specific intellectual fields which have, as one of their main goals, to formalize several aspects of our multifaceted reality.

Consider a simple example to illustrate our idea. Let us assume that someone is experiencing mental and emotional "singularities" during his/her daily life, which prevent him/her living a better life.[6] Nobody would expect this person to go to see a cosmologist in order to improve his/her quality of life, because such a scientist claims to have a more precise and global understanding of how reality works, at the physical (micro and macro) level. On the contrary, what is expected is that this person goes to the psychologist (resp. psychiatrist) because (implicitly) the formal models and structures employed in psychology (resp. psychiatry) are fundamentally more appropriate for describing mental and emotional phenomena of human beings.

This example can be generalized to dozens of scientific disciplines, and in this way one can support the unique importance and the kind of *formal exclusivity*

---

[6]The term "singularity" is used here to denote an irregular phenomenon, which could be replaced from a medical or psychological perspective by the word "pathology."

that each single scientific discipline possesses. Each of them offers a structurally valuable and seminal view of reality that has evolved through many years of intellectual development. If we want to find a suitable global description of the reality as a whole, we should integrate (gradually) the most essential aspects and principles of all the authentic scientific fields existing today.

One of the strongest conclusions of the intuitionist and the (classic) cognitive school is that the reality of mathematics occupies at least the cognitive (i.e., mental) dimension in our universe. Moreover, the next question to ask is to what extent the reality of mathematics includes the physical dimension. In order to enlighten this issue, let us imagine a wooden stick, used for playing drums, and let us denote this object by D. Let us assume that we use all the results of modern physics for giving a spatio-temporal description of D in a four-dimensional setting (3 spatial dimensions and 1 temporal dimension). Here, we can imagine the usage of all the most sophisticated equations and mathematical frameworks developed in quantum mechanics, quantum field theory, relativity, among others, for describing the general behavior of D at a temporal basis, e.g. its position, its (general) texture, and its weight at a given time t. Even more, it could happen that the equations needed for such a description require thousands of lines to be written, which is completely fine and in coherence with our present discussion. Let us call $\mathcal{M}ath(D)$ the mathematical structure described by the whole spatio-temporal description of D until the present time; i.e. the formal logic(s) used, the specific language, the formulas describing the corresponding properties, and the concrete collection of mathematical models involved.

So, $\mathcal{M}ath(D)$ can be considered the most precise formal model of D possible. $\mathcal{M}ath(D)$ is clearly a mathematical object. Moreover, let us assume that we focus our attention on understanding more deeply all the mathematical properties of $\mathcal{M}ath(D)$, from the most elementary one until the most sophisticated one, from the most insignificant one until the most valuable, from the most local one until the most global. Such an inquiry about $\mathcal{M}ath(D)$ is basically a quest for getting a better understanding of its mathematical nature and therefore about the character of its mathematical realm, or the reality of its mathematical structure. Nonetheless, by doing such a mathematical exploration, we will immediately get information about the potential physical evolution of D in time. For instance, we will know in more detail its texture, color, and weight in the near future;[7] which are basically physical properties. Essential aspects of the reality of a mathematical object like $\mathcal{M}ath(D)$ were fundamental to ground essential aspects of the physical reality of D. More generally, the same thought experiment can be done by replacing D with any object suitable to be studied in modern physics. Thus, one of the main conclusions of this potentially huge collection of thought experiments is that the reality of

---

[7]Here, we assume implicitly minimal conditions of a moderate isolation for D in the sense that there are no external influences considerably changing its physical configuration. This can be originated, for example, by an external agent who interacts with it.

mathematical structures and properties goes beyond a cognitive realm and touches quite fundamentally the physical reality.

4.3   Towards the Mathematics of Reality

4.3.1   An Initial Taxonomy for the Size of Phenomena in Nature

Here, we will do an initial rough classification of the size (or *level*) of the (minimal) space that an event in nature requires to happen. For that, we take as referential framework the human being; i.e., the visual field of the naked human eyes. Explicitly, let $\mathcal{E}$ be an event happening in the universe during a period of time starting at t_1 and finishing at t_2 related to a fixed (human) observer (on earth). We say that $\mathcal{E}$ is a *mecro* event if, roughly speaking, the naked human eyes can see, perceive, and differentiate the essential structural movements happening in $\mathcal{E}$ between t_1 and t_2. Any event $\mathcal{E}$ of a smaller magnitude, essentially imperceptible to the naked human eyes will be called a *micro* event. Lastly, any event of bigger size and (again) imperceptible/just barely perceptible to the naked human eyes is a *macro* event.[8] Typically, most of the phenomena studied in quantum mechanics, chemistry, and in cellular biology are micro events. Meanwhile, a lot of phenomena studied in anthropology, psychology (and in general in the social sciences) are mecro events. Finally, a huge amount of phenomena studied in cosmology and astronomy are macro events. It is important to mention here that this rough spatial taxonomy is slightly different from the standard distinction used in physics in terms of microscopic and macroscopic scales. Effectively, a lot of macroscopic events (for example, the ones encompassing ca. 1 cubic micra) are considered in our taxonomy as micro events, while others can be considered as mecro events (e.g. the ones encompassing ca. one cubic meter) [21].

4.3.2   The Ontological Role of Mathematics Within the Existing Realm

Based on the former considerations, which can be seen as an ontological restriction for the (meta-)physical scope that mathematics plays within nature; our claim is that *mathematics plays an existent structural role within the (meta-)physical reality*.

---

[8]Based on more precise estimations coming from the visual science/optics, one could give specific spacial thresholds for the intervals where each level is located (e.g. in terms of the amount of cubic meters of the space encompassed by the event). However, for the purpose of our presentation, we will omit this.

In order to understand our claim more precisely, we need to be more explicit in what "mathematics" means in this context. Before doing that, we assume the following modified version of Tegmark's external reality hypothesis [23]: there exists an external reality that is strongly influenced by humans (and human minds), but ontologically not limited to their existence. So, let us define "mathematics" beyond the existent syntactic and semantic formal frameworks used by working mathematicians and physicists, (e.g., ZFC set theory, Peano arithmetic, (non-)standard analysis, differential and algebraic geometry, among others), namely, as one fundamental ontological dimension of the external reality.

For example, assuming the consistency of ZFC, we claim the entities in nature formed by collections of well-defined objects possess an existent and coherent set-theoretical ontological dimension. Similar phenomena would hold for arithmetic (number theory), algebraic topology and geometry, graph theory, and most of the other relevant areas of mathematics, relative to the corresponding entities in nature being closer to the semantic origin of such theories.

After studying in more detail the lines of argumentation of Tegmark (for example, in [23]), one can see that our claim is moderate and represents a less absolute version of most of his qualitative and quantitative theses. In particular, two simple meta-conclusions that general relativity and quantum mechanics can teach us are that nature at the micro- and macro-level can be described with specific mathematical models that allow us to make precise predictions about spatio-temporal properties of physical objects like, for example, planets, stars, or atoms.

Notwithstanding, when we move to an intermediate degree into the (meta-) physical realm (i.e., to the mecro-level, namely, to a human-scale level) we encounter strong qualitative limitations on our search for the same kind of purely mathematical predictive models. We will describe such constraints in more detail in the next section.

A second primary claim, closely related to the first one, is that *the external reality is at least as precise as Mathematics.* In other words, like in mathematics, any minimal change that one makes to any formal structure (e.g., topological space, group, graph), is permanently saved into such structure, (e.g., a new topological space obtained after removing a point from another one, a new group obtained defining a new binary operation on the underlying set, a new graph obtained by adding an extra node and edge), so is the case for any minimal modification that occurs to any entity in nature at every degree (i.e., macro-, mecro- and micro-level). This is a straightforward consequence emerging from the fact that throughout the centuries, mankind has gradually seen how sophisticated and meta-physical processes of our world, which, in principle, have not shown any kind of regularity or law of development could be understood, described, and (structurally) modeled by very accurate theoretical frameworks, e.g. quantum mechanics, game theory, information theory, statistical mechanics, and (general and special) relativity.

In addition, another fundamental consequence of the former claim is that, at least from a theoretical point of view, any mecro-phenomenon as the ones studied in anthropology, ethnology, sociology, jurisprudence, or political sciences can be described with at least mathematical precision. In fact, what usually happens is that the particular phenomenon (being the object of study) is highly sophisticated and involves the integration various variables and initial data that needs to be at disposition.

From the standard way in which the former disciplines are taught, we see that it is not always the case that the corresponding researchers have the minimal mathematical training to be able to spend significant time of their research looking for such formalizations.

4.4   The Nature of a Formal Model for Reality at a Mecro-Level

Let us assume that we have a formal model for the whole reality. Specifically, it means that we have developed a mathematical-physical theory of everything, a finite collection of concepts and relations involving those concepts (e.g., equations) such that we can model any phenomenon of reality at any scale (macro-, mecro- and micro-scale). In addition, modeling in this context should be understood as predicting in advance the "state" of a subject in space and time. Secondly, state refers to several possible constituent dimensions of this subject, like the spacial, physical, chemical, and psychical state. In particular, such a formalism will anticipate the physical coordinates of any particular person at any time in the future. Let us assume that a person, let us say Michael, starts at 6 am to compute his spacial coordinates for the same day at 6 pm. Let r (hours) be the exact time when Michael realizes his spacial position.

The question is, what is the minimal possible value for r? Let us assume that "Michael realizes" also means that he knows roughly at r hours the coordinates of his position at 6 pm related to his position at r hours. Let us assume that Michael is, physically speaking, free enough to choose (as a "normal" human being) his immediate spacial place in a conscious self-aware manner. Summarizing, at r hours Michael knows, roughly, the coordinates of his current position as well as the coordinates of his position at 6 pm and, at the same time, he is free enough to choose the coordinates of his spacial location at 6 pm.

So, he can authentically choose at r hours if he wants to occupy the predicted position at 6 pm or a new one. Now, let us assume that Michael is able to change the position of his center (i.e., mass center) by more than a centimeter (in any spatial direction) in less than a second.

Under these assumptions, we deduce that in order to maintain coherently both hypotheses (i.e., the practical validity of the predictions of our formal theory of everything and the concrete and actual spacial freedom of movement of Michael), we necessarily infer that r hours is a time less than a second before the "predictable" time of 6 pm.

We can clearly modify this thought experiment to show that (in practical terms) in such a theory of everything the more accurate the (spacial) prediction is, the less "prediction" (in time) it is. In other words, when the free subject (e.g., Michael) realizes that he is able to choose any human possible spacial location besides the predicted one, he should be already on the time corresponding to that prediction. Thus, the prediction time is essentially, in practical terms, equal or posterior to the time of "realization." We will call this phenomenon *the Unpredictability Principle of the Mecro-level* (resp. *the Unpredictability Principle of Natural Human Will*).

The former thought experiment shows that any genuine "theory of everything," being able to "explain and foresee," in particular, mecro-level phenomena; should possess new formal and descriptive principles, which would allow us to make "predictions," but more from a qualitative perspective.

We can also conclude that one of the natural phenomena that causes most foundational challenges to any kind of theory that we have within any scientific discipline so far, is what we call *natural human will*; the seminal, original, natural cause of any (spacial) movement of every mentally healthy human being. Here, we use the word "natural," because as a tree is part of nature, so too is such cause of human movement. Hence, natural human will is a very concrete entity of reality whose ontological nature requires formal structures going beyond the (classic) mathematical realm.

In addition, this unique entity (i.e., natural human will) offers a (sort of) straightforward bound for one of our former thesis stating that mathematics performs just a structural role in reality, without being able to fill it completely. This is mainly due to the strong explanatory gap that human will clearly establish.[9]

4.5 The Singularity and the Continuous Model of our Spacetime: *Pseudo-Sculpting Irrational Computable Numbers*

The view that our three-dimensional space has structural similarities, quantitatively speaking, with the mathematical object $\mathbb{R}^3$ (respectively between "time" and $\mathbb{R}$), is one of the most used spacial (resp. temporal) models that the (standard) working mathematician and some physicists use when they want to obtain deeper intuitions and more precise formalizations about special phenomena in (mathematical) analysis, topology, geometry, and mathematical physics (among others). Informally, one can "zoom" any spatial situation and object until "infinity" in order to obtain more precise information about it. This is something so standardized that most of the amateur and professional mathematicians no longer consciously realize is simply a kind of working principle.

Furthermore, during the last decades we have seen an exponential growth in the processing speed of computer programs and artificial processes, leading to some academicians to think that it is plausible that there will be intelligent machines

---

[9]For further information about (free) will from several perspectives see, for instance, [14] and [5].

which are able to construct even more intelligent machines, each time in at most half of the time of its predecessor and each one having more sophisticated abilities for solving spacial tasks, which generate at some point the so called singularity [9].

Let us assume the former two hypothesis; we will consider a formal thought experiment as follows:

4.5.1   Continuous Syntactic Notation

Let a be a non-negative real number. Let us write a in binary representation

$$a = \sum_{i=-n}^{\infty} a_{-i} 2^{-i},$$

for some non-negative integer n and $a_{-i} \in \{0, 1\}$ for all i. In order to have a unique representation, we will assume that there are infinitely many i such that $a_{-i} = 0$.

We want to represent each a with a bounded two-dimensional symbol which codes in a bijective way its whole binary representation. Let S be the hyperrectangle (box) $[-1, 1] \times [0, 1] \times [0, 1]$. Thus, we will represent the entire part of a (i.e., the part without decimals) as follows:

$$I_a = \{[-1, 0] \times [1/2^k - 1/2^{k+2}] \times [0, 1] \bigcup [-1/8, 1] \times [0, 1] \times [0, 1] :$$

$$k = 1, 2, \cdots, \lfloor a \rfloor = \sum_{i=-n}^{0} a_{-i} 2^{-i}\}.$$

Thus, each of the digits of the entire part of a is codified by a particular (horizontal) box. For the decimal part $\sum_{i=1}^{\infty} a_{-i} 2^{-i}$ we do an analogous procedure with the possibility of having infinitely many (vertical) boxes:

$$D_a = \left\{ [1/2^i - 1/2^{i+2}] \times [0, 1] \times [0, 1] : i \in \mathbb{N}_{\geq 1} \wedge a_i = 1 \right\}.$$

For a negative number $-b$ we just add the (horizontal) box $[-1, 0] \times [3/4, 1] \times [0, 1]$ to the notation of b in order to define its *syntactic continuous notation*. The notation of the zero will be represented for simplicity with the whole box S. A number a written in this way is denoted schematically as $\lfloor a \rceil$.

This notation has the advantage that it codes so much information about a number as the classical binary notation, but such a representation could be "formally written" in a "bounded space."

Assuming one could explicitly "write" in this continuous notation "real" numbers in the physical world, one could completely decode the corresponding number from its morphological representation, which is unique and geometrically characterizes each of the binary digits of the corresponding number. In this case, the exact way in which the morphological symbols of the notation are "written" has a structural and direct influence on the meaning of the corresponding number.

Remark 4.1 The two-dimensional version of this notation consists for positive numbers exactly of the former construct, but omitting the last component, namely, (a, b) belongs to the $2D$-*continuous notation* of a number α, if and only if $(a, b, 0)$ belongs to the $3D$-continuous notation of α. For negative numbers one can easily generate a slight variation of the three-dimensional one.

4.5.2   "Singular" Sculptures of Irrational Computable Numbers into a Continuous Physical Realm

Let A be a computable real number, informally, a number such that there exists a computable function with the property that for any accuracy measure $\epsilon > 0$, it is able to compute a rational number r such that $|A - r| \leq \epsilon$ [1]. For example, $\pi, e, \sqrt{2}$ are computable numbers.

   Let us assume that $10\,cm$ correspond to one unity of measure; i.e., the (physical) distance between 0 and 1. Let us also fix a computable irrational number α; the computer C_s that originates ("hypothetically") the singularity; an initial computer program (and the corresponding hardware) being able to generate recursively the binary digits of α and translate them into the (3D) continuous notation; and an initial 3D Printer $P_A(12)$ which is able to print in the $3D$-space, first the integer part of α, and then the rational number coinciding with the decimal part of α in exactly the first 12 decimal digits (and having zeros in all the others binary digits).

   With this initial setting (and potentially enough resources at hand) C_s is able to produce, at exponential speed, improved versions of $P_A(12)$. Specifically, let us say that $P_A(12n)$ (for $n \in \mathbb{N}$ arbitrarily big) can be produced in around 10^{-n} seconds. This is possible, if we are near enough in time and space to the singularity (event), since there the speed and efficiency of artificial devices increase without limit, and, on the other hand, we are assuming that our spacetime has a continuous nature.

   In conclusion, after the singularity occurs, and assuming that we generate the former initial setting, we have obtained a concrete 3-dimensional sculpture of α within a box of $20\,cm$ of length, $10\,cm$ of width, and $10\,cm$ of height. Under these "singular" conditions, we could be able to generate physical versions of abstract mathematical entities as the number π, or e.

   Moreover, due to the "singular" nature of C_s, which exponentially doubles its computation speed, we see that given a Turing machine p, C_s would be able to determine within a finite amount of time if p halts or not. Effectively, we partition one unit of time (e.g., a second), modeled by a close real interval, into an infinite amount of intervals I_n of length $1/2^n$ (seconds) and in this period of time C_s can simulate the n-th computation of p. So, after a second, C_s is able to know if p halts (or not). Therefore, C_s would be able to do an infinite sum of the corresponding probabilities involving only the lengths of the halting Turing machines in a finite amount of time. So, C_s could print (e.g., compute) Chaitin Omega number [7], which is a contradiction. In conclusion, both assumptions cannot be simultaneously true.

Both assumptions made above are also strongly questionable for additional reasons. Effectively, the singularity as a whole seems to come more from a science-fiction scenario without any strong qualitative evidence for it. Besides, if we talk about machines replicating human intelligence in a (strictly speaking) general (global) way, then they should be able to show human-like performance, at least in logic-mathematical, linguistic, kinesthetic, musical, interpersonal, and intra-personal dimensions of human intelligence [12]. Until now, our most efficient artificial devices perform, in the best cases, intelligent behavior in (restricted) sub-activities regarding 2–3 of the former dimensions.

Thus, one can see in this aspect a strong "influence" of the film and the commercial industry, creating fantastic and sub-real scenarios where robots perform human-like abilities at levels far beyond the authentic capacities of the real robots constructed in the industry. In addition, some advertisement programs offering AI products create a cognitive effect that make it tricky to distinguish the thin border between the description of an ("objective") AI tool obtained by AI researchers and engineers in a laboratory and the devices shown in some science-fiction films during the last decades.

Returning to our first assumption—the "continuity" of our spacetime. On this point, there is theoretical evidence coming from string theory supporting the thesis that our spacetime can be neither discrete nor continuous [2]. In particular, under some fixed small numbers l_P (Planck Length) and t_P (Planck Time) [6, 11, 20] it makes no sense to talk about a spacial distance smaller than l_P; and a period of time smaller than t_P. Although, it is also true that a lot of formal models for several physical phenomena are based essentially on continuous aspects of the real numbers, since each of them has a grounding (standard) mathematical framework. This continuous spacial model that most mathematicians and physicists use on a daily basis in their research has been, at the same time, fundamental for the further formal development of physics and mathematics. Hence, there seems to be genuine aspects of this approach which are structurally coherent with some (corresponding) external aspects of reality. The important question at this point is, how to develop a finer *conceptual filter*, which allows us to distinguish between the physical and the meta-physical (e.g., mental) components of our models?

Due to the clear and large evidence given by the effectiveness of mathematical methods and models in physics, one can say that, at least at the meta-physical (e.g., mental) level, our (continuous) models of space, time, and spacetime have a coherent ontological existence.

4.6 Final Remarks and Conclusions

We have seen that entities of a mathematical nature appear naturally as structural quantitative commonalities of a particular collection of genuinely possible physical scenarios and, therefore, objects having a typical mathematical essence seem to be structurally located in different scenarios simultaneously. So, our (external) reality appears to possess a *predictive mathematical entanglement*; i.e., if a pair of physically independent scenarios share ("isomorphically") initial structural commonalities from a mathematical point of view, then any formal new fact (i.e., formal prediction) of one scenario described by using essentially mathematical rules of inference can be instantly translated to the other one. Here, the hypothesis that both scenarios fulfill is strong enough to include their total (meta-)physical "history" until the spatio-temporal point of consideration.

Contrastingly, the (unreasonable) effectiveness of mathematics for the description of physical phenomena [13, 24]; jointed with the fact that the micro- and macro-events of our universe have been (structurally) modeled by the mathematical frameworks used in modern physics as a whole [22, 23], constitute initial sufficient support for stating that our reality has (at least) a mathematical precision at the mecro-level also, and, therefore, at any level. However, natural will offers a concrete phenomenon of an object of our existential realm whose formal description requires new principles beyond the (usual) mathematical domain; and whose predictive models should contain a permanent uncertainly inside.

We have seen that under "singular" and "continuous" assumptions of our spacetime, we could potentially "sculpt," in an explicit way sophisticated mathematical structures like computable irrational and non-computable numbers. However, the search for the best formal (e.g., mathematical) structure that can globally describe our spacetime at any level is one of the most fundamental questions of modern physics in its quest for unifying our best physical theories.

Finally, it is fundamental to construct a more natural and renewed logical and semantic framework for general mathematics, which allows us to generate *explicit models* that do not depend on any kind of consistency test; and, at the same time, that can support more accurate formal descriptions of (meta-)physical phenomena.

Acknowledgments The author thank Ulf Krumnack for all the valuable discussions on these topics and for all the support and kindness throughout the years.

References

1. Aberth, O.: Analysis in the computable number field. Journal of the ACM (JACM) **15**(2), 275–299 (1968)
2. Arcani-Hamed, N.: Space-Time is doomed. What replaces it? Messenger Lectures, University of Cornell (2010)
3. Barendregt, H.P., et al.: The lambda calculus, vol. 3. North-Holland Amsterdam (1984)

4. Benacerraf, P.: What numbers could not be. The Philosophical Review **74**(1), 47–73 (1965)
5. Bergson, H.: Time and free will: An essay on the immediate data of consciousness. Routledge (2014)
6. Bernal, A.N., Lopez, M., Sánchez, M.: Fundamental units of length and time. Foundations of Physics **32**(1), 77–108 (2002)
7. Chaitin, G.J.: Algorithmic information theory. IBM journal of research and development **21**(4), 350–359 (1977)
8. Chalmers, D.: Naturalistic dualism. The Blackwell Companion to Consciousness pp. 359–368 (2007)
9. Chalmers, D.: The singularity: A philosophical analysis. Journal of Consciousness Studies **17**(9–10), 7–65 (2010)
10. Fraenkel, A.A., Bar-Hillel, Y., Levy, A.: Foundations of set theory, *Studies in Logic and the Foundations of Mathematics*, vol. 67. Elsevier (1973)
11. Garay, L.J.: Quantum gravity and minimum length. International Journal of Modern Physics A **10**(02), 145–165 (1995)
12. Gardner, H.: Reflections on multiple intelligences: Myths and messages. Phi Delta Kappan **77**(3), 200 (1995)
13. Hamming, R.W.: The unreasonable effectiveness of mathematics. The American Mathematical Monthly **87**(2), 81–90 (1980)
14. Kane, R.: A contemporary introduction to free will (2005)
15. Kunen, K.: Set theory an introduction to independence proofs, *Studies in Logic and the Foundations of Mathematics*, vol. 102. Elsevier (2014)
16. Lakoff, G., Núñez, R.: Where Mathematics Comes From: How the Embodied Mind Brings Mathematics into Being. Basic Books, New York (2000)
17. Mendelson, E.: Introduction to Mathematical Logic (Fifth Edition). Chapman & Hall/CRC (2010)
18. Núñez, R.: Conceptual metaphor, human cognition, and the nature of mathematics. The Cambridge handbook of metaphor and thought pp. 339–362 (2008)
19. Núñez, R., Lakoff, G.: The cognitive foundations of mathematics. Handbook of mathematical cognition. Psychology Press (2005)
20. Padmanabhan, T.: Physical significance of Planck length. Annals of Physics **165**(1), 38–58 (1985)
21. Reif, F.: Fundamentals of statistical and thermal physics. Waveland Press (2009)
22. Tegmark, M.: The mathematical universe. Foundations of Physics **38**(2), 101–150 (2008)
23. Tegmark, M.: Our mathematical universe: My quest for the ultimate nature of reality. Penguin UK (2014)
24. Wigner, E.: The unreasonable effectiveness of mathematics in the natural sciences. In: Philosophical Reflections and Syntheses, pp. 534–549. Springer (1995)
25. Zalamea, F.: Filosofía Sintética de las Matemáticas Contemporaneas. Editorial Universidad Nacional de Colombia, Bogotá (2009)
26. Zalamea, F.: Synthetic philosophy of contemporary mathematics. Urbanomic (2012)

Chapter 5
The Physical Numbers

5.1 Introduction

Nowadays, the foundations of the most outstanding formal frameworks in theoretical physics like quantum mechanics, general relativity, and string theory, include mathematical frameworks like the theory of (operators over) real and complex Hilbert spaces, differential geometry, global and infinite-dimensional analysis, and De Rham cohomology, [10, 18, 29, 33]. Almost all of these theories use as seminal conceptual tools the mathematical notions of the real and complex numbers. The canonical and most common objects of study in physics are finite configurations of objects in nature. However, why do we use such "infinite" and meta-physical structures (e.g., the real and the complex numbers) in order to describe concrete (finite) physical phenomena? Is it necessary to employ such "complex" mathematical notions in theoretical physics, which are not even completely understood in purely mathematical and logical terms?

In order to offer valuable answers to these questions, it is necessary to adopt an interdisciplinary approach. Effectively, the quantitative aspects of the formal frameworks currently used to describe physical phenomena involve a closer study into the fields like philosophy and the foundations of mathematics (e.g., set theory, first- and higher-order logic), cognitive sciences (e.g., formal conceptual creation of the human mind), and, of course, philosophy and foundations of physics (e.g., the structural nature of the (external) physical reality).

In a global research program finding conceptually simpler foundations for physics, it seems natural to start with two of the most omnipresent mathematical structures used (as a conceptual tool) in physical research—the real ($\mathbb{R}$) and the complex ($\mathbb{C}$) numbers.

From a set-theoretical point of view, it is possible to reduce the formal construction of the complex numbers to the construction of the real numbers, which can be reduced, at the same time, to the mathematical construct of the rational numbers. Furthermore, the latter notion can be defined in terms of the integers. Finally, this

© Springer Nature Switzerland AG 2020
D. A. J. Gómez Ramírez, *Artificial Mathematical Intelligence*,
https://doi.org/10.1007/978-3-030-50273-7_5

collection of whole "positive" and "negative" numbers can be formalized in terms of the "grounding" concept of the natural numbers (usually denoted by $\mathbb{N}$) [5]. In conclusion, we will start our re-foundational enterprise with a concrete investigation on the formal, physical, and cognitive "essence" of the natural numbers, in order to be able to give more concrete insights into a conceptually simpler foundational framework for modern physics.

5.2   Overview

One of the most ancient questions in the foundations of sciences is the one concerning the ontological and formal nature of what we call "the natural numbers," namely, the most intuitive and, at the same time, intelligible quantitative structures that we have face since our early years of life when we were "counting" objects of several natures around us. In addition, throughout hundreds of years of investigations going back to the Greek school, we have achieved sophisticated and elegant formalizations of these special kind of numbers as, for instance, the (first- and higher-order) Peano axioms for the arithmetic [25, Ch. 3]. How "near" are these axiomatizations to describing the close and structural relation between the non-embodied quantitative properties of the physical realm, our corresponding cognitive abstractions about those quantitative properties of nature, and what we call "the natural numbers?" Moreover, why does it seem so normal to talk about an initial natural number, but not about a final natural number?

In this interdisciplinary investigation, we use a multifaceted perspective in order to look deeper into the formal and physical core of the "numbers" used in arithmetic to clearly differentiate entities with a physical support (e.g., small numbers) and the ones with a strict meta-physical (e.g., mental) support. Furthermore, we propose an alternative way to generate these numbers in terms of formal partitions of the physical realm. Additionally, we observe that these new kind of numbers are formally enough to derive logically (within strict finitism) the mathematics needed in General Relativity and Quantum Mechanics, and, therefore, the expressive power of this new system is strong enough for developing foundational theoretical physics. Finally, we offer a bottom-up meta-principle for studying classic problems in number theory by means of solving them firstly within this system of physical numbers. This is done in order to obtain an initial "physical" test for the veracity of sentences in number theory.

5.3   The Natural Finiteness of the Universe

We are surrounded by a finite and bounded universe; we have a bounded body composed of a finite collection of atoms and with a, finite weight. Since we were born, we have seen a finite number of persons, trees, books, animals, cities, stars, and in general, physical objects. We have also spoken a finite number of words; we have made a bounded amount of physical movements with our body; we have understood a finite number of theories describing finitely many phenomena of our world and each theory consists of a finite number of symbols with a finite number of subjective meanings (i.e., meanings that a person in a finite amount of time could generate). In particular, we use a limited amount of morphological, syntactic and semantic tools and concepts in order to create a notion of "(in)finite(ness)."

In addition, we have an intuitive notion of "number of elements" of a collection of objects. This first notion of quantity is based on our physical experience with objects in our world (e.g., with collections of plants, toys, fruits, persons, or animals). After having this concrete experience with, for example, 3, 4, or 5 of such objects, we infer "mentally" that it is completely plausible to consider having also such a concrete experience with 6, 7, or 8 or "more" objects. Let us see in a coherent mind-world relation the plausible possibilities of interpretation of the word "more."

First, let us assume that there exists a quantum of matter; i.e., minimal physical particles such that all the elementary particles and the whole physical reality consist of these physical quanta, and they constitute the seminal physical components or fundamental "bricks" of the whole physical realm (for a related discussion see, for example, [13, 26] and [3]).

Let us define ω as the number of quanta (i.e., minimal spatial "blocks") of the physical reality "now" (with respect to the reader of this book as an observator with a concrete spatio-temporal position).[1] So, ω is a dynamic spatio-temporal formal concept, which could vary quantitatively over time.[2] By "physical reality" we mean the whole existing physical world beyond the observable world [31]. So, ω would be at least 10^{78}, since it is estimated that there are at least 10^{78} atoms in the observable universe.[3]

If we want to maintain coherence between our mental understanding of the word "more" before and the physical plausibility of really adding to this "more" amount of objects, we see that "more" could be at most ω, since each (physical) object consists of at least a physical quantum. Therefore, the maximal physically plausible group of things is the collection having as "things" the quanta in the whole physical reality. This group has ω number of elements.

---

[1] Note that the value of ω depends on the exact time when the reader is taking it into consideration. For example, "now" can mean for the author of this book 30.10.2018 at 12:23:03 CET.

[2] It is worth mentioning that the fact that ω could be finite (or not) is an open question in modern physics, and the answer depends on the exact value of the curvature of the universe [6]. However, some evidence suggests that ω should be finite [8, 11]. Finally, it is worth mentioning here that most of the concepts and frameworks developed in further sections (with the exception of Sect.5.7) can be developed without establishing the (non-)finiteness of ω.

[3] The exact number emerging from this kind of calculation tends to vary constantly due to the improvement of our technological devices and due to the appearance of new ways of doing the calculations. For example, in 2017 the British physicist Antonio Padilla deduced using standard (well-known) facts (in physics) that the amount of particles in the visible universe is approximately 3.28×10^{80}.

From this, we can conclude that when we consider a group of things with a number of elements strictly "bigger" than ω, we are generating a "mental" concept being expressed as a semantic combination of former concepts as ω and the concept "bigger," via, for example, a metaphorical process [16, 20]. However, such a huge group of things like $\omega + 7$, $\omega + \omega$, and ω^ω would be concrete examples of objects with an essentially mental or meta-physical nature expressed physically through a (particular) written and spoken language. In fact, these expressions can be considered as examples of formal conceptual blending [9, 12, 15]. Specifically, if we combine, for example, the concepts of ω and 7 to generate a new concept $\omega + 7$, which has semantic similarities to ω and 7 as numbers, we attach to the symbol $\omega + 7$ the property of "being a quantity." In addition, due to the fact that we use the symbol $+$ on the blend, we also attribute the property of "being bigger" than ω to the conceptual fusion $\omega + 7$ since one could potentially cancel out in both sides of $\omega + 7 > \omega$, the number ω, and then obtain the true statement $7 > 0$. The former process is analogical, since, we are used to "canceling" similar terms on both sides of an inequality $a + b > a$. However, in order to do that, we need to guarantee that both numbers $a + b$ and a are well defined.

In our case, w is a number which has a support in the physical world: i.e., there is a collection of physical elements (physical quanta) having exactly ω elements. So, we can consider ω to be a "physical" number, however, $\omega + 7$ is formally just a mental construction. Specifically, $\omega + 7$ is a conceptual blend of two physical numbers.

Thus, we are seeking to compare from a quantitative perspective one object, e.g. ω, with a physical collection of elements as canonical quantitative support, and a "mental" number, i.e. $w + 7$. So, how could we compare quantitatively a physical object with a mental one?

5.4   Our Classic Intuitions About the "Natural Numbers"

We have an intuitive understanding of the well-known "natural numbers," starting with the familiar concept of "1" as a formal name abstracting the idea of "having one object," from which we can see a multitude of examples in our world such as, our earth, our moon, each single person, and, in general, each single physical entity that we can distinguish as unique.

After that, we continue considering the concept of "2," also based on the number of elements of intuitive examples such as the parents of a person, our hands, feet, eyes, and male and female specimens of certain species of animals.

Later, we continue with the number three, being intuitively supported by, for instance, the dimensions of our visible world, i.e. the maximal numbers of pencils that we can put together such that each pair forms a right angle. A less common known collection of examples, but quite natural in anatomy are the layers of skin (epidermis, dermis, and hypodermis) and the general venous circulatory systems (systematic, pulmonary, and portal).

Going further, we can find some "natural" examples of collections of objects in our world having 4, 5, 6, 7, 8, 9, 10, 11, and 12 elements. For most of the "small" numbers we can actually form a collection of objects of the same "type" having exactly the corresponding number of elements. Moreover, in order to be sure that a particular group of objects has a specific amount of elements we count explicitly its elements; i.e., we start with any element of the collection and label it as "number one," and then we choose another element and give it the mark "two." Lastly, we finish when the last object has been labeled and we take this label as the "number of elements" of the collection.

In addition, one can visually perceive more or less in an intuitive way that some groups of objects have a certain amount of elements. But, the bigger the amount of elements, the harder the intuitive perception is.

Effectively, at some point we can neither do the explicit counting, nor have an intuitive visual perception of the quantity of elements of the set. Therefore, we are used to rely on the counting performed by someone else and expressed as a finite combination of words, who expresses the numerical abbreviations given by our decimal system. For example, when we read the information "the biggest city in the world has more than 24,000,000 people," we just use the abbreviation "24,000,000" in order to form an approximate idea of the "real" amount of people represented by that abbreviation, without having counted or even seen such a huge collection of human beings.

We stop the process of counting, due to practical constrains and we justify that by arguing that "one could continue this process again and again." But, what does it really mean? And, what are the concrete bounds of the expression "again and again?" In other words, exactly where and when does one stop this counting task? Let us consider the following situation: Suppose that one constructs a house with 12,000,000 square kilometers of extension, full of rooms and doors. Now, you receive the task of opening each door and counting how many doors there are. When you start to perform the task and after opening, let us say, the first 1000 doors, you could infer that, although it could take a lot of time, at some point you will find the last door to open. In fact, for any spatio-temporal physical process involving just a bounded amount of time, we will conclude that, in the case of performing a kind of sequential process, we should arrive at a last step. So, why is there not a final natural number? Why can the process of counting not finish at all? And, what does it mean to say that the natural numbers are "natural," if most of the "natural" processes finish at some point?

These questions are as fundamental as the ones regarding the reason why we should start with a unity of quantity if we want to count objects. In general, (the) beginning and (the) end; first and last object; and initial and final stage are fundamental components of natural phenomena, so why does it seem to be a tendency towards an overestimation of initial stages, and, at the same time, an underestimation of the final ones?

5.5 Additional Cognitive Considerations

One of the central reasons for "over-usage" of mental formal structures like the natural, rational, and real numbers in mathematics and physics is an "oversimplification" of the process of counting. In fact, let us assume that we need to demonstrate with genuine photos that there exist at least one hundred thousands pair of shoes on earth, which have not been taken in photos already. Then, it would be considerably easier to provide initial evidence for the starting quantities like $1, 2, 3, 4, 5, 6, 7$, and 8 pairs of shoes.[4] However, getting photographic evidence of the pairs numbered $99, 998, 99, 999$, and $100, 000$ would clearly be another story. Moreover, the searching and finding task of each new pair of shoes would be a unique challenge, not equal to any of the other pairs. So, just by saying that one obtains all possible formal quantities (e.g., numbers) by "adding" one, turns out to be an extreme simplification of the quantitative structure of nature.

On the other hand, from the perceptual point of view, our minds seem to believe that they can "understand" what the quantities behind $1, 2, 3, 4, 5, 6, 7, 8, 9, 10, 11, 12$ are. However, the more one increases the number, the less perception (some) people seems to have about the genuine (physical) quantitative properties of such number. The history of science has shown that we are able to grasp a genuine understanding of such big numbers as the numbers of atoms in the universe. So, what could limit us in our quest for perceiving and understanding completely the whole spectrum of (finite) (physical) numbers?

Finally, it is important to mention the fact that within the spectrum of (finite) numbers there are a huge collection of interesting and sophisticated natural phenomena, which are not (in any sense) simpler than the ones being modeled by "infinite" numbers (e.g., large, inaccessible cardinals). In fact, one of the main reasons why we lack more and better (mathematical) models for a lot of physical phenomena is that such phenomena usually depend on intricate and sophisticated finite interrelations of (physical) variables and (physical) contexts that cannot be "caught" by the (per definition) non-finite mathematical frameworks used during the "formalization" processes.

5.6 The Physical Numbers

5.6.1 First Approach: Counting Physically

The former considerations were made in order to introduce a new refined structural form of the so-called natural numbers (i.e., the collection of *Physical Numbers*), which have a concrete *physical support*, at least by groups of physical objects (e.g.,

---

[4]Since, one could take photos of one's own pairs of shoes.

quanta) having these numbers as amounts of elements. So, the physical numbers start with one (i.e., the "cardinality" of a physical quantum) and finish with ω (i.e., the amount of physical quanta of the whole universe).

Next, let us review the concept of "counting" within the former setting. Let us assume that we could develop a process for counting quanta of the physical reality. We would use our classical decimal system approach in order to do that. Let us suppose that we start with a fixed quantum of matter (such as the number "one") and we give to it the symbol α as an internal label, in order to know that it has already been counted. In the next step, we look for a new quantum and we give it the internal label $\alpha + 1$. After that, we look for a new one and we give it the internal label $\alpha + 2$. We continue in this way until we finish the *counting* with the last quantum ω of the physical reality. Thus, ω is the last physical number, and we are using the Greek letter "ω" as a tag for it.

Additionally, when we add two natural numbers a and b, we are assuming intuitively that both of them exist independently of one another. In other words, we suppose that each of these numbers is existentially supported by a collection of objects having exactly a (resp. b) elements and that these collections are basically "disjointed," namely, they have no common objects. For example, when we want to obtain any kind of natural support of the meaning of the expression $5 + 7 = 12$, we are mainly assuming that we have a collection of objects A and B with 5 and 7 elements, respectively, such that each element belongs exclusively to only one of the collections; and that the addition of these two numbers is simply the cardinality of the collection formed by joining together all the objects in A and B. The same happens with a pair of numbers for which we can imagine similar collections of objects. But, what happens if one of these numbers exhausts, by definition, all the possible gatherings of physical objects in the whole universe, and for example, has as its support the maximum thinkable collection of physical objects in the cosmos, i.e., ω? Then, in this case, the expression $\omega + 1$ would be just a mental construct, namely, a kind of formal conceptual blending. The reason is that we are simply seeking to create meaning for a new concept obtained by combining the concepts of ω and *one* by means of the concept of "addition," in a similar way as for the addition of 5 and 7. So, in order to do that, we imagine collections of ("separate") mental objects (having a physical imagined nature), which we put together to obtain support for the resulting addition, which, in the end, is strictly mental. Shortly, $\omega + 1$ will be the morpheme that we use to denote a conceptual blending of the concepts ω and 1.[5]

For the sake of a better semantic precision, we can say that $\omega + 1$ is a *mental number*, since we construct it by doing a mental analogy with the way in which we intuitively add smaller natural numbers. With this terminology we can classify the classic natural numbers into a finite amount of physical numbers and also a finite amount of mental numbers, namely, all the (mental) numbers considered in scientific journals overcoming, from a conceptual point of view, the quantitative nature of ω.

---

[5]Further information about conceptual blending can be found in Chap. 7.

This refinement in the terminology is fundamental when we want to use mathematical theories for modeling and describing physical phenomena.

Here, it is worth mentioning that the "seminal formal tool" for generating the natural numbers, namely, the possibility of "adding one again" (e.g., with a successor function [25, Ch. 3]) vanishes in this context. Effectively, let us assume that we have a physical number β, supported by a physical collection of elements S_β of the corresponding size. In order to be able to "add a new number" to β, we need to be able to find an extra physical particle lying outside S_β. This can only be possible if S_β is not the whole existing physical realm.

Specifically, "physical counting" has an end, which gives us a concrete and accurate understanding of the whole process of counting as part of the physical ("tangible") phenomena.

5.6.2   Second Approach: Partitioning Physically

Previously we have used a physical version of the addition operation as a fundamental binary relation for defining the physical numbers. Now, we want to adopt a new posture taking a kind of "physical partition" (or "physical division") as a seminal operation.

In fact, we will start by assuming a very natural fact as the initial premise, namely, the *The Integrative External Reality Hypothesis (IERH)*: An external (physical) reality exists in which all humans are (physically) embedded.[6]

We start with the initial physical number (or ph-number) α, which is physically supported by *The External (Physical) Reality (EPR)* guaranteed by the IERH. Now, we informally obtain a physical support for the next ph-number just by partitioning the EPR into a strict sub-collection of physical entities and its complementary part. For example, let F be the collection of all Fermions in the EPR and F^+ the collection of all additional types of particles and entities in EPR, e.g., bosons. In the next step, we choose exactly one of the former sub-collections, for instance, F, and we bisect it again, e.g., into the collection G formed by the gluons, and the one formed by the other types of Fermions, (e.g., photons). So, we obtain a finer partition of EPR into more sub-collections.

We continue in the same fashion until we finish with a maximal partition describing a physical support for ω.

We note that in this case *the fundamental grounding support* for any physical number, formed by the union of all the sub-collections, is always the same, namely, the External (Physical) Reality. On the other hand, in the former approach we did

---

[6]This hypothesis is from a physical perspective very similar to its predecessor the External Reality Hypothesis (ERH) [30], (for further discussions see [4]), with the additional element that the role of humans into the existence of the physical realm is (from a cognitive and a phenomenological perspective) more preponderant in IERH. However, regarding our present discussion, we can assume essential quantitative similarities between them.

not choose a common grounding physical base for any of the ph-numbers, except for the last one.

A fundamental feature of the physical supports of the ph-numbers is that they are (potentially) time-dependent; i.e., membership to one of the sub-collections could depend on the specific time where the (counter) observer is doing his/her considerations. In this way, the formal nature of the ph-numbers is structurally related to the current state of the EPR.

5.7 Towards a Formalization of the Physical Numbers

In this section, we present the first formal framework to describe in an axiomatic way the quintessence of the physical numbers. Due to their seminal nature they are a new kind of entities which are constituted as formal/natural blending of physical and abstract (mathematical) structures. So, let us show this fusion more explicitly.

First, we will define primitive kinds of entities and relations among them, which should be seen as the analogous and refined version of the notions of set and the membership relation in the Zermelo–Fraenkel-Choice set theory [25]. Moreover, let us assume implicitly that we have constructed a refined logical formalism with features coherent with the new cognitive foundations program (see Chap. 3). So, we will make more explicit some of the features that such a formalism should possess without going deeper into the details, since this will be the subject of research of subsequent studies in AMI.

Second, we will write our foundational axioms in a setting similar to many-sorted logic [24]. Specifically, we will quantify over objects having a purely physical nature (e.g., physical sub-spaces), as well as objects having a mixed formal/physical nature (e.g., the ph-numbers).

Third, E_R will denote the External (Physical) Reality. Additionally, let us denote by A any well-defined sub-space of E_R (e.g., the current collection of all photons; stars; planets; the Milky Way; the earth, among others[7]). We will use the symbol $\mathbb{F}$ to denote the ph-numbers. In this context, we can talk about a *physical membership relation* among the *physical elements* of $\mathbb{F}$. So, let us denote this ph-relation by $\in_p$. Moreover, let us fix a physical sub-space A (this can be denoted by $A \subseteq_{ss} E_R$). In general, if A and B are physical sub-spaces, then we denote the fact that A is physically contained in B by $A \subseteq_{ss} B$ (i.e., each quantum (of any kind) of A belongs to B). $P_a(A)$ denotes the space of (all potential) finitely constructed partitions of A; i.e., "physically-disjointed" finite collections of sub-spaces of A, whose physical union recovers A again. Note that $P_a(A)$ is a formal/physical entity, where any partition of A can be found, but not where all the partitions of A are simultaneously

---

[7]Note that any of these physical sub-classes can be seen as objects which depend on time, specifically, they depend on the time when they are considered. For instance, the collection of stars now will be different from the collection of starts some million years ago.

located.[8][9] If E is a fixed partition of A, then we write $E \in_f P_a(A)$. In addition, the fact that a physical sub-space e belongs to a particular partition E will be denoted by $e \Subset E$. If a partition E consists exactly of the sub-spaces $A_1, \cdots, A_m$, then we denote this fact more explicitly by $E = [\![A_1, \cdots, A_m]\!]$. Let $P_1 \in_f P(A_1)$ and $P_2 \in_f P(A_2)$ be partitions whose physical domains are disjoint (i.e., A_1 and A_2 has no quantum in common), then we denote the new partition formed by integrating both of them by $P_1 \uplus_{par} P_2$ with physical domain $A_1 \uplus_f A_2$, i.e., the physical union of both physical sub-spaces A_1 and A_2. If $e \Subset P_1$, then we denote by $P_1 \setminus_{par} [\![e]\!]$, the partition of the physical sub-space $A_1 \setminus_f e$, consisting of all the physical sub-spaces of P_1 except e.

Fourth, the fundamental partial "physical" functions will be the following: The *physical cardinal* of a partition of A, $C_A : P_a(A) \to \mathbb{F}$; the *physical successor* $s_p : \mathbb{F} \twoheadrightarrow \mathbb{F}$; the *physical addition* $+_p : \mathbb{F} \times \mathbb{F} \twoheadrightarrow \mathbb{F}$; the *physical multiplication* $*_p : \mathbb{F} \times \mathbb{F} \twoheadrightarrow \mathbb{F}$; and the *physical quotient* between ph-numbers, $\div_p : \mathbb{F} \times \mathbb{F} \twoheadrightarrow \mathbb{F}$. The corresponding relations are $=_p$, $<_p$ and $\leq_p$ contained in $\mathbb{F} \times_p \mathbb{F}$.

Fifth, the "initial" physical number corresponding to the simplest partition containing just one sub-space, i.e. the whole corresponding sub-space in consideration, will be denoted by α, i.e. $C_A([\![A]\!]) = \alpha$, for any physical sub-space A. This physical number corresponds intuitively to the natural number zero (see Sect. 5.12). The "last" physical number corresponding to the maximal physical partition of the whole external reality into all its physical quanta (E_R^{max}) will be denoted by ω, i.e., $C_{E_R}(E_R^{max}) = \omega$. Similarly, for any $A \subseteq_{ss} E_R$, one can define the maximal partition relative to A, (i.e., A^{max}) in the same way, and the corresponding maximal (or final) physical number relative to A, by $\omega_A := C_A(A^{max})$. So, $\omega = \omega_{E_R}$. Note that intuitively a physical number coming from a partition that classically possesses $n + 1$ physical sub-spaces ($n + 1$ understood as a natural number), will simulate the natural number n (see Sect. 5.12).

Now, we describe all the axioms describing (implicitly) the formal structure of the physical numbers.

---

[8]The last constraint is due to physical considerations, since the amount of energy required for reproducing all the possible sub-spaces of A (which is required to generate all its partitions) increases exponentially and in a lot of cases is simply not available, e.g. $A = E_R$.

[9]To better illustrate the idea behind this concept, let us think about the entity T consisting of a person J, a white board, and a marker. This object can be thought of as "the collection of all (potential) possible written (short) expressions of thoughts of J." In fact, one could (potentially) find in T any possible (short) written expression of thoughts of J, just by asking J. However, T does not consist, at any time, of the collection of all the possible (short) written expressions of thoughts of J altogether.

5.7.1 Axioms Defining the Initial Physical Number

Here we define the seminal properties of the initial physical number α.

$$(\forall A, B \subseteq_{ss} E_R)(C_A(\llbracket A \rrbracket) =_p C_B(\llbracket B \rrbracket) =_p \alpha).$$

And the condition characterizing the fact that only the trivial partitions correspond to the initial number

$$(\forall A \subseteq_{ss} E_R)(\forall P \in_f P_a(A))(C_A(P) =_p \alpha \leftrightarrow P =_p \llbracket A \rrbracket).$$

5.7.2 Axioms for the Final (Global and Relative) Physical Number

The following axiom assures that ω_A is the biggest physical number relative to partitions in a physical sub-space $A \subseteq_{ss} E_R$.

$$(\forall A \subseteq_{ss} E_R)(\forall P \in_f P_a(A))(C_A(P) \leq \omega_A \wedge \omega_A \leq \omega)$$

Now, we assure that only the maximal relative partitions produce the maximal relative physical numbers

$$(\forall A \subseteq_{ss} E_R)(\forall P \in_f P_a(A))(C_A(P) =_p \omega_A \leftrightarrow P = A^{max}),$$

where the symbol $=$ among partition simply means that both partitions are exactly the same, i.e. they consist of exactly the same physical sub-spaces.

5.7.3 Axioms the (Physical) Equality

We define the (physical) equality among physical numbers, implicitly through a finite recursion approach. So, the initial step was already defined on the former axioms structuring α by defining (physical equality) for trivial partitions.

Now, based on that we define (physical) equality more generally

$$(\forall a, b \in_p \mathbb{F})((a \neq_f \alpha \wedge b \neq_p \alpha) \rightarrow (a =_p b \leftrightarrow ((\forall A, B \subseteq_{ss} E_R)(\forall P_1 \in_f P_a(A))$$

$$(\forall P_2 \in_f P_a(B))(\forall A_1 \Subset P_1)(\forall B_1 \Subset P_2)((C_A(P_1) = a \wedge C_B(P_2) = b) \rightarrow$$

$$(C_{A\setminus_f A_1}(P_1 \setminus_{par} [\![A_1]\!]) = C_{B\setminus_f B_1}(P_2 \setminus_{par} [\![B_1]\!]))))).$$

The following axiom guarantees the non-triviality of the equality:

$$(\forall A \subseteq E_R)(\forall P \in_p P_a(A))(\forall A_1 \in P)(C_A(P) \neq_p \alpha \rightarrow$$

$$C_A(P) \neq_p C_{A\setminus_f A_1}(P \setminus_{par} [\![A_1]\!]) \wedge C_A(P) = C_{A\setminus_f A_1}(P \setminus_{par} [\![A_1]\!]) + s(\alpha)).$$

5.7.4   Partitioning Axiom

This axiom describes the quantitative coherence of physical numbers when one adds (resp. suppresses) sub-spaces to partitions:

$$(\forall A', A'' \subseteq_{ss} E_R)(\forall P', Q' \in_f P_a(A'))(\forall P'', Q'' \in_f P_a(A''))((\exists B' \in P')(\exists B'' \in P'')$$

$$(\exists C', D', C'', D'' \subseteq_{ss} E_R)(B' = C' \uplus_f D' \wedge B'' = C'' \uplus_f D'' \wedge Q' = (P' \setminus_{par} [\![B']\!]) \uplus_{par} [\![C', D']\!]$$

$$\wedge Q'' = P'' \setminus_{par} [\![B'']\!] \uplus_{par} [\![C'', D'']\!]) \rightarrow (C_{A''}(Q'') =_p C_{A'}(Q') \leftrightarrow C_{A'}(P') =_p C_{A''}(P''))).$$

5.7.5   Retraction-Extension Axiom

This axiom guarantees the quantitative immutability of the physical numbers regarding the inner size of the sub-spaces of partitions

$$(\forall A \subseteq_{ss} E_R)(\forall P \in_f P_A(A))(\forall B \in P)(\forall C \subseteq_{ss} E_R)(C \subseteq_{ss} B \rightarrow$$

$$C_A(P) =_p C_{(A\setminus_f B)\uplus_f C}((P \setminus_{par} [\![B]\!]) \uplus_{par} [\![C]\!])).$$

5.7.6   Axiom for the (Physical) Successor Function

In this axiom we characterize the refinement of the classic successor function for the Peano arithmetic (see Sect. 2.3.3).

$$(\forall a, b \in_p \mathbb{F})(a =_p s(b) \leftrightarrow (\exists A', A'', A''' \subseteq_{ss} E_R)(\exists P', P'' \in_f P(E_R))(A' = A'' \uplus_f A'''$$

$$\wedge A' \in P' \wedge A'', A''' \in P'' \wedge C_{E_R}(P') =_p b \wedge C_{E_R}(P'') =_p a$$

$$\wedge P'' = (P' \setminus_{par} [\![A']\!]) \uplus_{par} [\![A'', A''']\!])).$$

5.7.7 (Physical) Addition Axiom

This axiom assures the existence of the (physical) addition of two physical numbers only in the cases that the physical quantitative constrains allow such an addition to exist.[10]

$$(\forall a, b, c \in_p \mathbb{F})(a +_p b =_p c \leftrightarrow ((a =_p \alpha \to b =_p c) \wedge (a \neq_p \alpha \to (\exists a' \in_p \mathbb{F})$$

$$(\exists A', A'', A''' \subseteq_{ss} E_R)(\exists P' \in_f P_a(A'))(\exists P'' \in_f P_a(A''))(\exists P''' \in_f P_a(A'''))$$

$$(A''' = A' \uplus_f A'' \wedge P''' = P' \uplus_{par} P'' \wedge s(a') = a \wedge C_{A'}(P') =_p a'$$

$$\wedge C_{A''}(P'') =_p b \wedge C_{A'''}(P''') =_p c)))).$$

5.7.8 (Physical) Multiplication Axiom

For the sake of clarity we divide this axiom in two parts by using an auxiliary multiplicative operation $*_p'$. So, let us first define axiomatically this operation

$$(\forall a, b, c \in_p \mathbb{F})(a =_p b *_p' c \leftrightarrow (\exists A \subseteq_{ss} E_R)(\exists x, y, z \in_f P(E_R))$$

$$(C_{E_R}(x) = a \wedge C_{E_R}(z) = c \wedge C_{E_R}(y) = b \wedge$$

$$(\forall w \in y)(\exists h_w \in_f P_a(w))(C_w(h_w) = c)) \wedge \uplus_{w \in y}(h_w) = x),$$

where $\uplus_{w \in y}(h_w)$ denotes the partition of E_R formed by the physical (disjoint) union of all fixed h_w. Informally, the operation $*_p'$ is the only one that is axiomatized in a way that the multiplication of physical numbers corresponding to partitions with n and m elements ($n, m \in \mathbb{N}$) produces a physical number with a partition of $n * m$ elements. This small intermediate step should be done in order to understand better central axiom structuring the (physical) multiplication:

$$(\forall a, b, c \in_p \mathbb{F})(a =_p b *_p c \leftrightarrow ((b =_p \alpha \to a =_p \alpha) \wedge (c = \alpha \to a =_p \alpha) \wedge$$

$$((b \neq_p \alpha \wedge c \neq_p \alpha) \to (\exists a', b', c' \in_p \mathbb{F})(s(a) =_p a' \wedge s(b) =_p b' \wedge s(c) =_p c'$$

$$\wedge a' =_p b' *_p' c')))).$$

---

[10]Remember all the physical considerations done in the former sections of this chapter.

5.7.9   (Physical) Quotient Axiom

This axiom defines the (physical) division naturally in terms of the (physical) product:

$$(\forall a, b, c \in_p \mathbb{F})(a =_p b \div_p c \leftrightarrow (c \neq_p \alpha \wedge b =_p a *_p c)).$$

From an ontological perspective one can say that *physical division* is a fundamental operation because it allows us to maintain "control" over the size of the ph-numbers, based on a physical fact (i.e., the quantitative structure of the physical partitions of sub-spaces of E_R). This type of formal regulation is harder to obtain when we consider operations which increase, quantitatively speaking, the size of computed numbers like addition and multiplication.

5.7.10   (Physical) Order Axioms

The following axioms establish the meaning of the (physical) order relation for the ph-numbers:

$$(\forall a, c \in_p \mathbb{F})(a \leq_p c \leftrightarrow (\exists b \in_p \mathbb{F})(a +_p b =_p c)).$$

$$(\forall a, c \in_p \mathbb{F})(a <_p c \leftrightarrow (\exists b \in_p \mathbb{F})(b \neq_p \alpha \wedge a +_p b =_p c)).$$

5.7.11   Refining the Peano Axioms

With the help of the former axioms one can see that the following axioms hold, which refines the classic Peano axioms (Sect. 2.3.3) in our multidisciplinary setting:

1. $(\forall a, b, c \in_p \mathbb{F})(a =_p b \rightarrow (a =_p c \rightarrow b =_p c))$
2. $(\forall a, b \in_p \mathbb{F})((a \neq_p \omega \wedge b \neq_p \omega) \rightarrow (a =_p b \rightarrow s(a) =_P s(b)))$
3. $(\forall a \in_p \mathbb{F})(s(a) \neq_p \alpha)$
4. $(\forall a \in_p \mathbb{F})(a \leq_p \omega)$
5. $(\forall a, b \in_p \mathbb{F})(s(a) =_p s(b) \rightarrow a =_p b)$
6. $(\forall a \in_p \mathbb{F})(a +_p \alpha =_p a)$
7. $(\forall a, b, d \in_p \mathbb{F})(a +_p s(b) =_p d \rightarrow a +_p s(b) =_p s(a +_p b))$
8. $(\forall a \in_p \mathbb{F})(a *_p \alpha =_p \alpha)$
9. $(\forall a, b, c \in_p \mathbb{F})((s(\alpha) <_p b \wedge s(\alpha) \leq_p a) \rightarrow a <_p a *_p b)$
10. $(\forall a, b, d \in_p \mathbb{F})(a *_p s(b) =_p d \rightarrow (a *_p s(b) =_p a *_p b +_p a))$

11. (Principle of (Physical-Finite) Induction) Let $D(x)$ be a well-formed formula (in our formalism[11]) with a single free variable. Then

$$D(\alpha) \rightarrow ((\forall a \in_p \mathbb{F})(a \neq_p \omega \rightarrow (D(s) \rightarrow D(s(a))) \rightarrow (\forall a \in_p \mathbb{F})(D(a))))$$

The former axioms represent genuine refinements of the classic Peano axioms in the sense that all of them are formally supported by physical partitions. Thus, in this way they are grounded on a brighter multidisciplinary foundation than the traditional ones.

5.8 Comparison with the Natural Numbers

In this section we will establish a formal relation between the ph-numbers and some subset of the natural numbers. The main intuition here is that the ph-numbers are "very similar" to a finite subset of the natural numbers, although their formal nature is, in some aspects, different. Here, we assume that the natural numbers start with zero, 0.

In fact, let us define recursively a partial function $\Phi : \mathbb{N} \nrightarrow \mathbb{P}$: Firstly, $\Phi(0) = \alpha$. Secondly, assume that $\Phi(n) = \beta$ is already defined for some $n \in \mathbb{N}$, and $\beta \neq \omega$. Thus, one partition B of E_R and one physical sub-space b of it exists, such that a non-trivial partition p_b of b (i.e., $p_b \neq [\![b]\!]$) exists; otherwise all the components of B would be quanta and so $B = \omega$.

So, we just fix a quantum of space in q lying in b and split b into the sub-spaces consisting of q, and all the other quanta in b different from q. So, if we denote the new resulting partition, including the former subdivision, by b', then we define $\Phi(n + 1) = c_{E_R}(b')$. Lastly, one can prove that the former number is independent of the partition chosen and that Φ possesses a partial inverse function $\Phi^{-1} : \mathbb{F} \rightarrow \mathbb{N}$. In addition, one can also observe that the classical operations on the natural numbers are compatible with the new physical operation over the range (of the natural numbers) that they make sense. For example, one can prove the following compatibility relation:

$$(\forall a, b, c \in \mathbb{F})(a *_p b = c \leftrightarrow \Phi^{-1}(a) * \Phi^{-1}(b) = \Phi^{-1}(c)),$$

where $*$ denotes the multiplication between natural numbers.

Thus, there is a close structural relationship between the classic natural and the physical numbers. Specifically, one can say that the ph-numbers are (a sort of) formal-physical refinement of the natural numbers, since they contain the

---

[11] Here, we implicitly assume that we have constructed an enhanced logical system being able to simulate the essential features of a first-order logic theory and following the requirements described in Chap. 3.

basic intuition that we have for "small" natural numbers and their arithmetic, and, simultaneously, they possess a more sound and compatible physical-formal nature.

5.9   Pragmatic Considerations

In this section we describe more concretely how to "count physically" within this numerical system. So, let us set some additional terminology. Let A and B be physical sub-spaces of E_R such that A is a physical sub-space of B (we denote this by $A \subseteq_p B$); i.e. any (physical) quantum of space in A also lives in B. Let us assume that there exists at least one quantum q which lives in B but not in A. So, we define the *external environment of A in B* as the sub-space of B consisting of all the quanta in B not living in A. This space will be denoted by $\mathscr{E}_B(A)$.

Now, one working principle in our physical numerical system is that when we are working with a physical number c, we always have a specific partition z and a sub-space A of E_R in mind such that $C_A(z) = c$. So, if we want to (potentially) "add physically" one element to z, we need to check that it is possible to choose a suitable sub-space $\gamma \in z$, together with a specific sub-space μ of γ such that $\mathscr{E}_\gamma(\mu)$ is physically well-defined; i.e., there exists at least one quantum in γ not living in μ. So, we can generate the partition z' being identical of z in all the sub-spaces excepting in γ. Namely, we replace γ by the sub-spaces μ and $\mathscr{E}_\gamma(\mu)$. Thus, $C_A(z')$ would be the desired physical number formed by "adding physically one." We could employ the natural notation $C_A(z') = c +_p \alpha$ to indicate the former relation.

5.10   Explanatory Scope of the Physical Numbers in Mathematics and Physics

Let us assume in this section that ω corresponds to a finite quantity; i.e., our universe has a finite number of fundamental particles at any time [11]. So, in [37], it is essentially shown that assuming strict finitism [23] some of the most relevant applied mathematical theories can be essentially derived. For instance, Calculus, Metric Space Theory, Complex Analysis, Standard Integration Theory, Hilbert Space Theory, and Semi-Riemannian Geometry are specifically reconstructed within a "strictly finite" semantic apparatus. This implies that the grounding mathematical frameworks needed for describing Quantum Mechanics and General Relativity can be recovered within this formal "finiteness" system.

In particular, the physical numbers, viewed as a natural structure within strict finitism, possess enough explanatory power for reconstructing and obtaining, respectively, (at least) some of the most basic mathematical theories and some of the most seminal physical theories available today.

5.11   (In)finiteness and Immensity

So far, we have developed enough ontological and pragmatical background for being able to have a closer and more detailed look into one of the most fundamental topics in (natural) sciences and mathematics, i.e., (in)finiteness. In fact, based on the concrete examples described in the last sections, we infer that in most of the cases our minds associate the notion of "finiteness" with collections of objects/ideas where we have a "conscious intuition/perception" that either we can "count" effectively all the (separated) members of such a collection or we can "visualize" them all together in a (mental) bounded scenario. Nonetheless, the more elements that we abstractly add to our initial quantity, the more challenging the corresponding counting/visualization is for our conscious mind. For instance, let us consider the generic statement: "The number n denotes a finite quantity," and let us denote it by $F(n)$. For us, it is very natural to accept the veracity of $F(1)$, $F(20)$, $F(300)$, $F(5000)$, and $F(7000000)$, because we have a lot of examples of collections of objects in nature with these amounts of objects; (a little bit) more demanding can be $F(8000000000)$ and $F(90000000000000)$, since the number of such desired collections start to decrease rapidly. An extreme case would be $F(3.28 \times 10^{80})$, because the only corresponding physical scenario offering a solid cognitive support for our "conscious understanding" of the veracity of such a sentence (i.e., $F(3.28 \times 10^{80})$) is the entire number of particles of our visible universe, which lies on the borders of what modern physics could accept as existent and simultaneously as visible from a micro perspective. Lastly, statements like $F(10^{90})$ and $F(100^{200})$ are, strictly speaking, only linguistic ones, whose veracity lies in a meta-physical dimension. Effectively, since the symbols 10^{80} and 100^{200} are only abstract mental creations without probable support in the (physical) universe, the adjective "finite" seems to not provide a scientifically provable/refutable property about them.[12]

So, when we consider numerical quantities close to ω, we can talk about *immense numbers*, since their corresponding semantic supports are so gigantic they are only comparable to the (discrete) cardinality of the whole (visible) universe. Moreover, due to their size, any physical partition $\rho \in_f P(E_R)$ of an immense (physical) number m (i.e., $C_{E_R}(\rho) = m$) should fulfill that any physical element x of ρ lies at

---

[12]However, the ((un-)conscious) mind could make a "fast metaphor" of the form $1, 2, 3$ *are numbers like* 10^{80} *and* 100^{200}, therefore if $F(1)$, $F(2)$, *and* $F(3)$ *are quite natural and intuitively true statements, then* $F(10^{80})$ *and* $F(100^{200})$ *should also be natural and intuitively true statements.* If the former "fast metaphor" is analyzed in a finer manner, one concludes that the semantic transfer between $1, 2, 3$, and 10^{80} and 100^{200} is, strictly speaking, not so robust as first thought. Effectively, there are plenty of external collections of objects in nature serving as semantic support for the numerical symbols "1,""2," and "3." Nonetheless, the symbols "10^{80}" and "100^{200}" are more sophisticated syntactic configurations of the atomic symbolic units "0,""1,""2," and "8" and the (explicitly and implicitly used) conceptual operations of "juxtaposition" and "exponentiation," whose concrete semantic support lies entirely on a phenomenological dimension; i.e., it is a mental construct.

the micro-level (see Sect. 4.3, Chap. 4). In fact, let us take the former statement for defining an immense number.

Let η be an immense number. Although η in the classic sense can be considered as just a finite natural number, when we do a closer analysis of the kind of explicit mental representations that we can create for it in terms of collections of physical objects, we encounter a structural difference in comparison with "smaller" numbers from our mecro-level environments. Explicitly, let us assume an ontological extension of the principle of structural coherence (see [7]), namely, that there exists a fundamental isomorphism (or structural equivalence) between the conscious and unconscious experience as a whole (i.e., (un-)consciousness) and the (biological) perceptual and non-perceptual physiological experience (i.e., (un-)awareness).[13] So, any physical support for η should have such a huge quantity of elements that the corresponding (isomorphic) mental counterpart should be present in the unconscious mind.

Contrastingly, let us analyze in more detail the notion of "infinity." From a perspective of cognitive psychology, Lakoff and Núñez suggest that essential semantic aspects of the notion of infinity are produced by the basic metaphor of infinity, which adds (potentially) a new step to a given sequential process and (metaphorically) produces a global final state producing the "infinite" conceptually [21, Ch.8], [32]. Now, with the former cognitive characterization of infinity, we infer that any kind of mental representation of any physical support for this metaphorically inspired notion should be, again, essentially unconscious. In fact, otherwise one could generate a three-dimensional (spatial) representation of a physical scenario whose cardinality is infinity; i.e., such that one can always add a new element ("step") without changing the quality of the scenario. But this is a contradiction, since in any bounded three-dimensional scenario formed with strictly distinguished elements adding a new one would immediately increase its cardinality (i.e., number of elements).

Taking a perspective based on cognitive linguistics, we see that the notion of infinity is associated to the formal negation of the concept of finiteness. Effectively, an entity is "infinite" if and only if *no* finite (sequential) process (i.e., with a beginning and an end) can capture its quantitative nature. So, assuming for a moment this semantic interpretation for "infinity," we see again that the mental counterpart of any object in nature serving as its physical support, should be unconscious.

For the two former characterizations of the notion of infinity we can also deduce straightforwardly that no physical partition (defining a physical number) can be used

---

[13]More specifically, we extend Chalmer's original correspondence [7, §7] to the physiological aspects of the body which have non-direct availability for global control, but indirectly influence the quality of the perceptual signals sent to central systems. In other words, for any non-central structure of the body affecting in any way the quality of the functional signals generating the whole network defining global control, there is a corresponding (isomorphic) non-central experiential structure. So, the sum of all such experiential structures configure what we could call the unconscious mind.

as a suitable physical support for it. Therefore, we conclude that from a purely physical perspective the notion of infinity has (semantic) support being purely meta-physical, which is not the case for the notion of immensity (as showed before).

However, due to the fact that for both notions there exists no conscious ("spatial") mental support,[14] there is a (conscious) tendency to identify (confuse) them. Such identification is, strictly speaking, an oversimplification, which could possess some degree of precision depending on the thematic framework where they are employed (i.e., identified) (see for instance [14]).

The philosophical question of why the assumption of the ideal (mathematical) notion of "infinity" seems to have concrete and pragmatic applications on the physical world can be enlightened in a new way following our former approach. Effectively, it is not the notion of infinity which ontologically provides useful insights about our finite nature, it is the notion of immensity which (disguised with a formal dress of "((no-)finiteness," i.e. infinity) provides the genuine existential basis from where emerges its apparent pragmatic scope. In other words, the authentic foundational notion grounding the (usually meta-physical) concept of infinity is the one of immensity, which, as seen before, possesses a more solid existential framework supported by physics.

5.12 Counting as Partitioning

The classic and well-known activity of "counting" as one of the most practiced (mental) activities used for obtaining coherent semantics about (natural) numbers and for doing (mental) operations between them, can be reduced, from a cognitive perspective, to the act of generating partitions in a sequential manner and "avoiding" the last component of the corresponding partition. Effectively, our basic intuitions of the number "one" are often based on examples where we perceive a specific entity X as a unity (e.g., a mango, a tree, a person). In this case, what we are doing is simply a partition of E_R in two sub-spaces, namely, X and $\mathscr{E}_{E_R}(X)$, where we focus our conscious attention to X and we, in some sense, "ignore" (temporally) $\mathscr{E}_{E_R}(X)$. Further, for the number two, usually we choose an additional sub-space Y of $\mathscr{E}_{E_R}(X)$,[15] corresponding to a second object, that, in a lot of cases, possesses strong (geometrical) similarities with X and we continuing our "counting" process simply by refining the initial partition to the partition consisting of X, Y and $\mathscr{E}_{E_R}(X \cup_{ph} Y)$, where $X \cup_{ph} Y$ denotes the *physical union of X and Y* (i.e., the space consisting of X and Y) and finally we "ignore" (quantitatively speaking) the last sub-space $\mathscr{E}_{E_R}(X \cup_{ph} Y)$. Note that in the whole process the conscious attention moves

---

[14]Namely, one cannot generate conscious mental (three-dimensional) representations of physical entities supporting any of them.

[15]In general, we choose this second object physically disjoint from the first one, namely, X.

temporarily from the initial space (X) until the penultimate (Y), with the implicit assumption that the "counted" entities belong to different spaces of the partition.

This method has a lot of practical advantages materialized in the efficient way in which we are able to manipulate very large numerical quantities by means of quite fast artificial artifacts. However, it also possesses considerable limitations because it is basically restricted in a spatio-temporal way to the conscious mind, which in several respects is weaker than the unconscious one; e.g. regarding perceptual processing [2, 22].

In conclusion, the former remarks suggest the possibility of looking for additional and more effective ways for understanding quantitative properties about numbers beyond counting. One instance of this could be obtained by gradually subitizing partitions with a stronger usage of the unconscious mind (for larger numbers) and based on a choice of suitable visual fields that can maximize the size of the numbers to be perceived. This kind of generic subitizing procedure is supported for initial (e.g., not so large) numbers by the structural relations between our mental (e.g., neurobiological) understanding of numbers and our (cognitive) spatial representations [19].

5.13   Towards Physical Number Theory

We propose explicitly a new direction to stating and, subsequently, to solving problems in number theory, namely, to state them over the physical numbers, and then to try to solve them in this new numerical system. In fact, when we work with the ph-numbers, we are embedded in a physical-finite setting where we have not only formal but also more physical tools at our disposal. By contrast, as shown in the former section, the logical-deductive power of frameworks dealing with structures like the ph-numbers is powerful enough for reconstructing other mathematical areas very closely related with number theory as real and complex analysis, among others.

Is it not peculiar that in elementary number theory, although most of the problems are easy to state, it turns out that their solutions are, in a lot of cases, not so easy to find at all? could not the cause of this formal phenomenon be a "grounding singularity" (e.g., lack of precision) on our formalizations of the seminal structures used here, namely, the natural and integer numbers? (see, for example, [1, 17, 36]).

We suggest within this chapter a specific way of solving more conjectures in (elementary) number theory; i.e., to obtain a more accurate formal notion of the corresponding and most intuitive numerical system used here. This is instead of trying to develop more sophisticated conceptual frameworks based on the notion of the natural number, which should be better understood from a foundational bottom-up point of view.

5.14   Conclusions

Any intuition and subsequent formalization that we have formed throughout the centuries regarding axiomatic and structural properties of those entities called "natural numbers" are essentially based, and start and finish, in our ((meta-)physical) interaction with the environment. Simultaneously, they are also supported by concrete examples of quantities living in our physical realm. Therefore, it is worth finding a more concrete semantic basis and genuinely physical foundations for these "natural" quantities, which are so close to our daily cognitive experiences with the world, and, at the same time, so abstract and timeless for us.

Hence, one of the central reasons for the high complexity that the solutions of quite simple arithmetic problems have, e.g. the Goldbach's conjecture [28, 35], is that we should first develop a more powerful, elegant, and natural axiomatic system closer to our physical realm, instead of building more sophisticated and strongly meta-physical formal frameworks grounded mostly in a clearly platonic axiomatization [25, Ch. 3], [27, 34].

In addition, our physical numerical system is an explicit candidate of a fundamental quantitative concept which can facilitate the development of automated theorem provers, because the nature of the physical numbers is, by definition, a more tractable one from a purely computational point of view. The reason is that they are physically grounded, and, therefore, closer to being implemented than the natural numbers, which by definition cannot be even saved into the memory of any hardware ever, due to their infinite essence.

Finally, due to the fact that the numerical system proposed here is, in principle, compatible with strict finitism, we can deduce that the ph-numbers are rich enough for describing the essentials formal patterns of (many) number theoretical problems and for re-generating standard seminal mathematical and physical theories.

Acknowledgments The author would like to thanks Mayra Alejandra Marin Espinosa and Homero Cantera-Lopez for all their valuable suggestions.

References

1. Apostol, T.M.: Introduction to analytic number theory. Springer Science & Business Media (2013)
2. Bargh, J.A., Morsella, E.: The unconscious mind. Perspectives on psychological science **3**(1), 73–79 (2008)
3. Bernal, A.N., Lopez, M., Sánchez, M.: Fundamental units of length and time. Foundations of Physics **32**(1), 77–108 (2002)
4. Bernal, A.N., Sánchez, M., Soler Gil, F.J.: Physics from scratch. Letter on M. Tegmark's "The Mathematical Universe". arXiv preprint arXiv:0803.0944 (2008)
5. Bloch, E.D.: The real numbers and real analysis. Springer Science & Business Media (2011)

6. Callahan, J.J.: The curvature of space in a finite universe. Scientific American **235** (1976). DOI 10.1038/scientificamerican0876-90
7. Chalmers, D.J.: Facing up to the problem of consciousness. Journal of consciousness studies **2**(3), 200–219 (1995)
8. Cornish, N.J., Weeks, J.R.: Measuring the shape of the universe. Notices of the AMS **45**(11), 1463–1471 (1998)
9. Coulson, S.: Semantic leaps: Frame-shifting and conceptual blending in meaning construction. Cambridge University Press (2001)
10. Dijkgraaf, R., et al.: The mathematics of string theory. Gazette des Mathématiciens **106** (2005)
11. Ellis, G.F.: The shape of the universe. Nature **425**(6958), 566–567 (2003)
12. Fauconnier, G., Turner, M.: The Way We Think. Basic Books (2003)
13. Garay, L.J.: Quantum gravity and minimum length. International Journal of Modern Physics A **10**(02), 145–165 (1995)
14. Gardiner, A.: Understanding infinity: the mathematics of infinite processes. Courier Corporation (1982)
15. Goguen, J.: Mathematical models of cognitive space and time. In: D. Andler, Y. Ogawa, M. Okada, S. Watanabe (eds.) Reasoning and Cognition: Proc. of the Interdisciplinary Conference on Reasoning and Cognition, pp. 125–128. Keio University Press (2006)
16. Grady, J.: Primary metaphors as inputs to conceptual integration. Journal of Pragmatics **37**(10), 1595–1614 (2005)
17. Guy, R.: Unsolved problems in number theory, vol. 1. Springer Science & Business Media (2013)
18. Hartle, J.B.: Gravity: An introduction to Einstein's general relativity. Addison-Wesley (2003)
19. Hubbard, E.M., Piazza, M., Pinel, P., Dehaene, S.: Interactions between number and space in parietal cortex. Nature Reviews Neuroscience **6**(6), 435 (2005)
20. Lakoff, G., Johnson, M.: The metaphorical structure of the human conceptual system. Cognitive science **4**(2), 195–208 (1980)
21. Lakoff, G., Núñez, R.: Where Mathematics Comes From: How the Embodied Mind Brings Mathematics into Being. Basic Books, New York (2000)
22. Marcel, A.J.: Conscious and unconscious perception: Experiments on visual masking and word recognition. Cognitive psychology **15**(2), 197–237 (1983)
23. Mawby, J.: Strict finitism as a foundation for mathematics. Ph.D. thesis, University of Glasgow (2005)
24. Meinke, K., Tucker, J.V.: Many-sorted logic and its applications. John Wiley & Sons, Inc., New York, NY, USA (1993)
25. Mendelson, E.: Introduction to Mathematical Logic (Fifth Edition). Chapman & Hall/CRC (2010)
26. Padmanabhan, T.: Physical significance of Planck length. Annals of Physics **165**(1), 38–58 (1985)
27. Paris, J., Harrington, L.: A mathematical incompleteness in peano arithmetic. Handbook of mathematical logic **90**, 1133–1142 (1977)
28. Richstein, J.: Verifying the Goldbach conjecture up to 4×10^{14}. Mathematics of computation **70**(236), 1745–1749 (2001)
29. Takhtadzhian, L.A.: Quantum mechanics for mathematicians, vol. 95. American Mathematical Soc. (2008)
30. Tegmark, M.: The mathematical universe. Foundations of Physics **38**(2), 101–150 (2008)
31. Tegmark, M.: Our mathematical universe: My quest for the ultimate nature of reality. Penguin UK (2014)
32. Ueno, Y., et al.: The basic metaphor of infinity and the concept of a point. The Academic Reports, the Faculty of Engineering, Tokyo Polytechnic University **28**(1), 120–128 (2005)
33. Von Neumann, J.: Mathematical foundations of quantum mechanics. 2. Princeton university press (1955)
34. Wang, H.: The axiomatization of arithmetic. The Journal of Symbolic Logic **22**(02), 145–158 (1957)

35. Wang, Y.: The Goldbach Conjecture, vol. 4. World scientific (2002)
36. Wiles, A.: Modular elliptic curves and Fermat's last theorem. Annals of mathematics **141**(3), 443–551 (1995)
37. Ye, F.: Strict finitism and the logic of mathematical applications, vol. 355. Springer Science & Business Media (2011)

Chapter 6
Dathematics: A Meta-Isomorphic Version of "Standard" Mathematics Based on Proper Classes

6.1 Introduction

At the beginning of the twentieth century, there was a particular interest among mathematicians and logicians in finding a general, coherent, and consistent formal framework for mathematics. One of the main reasons for this was the discovery of paradoxes in Cantor's Naive set theory and related systems, e.g., Russell's, Cantor's, Burati-Forti's, Richard's, Berry's, and Grelling's paradoxes [3, 4, 6, 12, 14], and [11]. In particular, Russell's paradox offered one of the strongest motivations for developing new and more restricted set-theoretical frameworks. Specifically, the seminal works of E. Zermelo [16], A. Fraenkel [5], J. von Newmann [15], P. Bernays [1, 2], R. Robinson [13], and K. Goedel [7–9] allow for the construction of the most accepted and well-known logical formal frameworks of Zermelo–Fraenkel set theory with Choice (ZFC) [10], and more generally Von Newmann–Bernays–Goedel set theory (NBG) [12, Ch. 4].[1]

In the context of NBG set theory, the essential starting point was the intuitive idea that a new entity called a "proper class" should be formed from the general collection of all sets because this special collection was (in a certain sense) "too big." So, the general framework of NBG set theory is based on the primitive notion of class and the primitive relation of membership among classes. In addition, the notion of set is captured by restricting the classes to those who belong to at least another class. In this way one can guarantee with a suitable axiomatization that such classes remain small enough to prevent contradictory statements like Russell's paradox and to fulfill the main axioms of ZFC set theory required for constructing the most fundamental mathematical theories, e.g. analysis, (differential and algebraic) geometry, (abstract) algebra, and number theory.

---

[1] See Chap. 2 for a briefly summary of the fundamental pillars of NBG.

© Springer Nature Switzerland AG 2020
D. A. J. Gómez Ramírez, *Artificial Mathematical Intelligence*,
https://doi.org/10.1007/978-3-030-50273-7_6

In addition, an implicit working principle in NBG set theory is that small classes (or "sets") are more suitable objects to start and work with. On contrary, proper classes are too big and formally "too dangerous" to be able to ground any consistent and enough general mathematical theory.

In this chapter, we will show that these classic quantitative considerations about proper and small classes are tangential facts regarding the consistency of ZFC set theory. Explicitly, we will construct a logic theory D-ZFC (dual theory of ZFC set theory) strictly based on (a particular sub-collection of) proper classes with a corresponding special membership relation, such that ZFC and D-ZFC are meta-isomorphic frameworks. More specifically, for any standard formal definition, axiom, and theorem that can be described and deduced in ZFC set theory, there exists a corresponding "dual" version in D-ZFC and vice versa. In particular ZFC set theory is consistent if and only if D-ZFC is consistent.

From a cognitive perspective, the scope of this meta-result is deeper than first thought. In fact, it implies that a certain sub-collection of proper classes, which are objects that are understood by our minds in a qualitatively different way (in comparison with sets), can ground semantically (exactly) the same results and structures of (standard) mathematics that sets do. However, how could it be possible that objects that are, by definition, completely mental entities with no physical counterparts are able to play the role of "meta-equivalent" seminal bricks for the foundations of mathematics, which can be, in particular, seen as the language of physics? This issue will be further illuminated in the conclusions section.

Finally, we will do all the constructions of the dual axioms for Dathematics primarily for the sake of cognitive completeness in the presentation. In other words, after following all the dual constructions, the reader will be able to get a stronger and more pragmatic understanding of how the semantic pillars of Dathematics works. Our presentation is designed to give a faster understanding to a broader spectrum of readers, specially those not so much related to the standard foundational frameworks for mathematics.

6.2   Dual Notions and Axioms of Zermelo–Fraenkel Set Theory with Choice within NGB Set Theory

In this section, we will follow the treatment of E. Mendelson on the construction of the whole framework for classes and sets developed in NBF set theory [12, Ch. 4].

Von Newmann–Bernays–Gödel set theory is a special framework in the sense that it allows the existence of complementary classes, which can be seen as "dual" classes regarding the meta-class of all classes. Specifically, we will use the formal symmetry lying in the Axiom of the Existence of the Complement Class, which asserts that for any class X, there exists a (unique) *dual* class X^+ satisfying

$$(\forall a)(a \in X \leftrightarrow a \notin X^+),$$

where a varies over sets [12, Ch. 4, B4].

We will define dual notions of the main structural concepts of NBG set theory based on the former axiom.

Let us start with the dual notion of the membership relation $\in$, which we denote by ε.

This *dembership relation* is defined by the following axiom:[2]

$$(\forall A, B)(A \varepsilon B \leftrightarrow A^+ \in B^+).$$

In this case, we say that A is a *dember* (*delement*) of B.

For the dual notion of set, we analyze the corresponding dual formula:

$$M_d(X) :\Leftrightarrow (\exists Y)(X \varepsilon Y) \Leftrightarrow (\exists Y)(X^+ \in Y^+)$$

$$\Leftrightarrow (\exists Z)(X^+ \in Z).$$

What it means is that X^+ is a *sed* (dual set) if, and only if, its complement X^+ is a set, since Y varies over all classes, if and only if, Y^+ varies over all classes.

Now, let us prove that there is a "dual" theory of NBG set theory based on a special sub-collection of proper classes playing the dual role that sets play in NBG.

6.2.1   Dual Notion of Equality

The notion of equality for classes and its dual are exactly the same:

$$X =_d Y :\Leftrightarrow (\forall Z)(Z \varepsilon X \leftrightarrow Z \varepsilon Y)$$

$$\Leftrightarrow (\forall Z^+)(Z^+ \in X^+ \leftrightarrow Z^+ \in Y^+)$$

$$\Leftrightarrow (\forall W)(W \in X^+ \leftrightarrow W \in Y^+) :\Leftrightarrow X^+ = Y^+ \Leftrightarrow X = Y.$$

6.2.2   Dual Inclusion

The dual notion of inclusion, namely, *dinclusion* is defined as usual:

$$X \sqsubseteq Y :\Leftrightarrow (\forall Z)(Z \varepsilon X \rightarrow Z \varepsilon Y).$$

We express this by saying that X is a *subsed* of Y.

---

[2]In most cases, the name of the dual notions will be given by replacing (resp. adding to) the first letter of the original name with the letter "d," coming from "dual." For example, the dual of the membership relation is called "dembership relation."

6.2.3　Dual Proper Classes

The dual notion of proper class is called *d-proper class* and is given by

$$\neg M_d(W) \Leftrightarrow (\forall Y)(\neg(W \varepsilon Y)) \Leftrightarrow$$

$$(\forall Y)\neg(W^+ \in Y^+) \Leftrightarrow (\forall Z)(\neg(W^+ \in Z)),$$

where $Z = Y^+$ varies over all classes. So, W is a d-proper class if and only if W^+ is a proper class.

Informally, seds have very similar properties as sets, when replacing $\in$ by ε.

In addition, since one of the central notions of NBG set theory is the concept of set, we want to understand its behavior within the framework of the ε relation. So, we will focus our attention on the dual versions of the further axioms regarding sets.

6.2.4　Dual Axiom T

The dual version of the Axiom T, namely, the *Axiom T^+* coincides with the corresponding Axiom T due to the following reasons:

$$X = Y \Leftrightarrow X^+ = Y^+ \Rightarrow (\forall Z)(X^+ \in Z \leftrightarrow Y^+ \in Z)$$

$$\Leftrightarrow (\forall W)(X^+ \in W^+ \leftrightarrow Y^+ \in W^+)$$

$$\Leftrightarrow (\forall W)(X \varepsilon W \leftrightarrow Y \varepsilon W).$$

The last chain of equivalences holds due to the fact that W and Z vary over all classes, if and only if, W^+ and Z^+ do so too.

It is clear that

$$(\forall A, B)(A = B \leftrightarrow A^+ = B^+).$$

In conclusion, the Axiom T^+ states

$$X = Y \Rightarrow (\forall W)(X \varepsilon W \leftrightarrow Y \varepsilon W).$$

6.2.5　Dual Predicative Well-Formed Formulas

We denote sed variables (i.e., symbols that vary only over seds) by lower-case letters and classes by upper-case letters. So a *dual predicative well-formed* (dwf) formula is just a w. f. formula Φ, where all the bound variables are sed variables.

6.2.6   Dual Pairing Axiom

The dual version of the Pairing Axiom, namely, *Axiom* P^+ is the following:

$$(\forall x)(\forall y)(\exists z)(\forall u)(u\varepsilon z \leftrightarrow u = x \vee u = y).$$

It is equivalent to the sentence:

$$(\forall x)(\forall y)(\exists z)(\forall u)(u^+ \in z^+ \leftrightarrow u^+ = x^+ \vee u^+ = y^+).$$

x, y, z, and u vary over seds if and only if x^+, y^+, z^+, and u^+ vary over sets. So, the last expression is equivalent to

$$(\forall x^+)(\forall y^+)(\exists z^+)(\forall u^+)(u^+ \in z^+ \leftrightarrow u^+ = x^+ \vee u^+ = y^+),$$

where all variables appearing here are set variables. So, if we know that all symbols $\varXi^+$ vary over sets, then we could eliminate the symbols $(-)^+$ and obtain, in fact, just the classic pairing axiom of NBG. So, the Axiom P^+ is simply stating that $z^+ = \{x^+, y^+\}$, and we will denote this by $z = \langle x, y \rangle$. In other words, the Axiom P^+ states that for any seds x and y, there exists a (uniquely determined) sed z having as *denements* exactly x and y.[3]

6.2.7   Dual Null Set

For the *Axiom* N^+(*Null Sed*), let us first note that we can write the classic Axiom N in the following equivalent form:

$$(\exists X)(\forall Y)(\neg(Y \in X)).$$

Effectively, the empty set satisfies the former condition due to the fact that any proper class Z also fulfills $\neg(Z \in \varnothing)$.

However, a class X satisfying that any class Y does not belong to it would fulfill, in particular, the classic condition defining the empty set. Therefore due to the Class Existence Theorem [12, Prop. 4.4 Ch 4], both should be the same.

So, let us prove that the corresponding dual version of the former version of the Axiom T also holds. In fact,

---

[3]In this section we show explicitly the essential constructions and (in some sense similar) arguments due to achieving an axiomatic completeness in our presentation. However, in the next section we will prove a more general dualization result that requires only minimal technical requirements and can be applied far beyond the concrete axiomatization of NBG set theory.

$$(\exists X)(\forall Y)(\neg(Y\varepsilon X)) \Leftrightarrow$$

$$(\exists X)(\forall Y)(\neg(Y^+ \in X^+)) \Leftrightarrow$$

$$(\exists X)(\forall Y^+)(\neg(Y^+ \in X^+)) \Leftrightarrow$$

$$(\exists X^+)(\forall Y^+)(\neg(Y^+ \in X^+)) \Leftrightarrow$$

$$(\exists X^+)(\forall Z)(\neg(Z \in X^+)) \Leftrightarrow$$

$$(\exists X)(\forall Z)(Z \notin X^+).$$

The last sentence described the existence of a class whose complement is the empty set, which is true because the universal class V of all sets fulfills this property.

In this way, the *empty sed* is the universal class V. So, the *duniversal class* containing all the delements is the empty set.

6.2.8 Dual Unordered Pairs

We should define a unique value for $\langle X, Y \rangle$, where X and Y are any classes. So, we do this in the natural way:

$$Z = \langle X, Y \rangle :\Leftrightarrow Z^+ = \{X^+, Y^+\}.$$

Thus, the *unordered d-pair* is defined as the null sed if one of the classes is a d-proper class, and it is defined by the Axiom P^+ if both classes are seds. By definition, we get the equality $(\langle X, Y \rangle)^+ = \{X^+, Y^+\}$.

In addition, we define the *ordered d-pair* of X and Y, $\langle\!\langle X, Y \rangle\!\rangle$, as

$$\langle\langle X \rangle, \langle X, Y \rangle\rangle.$$

It is quite simple to prove that this notion fulfills the corresponding dual property that an ordered pair satisfies; i.e., two ordered d-pairs are equal if and only if the first and the second components coincide. Similarly, one defines ordered d-pairs with n components. Again, from this definition we can prove that $(\langle\!\langle X, Y \rangle\!\rangle)^+ = \langle X^+, Y^+ \rangle$.

6.2.9 Dual Axiom for the Existence of a Membership Relation

The axiom of the existence of the ε-relation states that

$$(\exists X)(\forall u)(\forall v)(\langle\!\langle u, v \rangle\!\rangle \varepsilon X \leftrightarrow u\varepsilon v).$$

It is equivalent to

$$(\exists X^+)(\forall u^+)(\forall v^+)(\langle u^+, v^+\rangle \in X^+ \leftrightarrow u^+ \in v^+).$$

So, this sentence shows the existence of a class whose complement is the $\in$-relation class. This holds since the complement of the $\in$-relation fulfills the statement above.

6.2.10 Dual Existence of Intersections

The axiom of the existence of *dintersections* of seds states that

$$(\forall X)(\forall Y)(\exists Z)(\forall u)(u\varepsilon Z \leftrightarrow u\varepsilon X \wedge u\varepsilon Y).$$

It is equivalent to the following statement:

$$(\forall X^+)(\forall Y^+)(\exists Z^+)(\forall u^+)(u^+ \in Z^+ \leftrightarrow u^+ \in X^+ \wedge u^+ \in Y^+),$$

where X^+, Y^+, and Z^+ vary over classes and u^+ varies over sets. The last statement is equivalent to the classic Axiom of the existence of the intersection class of two classes. Moreover, if we denote this new class by $Z = X \sqcap Y$, then it holds

$$X \sqcap Y = (X^+ \cap Y^+)^+.$$

Analogously, there is a notion of *dunion* of classes denoted by $X \sqcup X$ satisfying

$$X \sqcup Y = (X^+ \cup Y^+)^+.$$

6.2.11 Dual Notion of Complement

The notion of the *domplement* of a sed is given by the statement:

$$(\forall X)(\exists Z)(\forall u)(u\varepsilon Z \leftrightarrow \neg(u\varepsilon X)),$$

which is equivalent to

$$(\forall X^+)(\exists Z^+)(\forall u^+)(u^+ \in Z^+ \leftrightarrow u^+ \notin X^+)),$$

where again X^+ and Z^+ vary over classes, and u^+ varies over sets. As before, the former sentence is equivalent to the axiom of the existence of the complement class. If we denote this class by X^d, then

$$X^d = Z = (Z^+)^+ = X^+.$$

So, both notions coincide.

Note that due to the definition of equality, all the classes defined before are uniquely determined, which justifies the introduction of the new symbols.

6.2.12   Dual Existence of Domains of Classes

The sentence guaranteeing the existence of d-domains of classes is the following:

$$(\forall X)(\exists Z)(\forall u)(u\varepsilon Z \leftrightarrow (\exists v)(\langle\!\langle u, v \rangle\!\rangle \varepsilon X).$$

It is equivalent to

$$(\forall X^+)(\exists Z^+)(\forall u^+)(u^+ \in Z^+ \leftrightarrow (\exists v^+)(\langle u^+, v^+ \rangle \in X^+),$$

where X^+ and Z^+ vary over classes, and u^+ varies over sets.

This is equivalent to the Domain Existence Axiom. If we denote this new class by $Z = \mathbb{D}^+(X)$, then

$$\mathbb{D}^+(X) = (\mathbb{D}(X^+))^+,$$

where $\mathbb{D}(-)$ denotes the complement of a class.

It is easy to prove the last three dual versions of the Axioms of Class Existence [12, §1 Ch. 4], namely

$$(\forall X)(\exists Z)(\forall u)(\forall v)(\langle\!\langle u, v \rangle\!\rangle \varepsilon Z \leftrightarrow u\varepsilon X),$$

$$(\forall X)(\exists Z)(\forall u)(\forall v)(\forall w)(\langle\!\langle u, v, w \rangle\!\rangle \varepsilon Z \leftrightarrow \langle\!\langle u, w, v \rangle\!\rangle \varepsilon X),$$

and

$$(\forall X)(\exists Z)(\forall u)(\forall v)(\forall w)(\langle\!\langle u, v, w \rangle\!\rangle \varepsilon Z \leftrightarrow \langle\!\langle v, w, u \rangle\!\rangle \varepsilon X).$$

Furthermore, there is a natural dual notion of difference of sets, defined as d-difference of X and Y, i.e.,

$$X \curlywedge Y := X \sqcap Y^d.$$

6.2.13   Dual Class Existence Theorem

The general D-Class Existence Theorem (DCET) is the following:

Let $\Phi(w_\alpha, \cdots, w_l, X_1, \cdots, X_m, Y_1, \cdots, Y_n)$ be a dwf formula, where the only relation symbols allowed are $=$ and ε, the bound (seds) variables are exactly $w_\alpha, \cdots, w_l$, and the free variables occur among $X_1, \cdots, X_m, Y_1, \cdots, Y_n$. Then

$$\vdash (\exists Z)(\forall x_1) \cdots (\forall x_m)(\langle\!\langle x_1, \cdots, x_m \rangle\!\rangle \varepsilon Z \leftrightarrow \Phi(x_1, \cdots, x_m, Y_1, \cdots, Y_n)).$$

Now, there is a predicative well-formed formula Φ^+, corresponding to Φ, constructed in the following natural way: Φ^+ is obtained from Φ by replacing the relation symbol ε by $\in$ and replacing the bounded sed variables by bounded set variables with the same names.

So, it can easily be seen that the last sentence is equivalent to the following one:

$$(\exists Z^+)(\forall x_1^+) \cdots (\forall x_m^+)(\langle x_1^+, \cdots, x_m^+ \rangle \in Z^+ \leftrightarrow \Phi^+(x_1^+, \cdots, x_m^+, Y_1^+, \cdots, Y_n^+)),$$

which is a theorem due to the general Class Existence Theorem applied to the predicative wf $\Phi^+(w_\alpha^+, \cdots, w_l^+, X_1^+, \cdots, X_m^+, Y_1^+, \cdots, Y_n^+)$.

6.2.14   Dual Cartesian Product

The definition of dual Cartesian product of X and Y is the following:

$$(\forall x)(x\varepsilon X \boxtimes Y \leftrightarrow (\exists u)(\exists v)(x = \langle\!\langle u, v \rangle\!\rangle \wedge u\varepsilon X \wedge v\varepsilon Y)).$$

This class exists in virtue to the DCET and it holds

$$X \boxtimes Y = (X^+ \times Y^+)^+.$$

Similarly, one defines Cartesian products for more than two classes. The dual notion of relation is the concept of *delation*, namely, X is a (binary) delation if $X \sqsubseteq \emptyset^{[2]} := \emptyset \boxtimes \emptyset$. Based on this concept, we can directly define the dual notions concerning relations; e.g., X is an irreflexive delation of Y, $X Irr^+ Y$:

$$Rel^+(X) \wedge (\forall y)(y\varepsilon Y \rightarrow \neg(\langle\!\langle y, y \rangle\!\rangle \varepsilon X)).$$

Similarly, we can define $X Tr^+ Y$ (X is a transitive delation on Y), $X Part^+ Y$ (X *partially d-orders* Y), $X Con^+ Y$ (X is a connected delation of Y), $X Tot^+ Y$ (X *totally d-orders* Y), and $X We^+ Y$ (X *well-d-orders* Y). So, they also fulfill the corresponding dual properties, e.g., $X Tot^+ Y$ if and only if $X^+ Tot Y^+$.

In general, the following dual notions exist also due to the DCET:

6.2.15 Dual Notion of Power Class

By the DCET and the definition of equality, given a class X, there is a unique *Dower class* of X, denoted by $\mathbb{P}^+(X)$ containing as delements all subseds of X:

$$\vdash (\forall X)(\exists_1 Z)(\forall y)(y \varepsilon Z \leftrightarrow y \sqsubseteq X).$$

6.2.16 Dual Axiom U

The Axiom U^+ states that

$$(\forall x)(\exists y)(\forall u)(u \varepsilon y \leftrightarrow (\exists v)(u \varepsilon v \wedge v \varepsilon x)).$$

As usual, rewriting this statement with the classic notation we can see that it is equivalent to the corresponding Axiom U. We can easily prove that for any sed X

$$\sqcup X = (\cup(X^+))^+.$$

6.2.17 Dual Notion and Axiom of Sum Class

Similarly, by using the DCET, one proves the existence of a dual notion of Sum class, namely, for any class X there exists a class $Z = \sqcup(X)$ (*the D-Sum Class*) such that

$$(\forall y)(y \varepsilon \sqcup X \leftrightarrow (\exists v)(y \varepsilon v \wedge v \varepsilon X)).$$

6.2.18 Dual Axiom W

The dual version for the Axiom W (Power Set), i.e., the Axiom W^+ (*Dower Sed*) is the following:

$$(\forall x)(\exists y)(\forall u)(u \varepsilon y \leftrightarrow u \sqsubseteq x).$$

This sentence again holds because it is basically equivalent to the standard Axiom W (Power Set).

In particular, it is a formal computation to prove

$$\vdash P^+(\langle V, \langle V \rangle \rangle) = \langle V, \langle V \rangle, \langle V, \langle V \rangle \rangle, \langle \langle V \rangle \rangle \rangle,$$

where V denotes the *empty sed*, i.e., the universal class.

6.2.19   Dual Axiom S

It is a straightforward fact to state and to prove the Axiom S^+.

6.2.20   Dual Axiom R

The same holds for the Axiom R^+ in terms of a *univocal delation* $Un^+(X)$.

6.2.21   Dual Axiom of Infinity

The dual version of the Axiom I (Axiom of Infinity) is the Axiom I^+ (D-Axiom of infinity):

$$(\exists x)(V \varepsilon x \wedge (\forall y)(y \varepsilon x \;\rightarrow\; y \sqcup \langle y \rangle \varepsilon x)).$$

Informally, it states there exists a *(dual-)infinite* sed; i.e., a class whose complement is an infinite set.

6.2.22   Dual Axiom of Regularity

The dual version of the Axiom of Regularity (Axiom D) is the Axiom D^+:

$$\forall X (\exists W (W \varepsilon X) \;\rightarrow\; \exists y(y \varepsilon X \wedge \forall z(z \varepsilon y \;\rightarrow\; \neg(z \varepsilon X)))).$$

Again, it can be proved by using the corresponding form of the Axiom of Regularity.

6.2.23   Dual Axiom of Choice

Finally, there is a dual version of the Axiom of Choice, namely, the *Axiom of D-Choice*, stating that if x is a sed of pairwise *d-disjoint* seds, there exists a (d-choice) sed c containing exactly one denement of each of the seds of x. It is a direct consequence of (the corresponding classic version of) the Axiom of Choice.

Equivalently, we can explicitly state the dual form of Zorn's Lemma with the former terminology:

$$(\forall x)(\forall y)((y\mathrm{Part}x) \wedge (\forall u)(u \sqsubseteq x \wedge y\mathrm{Tot}u \rightarrow (\exists v)(v\varepsilon x$$

$$\wedge(\forall w)(w\varepsilon u \rightarrow w=u \vee \langle\!\langle w, v\rangle\!\rangle\varepsilon y))) \rightarrow (\exists v)(v\varepsilon x \wedge(\forall w)(w\varepsilon x \rightarrow \neg(\langle\!\langle v, w\rangle\!\rangle\varepsilon y)))),$$

which is equivalent to Zorn's Lemma, because all the variables vary over seds and then its dual expressions vary over sets. So, due to the dual properties of each of the delations expressed in the sentence, one obtains the corresponding classic form of Zorn's Lemma by reading "dually" this specific D-Zorn's Lemma.

Finally, D-NBG Sed theory will be defined as the first-order (sub-)theory (of NBG) having as logical axioms the standard 5 logical axioms of a first-order theory, as proper axioms we will consider the former dual versions of all the original axioms of NBG, and as logical rules (as usual) Modus Ponens and Generalization [12, Ch. 2 §3].

6.3   A More General Dualization Theorem

In the last section, we developed explicitly all the necessary basic (dual) notions and facts based essentially on the existence of a unique complementary class, and due to expository reasons as well. In fact, the fact that ZFC set theory is (one of) the most accepted foundational framework(s) for "modern" mathematics that compels us to present the minimal explicit results, since our work has direct implications on the identification of the seminal causes of the (in)consistency of such a framework. As the reader may suspect, the dual semantic and syntactic core properties lying behind most of the former statements do not depend on the existence of all the axioms of NBG (resp. ZFC).

Let L be a first-order language with equality and one binary operation symbol $\swarrow$. Let T be a first-order theory, including an axiom A_c guaranteeing the existence of a unique object, which plays the role of a formal "complement" with respect to $\swarrow$, i.e.,

$$(\forall a)(a \swarrow X \leftrightarrow a \nwarrow X^d).$$

Let us define a new binary relation symbol $\diagup$ by the sentence:

$$(\forall A, B)(A \diagup B \leftrightarrow A^d \swarrow B^d).$$

Let L^d be the language $L \cup \{\checkmark\} \setminus \{\checkmark\}$. Let Φ be a L-formula. We define the L^d-formula $\Phi^{(d)}$ by replacing in Φ every occurrence of $\checkmark$ by $\checkmark$.

With the former terminology, we can state our general Dualization Theorem.

Theorem 6.1 *Let Γ be a L-theory that includes the axiom A_c and let Φ be a L-sentence. Let $\Gamma^{(d)}$ be the corresponding $L^{(d)}$-theory consisting of the duals of the elements of Γ. Then, $\Gamma \vdash_L \Phi$ if and only if $\Gamma^{(d)} \vdash_{L^{(d)}} \Phi^{(d)}$. Furthermore, if M is a L-model of Γ, then the natural correspondence induced by the operator $(-)^{(d)}$ induces an isomorphism between M and $M^{(d)}$, where $M^{(d)}$ is exactly M as set and the interpretation of $\checkmark$ is given by means of the original interpretation of $\checkmark$ and the complements.*

Proof Let $H_1, \cdots, H_n$ be a L-proof of $\Gamma \vdash_L \Phi$, i.e., $H_m = \Phi$ and for any $i = 1, \cdots, m-1$, H_i is either an axiom of ZCF, an element of Γ, or it is a wff that can be deduced by a valid inference rule from the former $H_j - s$. It is straightforward to see that $H_1, \cdots, H_m$ is a L-proof of $\Gamma \vdash_L \Phi$, if and only if $H_1^{(d)}, \cdots, H_m^{(d)}$ is a $L^{(d)}$-proof of $\Gamma^{(d)} \vdash_{L^{(d)}} \Phi^{(d)}$. However, one can directly verify that the operator $(-)^{(d)}$ is its own inverse, and therefore it is an isomorphism.

6.4 Dathematics

Let us give the name "standard" (or set-theoretic) Mathematics to all formal mathematical theories that are grounded in ZFC set theory, for instance, Real and Complex Analysis, Geometry, Algebra, Number Theory, Topology, and Category Theory. We give the name *Dathematics* to the family of all dual versions of the (former) modern theories, where all subsequent concepts and theorems describing properties among them are expressed and grounded by D-ZFC.

Here, D-ZFC is, strictly speaking, a first-order logic dual sub-theory of NBG; i.e., in the same way that NBG is a conservative extension of ZFC. So too is D-NBG a conservative extension of (the corresponding theory) D-ZFC.[4] In particular, the fundamental objects of Dathematics are (a specific sub-collection of) proper classes (i.e., seds).

Furthermore, the empty sed is the universal class V, and the universal d-proper class V^+ is the empty set. So, in Dathematics, the quantitative properties, in a classical sense, are reversed. It is a natural meta-fact that (classic) mathematics and dathematics are (syntactically) meta-isomorphic; i.e., for any concept, theory, and conjecture in (standard) mathematics, there exists a symmetric d-concept, d-theory, and d-conjecture in dathematics with equivalent formal properties, and vice versa. For instance, we can prove the following syntactic meta-correspondence:

---

[4]The dual NBG first-order logic theory is just NBG itself considered with the former dual axioms (which are, in fact, theorems of the theory) and the former conventions about quantification over classes and seds, using explicitly only two binary relation symbols, i.e., $=$ and ε.

Theorem 6.2 *Let C be a conjecture in ZFC (seen as a sub-theory of NBG) given by a wff ϕ. Then there exists a corresponding dual conjecture C^+ in D-ZFC given by the dwf ϕ^+, such that C is provable in ZFC if and only if C^+ is provable in D-ZFC, i.e., $ZFC \vdash \phi$ if and only if $D - ZFC \vdash \phi^+$. Moreover, if P is a proof of C in ZFC, then the natural dual version of P in D-ZFC, namely P^+, is a proof of C^+ and vice versa. In other words, (standard) Mathematics and Dathematics are meta-isomorphic theories. In particular, they are equiconsistent.*

Proof The argument is similar to the one of the former theorem but with a small additional consideration. Effectively, let $P = \{P_1, \cdots, P_m\}$ be a proof of ϕ in ZFC. Then, based on the constructions done in Sect. 6.2, one can see that each P_i^+ is d-wff (i.e., a wff with respect to seds) and $P^+ = \{P_1^+, \cdots, P_m^+\}$ is a valid proof of ϕ^+ in the theory D-ZFC, which has the same inference rules of ZFC. The converse is straightforward. So, all (dual) well-formed statements that can be syntactically deduced from one theory can be (dually) mirrored into the other one. In conclusion, Mathematics and Dathematics are, in this sense, syntactically meta-isomorphic.

In particular, there exists a dual theory of the classical ZFC set theory, which can also be called *ZFC Sed Theory*.

6.5   Conclusions

The fact that set-theoretic Mathematics (based on ZFC [resp. NBG] set theory) and Dathematics are meta-isomorphic, and, in particular, one is consistent if, and only if, the other one is, together with the fact that the semantics for Dathematics are canonically given by "very big" objects (i.e., proper classes), shows that the cause of the Russell's paradox in Naive set theory is not only a matter of the "size" of the corresponding foundational objects (e.g., sets) but also lies within a deeper conceptual level in the formal framework in which sets are defined.

We have shown here that there is a formally identical version of standard Mathematics (i.e., Dathematics) structurally based on exactly the same type of objects that turn out to be avoided in NBG because of inconsistency issues, namely, proper classes. In particular, both formal frameworks are "equi-consistent" and both also simultaneously have, from a quantitative perspective, "diametrically opposite" seminal objects; i.e., sets and proper classes.[5]

---

[5]From a purely formal point of view, it would be worth exploring the construction of the dual versions of (real) analysis, topology, and number theory (to mention a few), using the original intuitions that we have about seds as some special type of proper classes, together with the former grounding framework developed here in terms of D-ZFC. Effectively, the initial conceptual bricks used for achieving this dual theory would be perceived cognitively in a new way in comparison with (our intuitions about) sets, which could enhance considerably our classic insights about mathematical phenomena.

Lastly, regarding the new cognitive foundations program (Chap. 3), this "singular" metamathematical phenomenon has a deeper meaning for our present understanding of the current foundational frameworks for mathematics (e.g., ZFC, NBG). In fact, the fact that our logic-deductive and semantic frameworks allow us to generate new meta-equivalent (formal) semantics (for Mathematics) with a "drastic" different pragmatic meaning indicates that our so-called semantics are, in fact, more syntactic constructions that do not "touch" transversely the cognitive essence of the corresponding (mathematical) objects in consideration.

It suggests that a more solid cognitive-semantic foundational framework for mathematics should be searched for also outside purely logic or classic metamathematical theories.

Acknowledgments The author wishes to thank Pedro Zambrano, Jose Manuel Gómez, Diana Carolina Montoya, and especially Diego Mejia for the useful suggestions during the elaboration of this manuscript. In addition, he would like to thank Jairo Gómez for the inspiration by teaching the beauty of formal consistent thinking. Finally, he thanks J. Kieninger for all the support and kindness. This work was supported by the Vienna Science and Technology Fund (WWTF), Vienna Research Group 12-004.

References

1. Bernays, P.: A system of axiomatic set theory. Studies in Logic and the Foundations of Mathematics **84**, 1–119 (1976)
2. Bernays, P.: Axiomatic set theory. Courier Corporation (1991)
3. Church, A.: The Richard paradox. The American Mathematical Monthly **41**(6), 356–361 (1934)
4. Curry, H.B.: Foundations of mathematical logic. Courier Corporation (1963)
5. Fraenkel, A.A., Bar-Hillel, Y., Levy, A.: Foundations of set theory, *Studies in Logic and the Foundations of Mathematics*, vol. 67. Elsevier (1973)
6. French, J.D.: The false assumption underlying berry's paradox. The Journal of Symbolic Logic **53**(04), 1220–1223 (1988)
7. Gödel, K.: The consistency of the axiom of choice and of the generalized continuum-hypothesis. Proceedings of the National Academy of Sciences **24**(12), 556–557 (1938)
8. Gödel, K.: The consistency of the axiom of choice and of the generalized continuum hypothesis with the axioms of set theory. Uspekhi Matematicheskikh Nauk **3**(1), 96–149 (1948)
9. Gödel, K., Feferman, S.: Kurt Gödel: Collected Works: Volume II: Publications 1939–1974, vol. 2. Oxford University Press (1990)
10. Jech, T.: Set theory. Springer Science & Business Media (2013)
11. Martin, R.L.: On grelling's paradox. The Philosophical Review **77**(3), 321–331 (1968)
12. Mendelson, E.: Introduction to Mathematical Logic (Fifth Edition). Chapman & Hall/CRC (2010)
13. Robinson, R. M.: The theory of classes a modification of von Neumann's system. The Journal of symbolic logic. **2**(01), 29–36 (1937)
14. Tait, W.W.: Cantor's Grundlagen and the paradoxes of set theory. Between Logic and Intuition: Essays in Honor of Charles Parsons (ed. G. Sher and R. Tieszen) pp. 269–290 (2000)
15. Von Neumann, J.: Mathematical foundations of quantum mechanics. 2. Princeton university press (1955)
16. Zermelo, E.: Untersuchungen über die grundlagen der mengenlehre. i. Mathematische Annalen **65**(2), 261–281 (1908)

Part II
Global Taxonomy of the Fundamental Cognitive (Metamathematical) Mechanisms Used in Mathematical Research

Chapter 7
Conceptual Blending in Mathematical Creation/Invention

7.1 Introduction

Over the last twenty years, conceptual blending has gained more and more attention and recognition as a fundamental ability of the human mind for merging concepts and ideas in a wide range of disciplines like linguistics, computer sciences, arts, literature, physics, and mathematics, among others [7, 18].[1][2][3]

In addition, J. Alexander has indirectly stressed an "omnipresence" of blending in pure mathematics. For example, he indicates a fundamental role of blending for obtaining such basic notions of modern mathematics, starting from the integer, rational, irrational, and imaginary numbers, and finishing with highly abstract concepts like generic points, motives, and k-theory [1].

Furthermore, the formal apparatus needed for obtaining more precise formalizations of conceptual blending started to be developed with the seminal work of J. Goguen and R. Burstall [13, 14] setting the basis of the theory of institutions, which allows us to find a global categorical framework for dealing with how to compare coherently semantic phenomena between theories expressed by different languages (i.e., signatures) through the use of a general notion of "abstract satisfiability," among others [5]. Subsequently, Goguen used this previous work in order to develop a general formalization of conceptual blending in terms of $3/2$−colimits [10, 11] and [12].

---

[1]Some of the results of this chapter can also be found in [4, 17] and [16]. However, all the meta-theorems and cognitive meta-generations presented here are an original product of the author, and in any particular case where secondary cooperation was valuable, this is explicitly acknowledged.

[2]For more detailed information see http://markturner.org/blending.html.

[3]In fact, there was a interdisciplinary research project called COINVENT (Concept Invention Theory), which aims to give a concrete formal (theoretical and computational) model of human concept creation based on Fauconnier and Turner's theory of conceptual blending [28].

© Springer Nature Switzerland AG 2020
D. A. J. Gómez Ramírez, *Artificial Mathematical Intelligence*,
https://doi.org/10.1007/978-3-030-50273-7_7

Some years later, T. Mossakowski and his collaborators were able to develop an entire computational tool called HETS (Heterogeneous Tool Set), which was grounded in something similar to a graph of institutions (described as languages and logics), as well as in the corresponding tools and translations [24].[4] HETS was also able to supply sound semantics (joint with the corresponding proof calculus) for heterogeneous specifications. In particular, the Common Algebraic Specification Language CASL [2] is supported by HETS. This language turns out to be useful for describing abstract concepts in classic mathematical areas such as abstract algebra and topology.

Particular versions of the former formalization obtained by Goguen in terms of colimits were able to be computed in HETS within a CASL specification. In fact, they have been able to model in a more accurate way how conceptual blending can "generate" concepts such as the (quasi-) complex numbers [8], simple versions of the integers, and the notion of prime ideals over commutative ring with unity [3].

Even by computing the specification of the prime ideal example as a blend, a new kind of commutative rings was suggested indirectly by the system HETS (i.e., the class of containment-division rings), which turns out to be equivalent to the notion of Dedekind domain in the finitely generated setting [15].

In addition, it was indicated in [15] that if a second-order version of the previous formalization is assumed, how one of the most fundamental notions of algebraic number theory (i.e., Dedekind domain joint with its prime spectrum) can be obtained as a blend of the notions of ideal over a Noetherian ring and the one of prime element on a quasi-monoid [15, §2].

On the other hand, the artificial intelligence group at the University of Osnabrück developed, over the course of several years, an outstanding (theoretical as well as computational) framework for analogical reasoning called Heuristic-Driven Theory Projector (HDTP), which is based on a specific form of monadic second-order anti-unification. It provides (structural) common information, "abstract analogies," of a source and a target domain described by first-order logic theories [27, 29].

Based on this work, it was shown in [21] how to generate the concept of abelian groups with elements of order at most two, the concept of commutative rings with unity with compatible divisibility relation, and a partial axiomatization of the integers as a blend of simpler concepts [21, §5]. In that case, the formalism of blending was based on sub-routines using analogical matching computed by HDTP and consistency checking [21, §3].

It is worth noticing that for almost all of the former case studies, the concepts involved were an approximation of standard mathematical definitions like the complex numbers and Noetherian rings. In each case, the conceptual decomposition involved single isolated and elementary notions and not entire collections of more sophisticated concepts belonging to the same mathematical area.

Here, we are going to prove explicitly how to generate fundamental and highly abstract concepts of Fields and Galois theory [20, Ch.5–6], as many-sorted first-

---

[4]http://hets.eu/.

order colimits of few elementary mathematical notions as the ones of abelian group, pointed abelian group, action of a group on a set, fixed points space, algebraic substructure, and distributive space. Some of these concepts are old and very well-known in the mathematical community and others are more implicitly used by the working mathematician and they emerge in a natural way during the process of generating such concepts of Fields and Galois theory through conceptual integration. An informal suggestion for considering these particular theories was first given in [3].

On the one hand, we obtain, strictly speaking, nine new meta-theorems describing how the former concepts (characterized as special objects of a specific co-complete category) are exactly represented as colimits of more elementary ones.

For example, let us state explicitly one of the (summarized) new meta-theorems of this work, whose formal proof is given by the whole collection of implementations described in the next sections:

The concept $\text{Aut}_F(E)$ *defined as the group of automorphisms of a field extension* E/F, *fixing the basis field* F, *viewed as a unique object of the co-complete category of many-sorted first-order theories with axiom-preserving signature morphisms, can be generated recursively by means of nine formal colimits (blends), starting from the concepts of abelian groups, pointed abelian groups, algebraic substructures, actions of a group on a set, and spaces of fixed points.*

Specifically, additional new meta-theorems will be explicitly stated as a result of the work developed in the following sections.

On the other hand, these meta-results start to suggest that this cognitively-inspired mechanism of syntactic and semantic combination should play a foundational role for bridging the cognitive gap that exists between the formal methods generally used in automated reasoning research and the ones used in formal cognitive sciences, trying to model the cognitive (actual) way in which a person states and solves a mathematical problem. Effectively, the work presented here can be seen as a kind of formal "symbiosis" between cognitively-inspired mechanisms (e.g., conceptual blending), formalized by means of a computationally-feasible framework (colimits of many-sorted first-order axiomatizations computed in HETS), which allows us to produce concretely mathematical concepts of a vastly abstract nature (e.g., the group of automorphisms of a field extension).

In addition, one of the main goals of this work is to start to enlighten and to offer evidence concerning the following fundamental questions in artificial intelligence and metamathematics: is it possible to meta-model and to meta-generate a considerably "huge" collection of mathematical theories (e.g., mathematical concepts and theorems) by modeling the most important cognitive processes discovered in cognitive sciences within the last decades, for example, analogical, inductive, abductive and case base reasoning, metaphorical thinking, and conceptual blending, among others?

We can see the examples developed in this work as seminal formal evidence that some of the current formalizations of one of these primary cognitive processes (i.e., conceptual blending) is, in fact, being able to formally generate complex mathematical concepts starting from very basic ones. Here, we can enlighten our approach "metaphorically" by saying that if our concepts were natural numbers, then we want to describe how some of our "natural" (conceptual) numbers can be generated from the "prime" numbers by means of combining them through a formal process called (conceptual) "product" (i.e., blending).

In addition, we suggest with this work a unique way of explaining, from a cognitive and computationally-feasible perspective and beyond the traditional and typical historic reconstructions, what the formal sources responsible for the generation of highly abstract mathematical concepts are. Effectively, the approach presented here is new in comparison with the standard historic reconstructions of the origins of Fields and Galois theories (see, for example, [6]).

7.2 Methods

7.2.1 Categorical Mathematical Concepts

We will specify our (categorical) *concepts* in many-sorted first-order logic [22].[5] Here, a (categorical) concept consists of the following components: a signature $\Sigma = (S, F, R)$, where S is a set of sorts, F is a set of functional symbols, each of them carrying a finite set of symbols of S specifying the n-tuple sort of the domain and the sort of the codomain (constants are just functional symbols with empty domain), and R is a set of symbols for relations with the corresponding m-tuple sort.[6]

A concept would have a (finite) set of sentences A in many-sorted first-order logic, called the axioms of the concept.[7]

Finally, an interpretation M of a concept is just a collection of sets M_S, functions M_F and relations M_R, with elements indexed by the corresponding sets S, F, and R, respectively; such that if M_g is an *interpretation* of $g : s_1 \times s_2 \times \cdots \times s_n \to s_m$, then $M_g : M_{s_1} \times M_{s_2} \cdots \times M_{s_n} \to M_{s_m}$.

Similarly for the interpretations of the symbols in R.

For simplicity, we define the class of models of a concept as big as possible; i.e., given the signature Σ and the finite set of axioms A, we define the class of models of the concept defined by (Σ, A) as the class $\mathbb{M}$ of all interpretations M,

---

[5]This is enough for the purposes of this chapter. However, later on we will give a more general notion of mathematical concept with a broader semantic scope and therefore it will be more suitable for the general purposes of our (cognitive) metamathematical quest.

[6]In most of the applications and results presented in this chapter we will use the word *concept* to mean *categorical concept*, in comparison with the more general notion of *(mathematical) concept* given later in this chapter. The main reason for this is simplicity during the presentation of the results. However, for the rest of the book the term "concept" essentially means "(mathematical) concept" in a broader generality.

[7]In general, a concept can have infinitely many axioms. However, the daily concepts used by the working mathematician have, in most cases, a finite number of axioms.

such that the interpretation of any axiom of A is, in fact, true over Σ, i.e. $\mathbb{M} \models_\Sigma A$. Here, we adopt the standard definition of satisfaction in many-sorted first-order logic. Furthermore, another way to express that is by saying that the class of models of a concept is just the dual of its set of axioms A into the class of all possible interpretations. This formalization allows us to say, by definition, that two concepts C_1 and C_2 are equivalent if their corresponding class of models coincide; i.e., if $\mathbb{M}_1 = \mathbb{M}_2$.

In conclusion, a concept for us consists of a triple $C = (\Sigma, A, \mathbb{M})$.

At this point, it is important to mention that due to practical considerations we do not allow infinite collections of axioms for defining (mathematical) concepts as, for example, within the framework of Formal Concept Analysis [9]. In fact, the working mathematician normally uses concepts with finitely many axioms and the implementations corresponding to the conceptual operation in consideration can be done more justly within a "finite" axiomatic setting.

Let us define the notion of morphism of concepts: if $C_1 = (\Sigma_1, A_1, \mathbb{M}_1)$ and $C_2 = (\Sigma_2, A_2, \mathbb{M}_2)$ are concepts, then a morphism $\phi : C_1 \to C_2$ is just a triple

$$\phi = (\phi_{\Sigma_1} : \Sigma_1 \to \Sigma_2, \phi_{F_1} : F_1 \to F_2, \phi_{R_1} : R_1 \to R_2),$$

such that the translation of the axioms of C_1 into C_2 induced by ϕ, i.e., $\phi(A)$, are deducible from the axioms A_2, that means $A_2 \vdash_{\Sigma_2} \phi(A_1)$.

It is a well-known fact that the collection of concepts with their morphisms forms a category $\mathbb{C}oncepts$, which can be also seen as the category of many-sorted first-order theories with axiom-preserving signature morphisms.[8] Moreover, in this category any V-shaped diagram, $\alpha : G \to C_1$ and $\beta : G \to C_2$ has a colimit [23].

For such a V-shaped diagram D, we will adopt the formalization followed in [3], namely, we use a simplified version of the formalization of Goguen ([10, 11] and [12]) for conceptual blending; i.e., we say that the colimit B of D is the *blending* of the concepts C_1 and C_2 regarding to (the identifications codified by) the concept G (through α and β). This simplification emerges from the fact that, in this particular context, the trivial order given by equality and defined on each set of morphisms between concepts seems to be the most natural and simple to be considered. So, in this specific case the notions of $3/2-$colimit and colimit coincides [10].

Let us use the notation $B = C_1 \vee_G C_2$, coming from the wedge sum of topological spaces (which is a particular case of a colimit construction), but adding the G as subindex in order to point out that the identifications of the two spaces is done among the generic space [26].

---

[8]Pragmatically speaking, we will show with plenty of examples that one can simulate almost any mathematical concept with this formalization, if one add enough sorts corresponding to power sets of other sets.

7.2.2 Structural Concepts

In order to obtain the fundamental concepts of Fields and Galois theory such as the ones of field, field extension, group of automorphisms of a field and $\mathrm{Aut}_F E$ (where E/F is a field extension) as a blend, we already presented the basic "structural" concepts required as conceptual "bricks" for starting to construct the whole "building" of this classic area of pure mathematics in Chap. 2.

We gave the explicit definitions of each of the structural concepts which are used, at least implicitly, in several areas of modern mathematics. More precisely, some of them are explicitly well-known concepts by the working mathematician (e.g., (abelian) group, action of a group on a set, and space of fixed points), whereas other ones are often implicitly used in modern mathematics without receiving, until now, any kind of special name (e.g., bigroup, pointed (abelian) group, and algebraic (bi-)substructure).

Finally, the concepts described in this section were obtained by trying to decompose the four former ones, coming from Fields and Galois theory, into their "minimal" conceptual building blocks. In particular, the concept of algebraic substructure appears as a simple natural solution to the question of decomposing the concept of field extension as a blend of the notion of field and a coherent additional concept. In fact, as it is later shown, this particular concept of algebraic substructure is a more general version of the notion of embedding, which seems to be structure-independent, since there is no specific intrinsic condition imposed on the corresponding binary operations, not even associativity. This process of formally decomposing (mathematical) concepts can also lead to the discovery of new seminal concepts, which allows us to understand more generally as well as more practically some of the classical concepts used by the working mathematician (e.g., embedding).

Here, it is relevant to give an intuitive account of the concrete form of a blending within this category of concepts in many-sorted first-order logic. Typically, one starts with two input concepts, defined by a finite collection of symbols for sorts, functions, relations, and a finite collection of axioms. Secondly, one identifies the symbols to be unified (blended) into the colimit, then, one "codifies" by hand this merging, by defining a suitable generic space and two morphisms from these spaces into the input spaces. These morphisms describe the way in which one wants to make the identifications. In the end, one obtains as colimit space, the collection of all the axioms from the two input spaces written using the "new" symbols induced by the generic space via the two corresponding morphisms. This is basically what HETS' command *combine* does.

7.3 Computations and Formal Proofs: Explicit Generation of Fundamental Concepts of Fields and Galois Theory

In this section, we use the system HETS in the language of CASL in order to compute the colimits of the former concepts, and we show explicitly some of the implementations. The reasons for this are as follows:

First, this kind of presentation not only introduces the reader in a direct manner to the HETS-CASL syntax and to the way HETS computes explicitly formal colimits, but it also shows that the CASL syntax can be formally understood at a basic formal level without any prior technical preparation. This fact allows researchers with an original formation in pure mathematics to potentially follow the formal arguments behind the code.

Second, one can see through the code lines the formal proofs of the mathematical statements lying behind. For example, the first three implementations described below show implicitly the mathematical proposition stating that (with the former formalization) the concept of a (mathematical) field [20, Ch.2, §1] can be expressed as an iterative colimit of the concepts of abelian group, pointed abelian group and, finally, the concept of distributive space.

Third, with this technical work we want to increase the level of mathematical rigor that this interdisciplinary research should require. For instance, classical works regarding the cognitive structural aspects of mathematical objects and theories like [19] and [1] emphasize mainly in psychological, linguistic, and historical considerations, however, there is a notorious shortage of metamathematical accuracy in their claims. Therefore, one of our purposes here is to start a solid formalization process and to set a clear metamathematical basis for claims concerning the cognitive modeling of highly sophisticated mathematical objects.

Finally, we show only explicitly the implementations for the colimits involved in the generation of the notion of field.[9]

7.3.1 Fields

We use this concept of bigroup (see Chap. 2) as an intermediate concept in order to obtain the notion of a field as a "fusion" of some of the former structural concepts.

So, let us show firstly how to obtain the concept of bigroup as a blend of the concepts of an abelian group and a pointed abelian group; i.e.,

$$BiGroup = AbGroup \vee_G PoinAbGroup.$$

---

[9]For the additional concepts the reader can find the implementation in the repository https://github.com/dgomezramire/FieldsGaloisBlendingGeneration.

We show how to specify the input spaces, the generic space, and the blending
morphisms:

```
logic CASL

spec GROUP=
     sort Element
     ops 0,f: Element;
         __ + __: Element * Element -> Element
     sort Elem2= { x: Element. not x = 0 }
%% Axioms of an Abelian Group with at least two elements.
         forall x : Element; y : Element . x + y = y + x
         forall x : Element; y : Element; z : Element
. (x + y) + z = x + (y + z)
         forall x : Element . x + 0 = x /\ 0 + x = x
         forall x : Element . exists x' : Element . x' + x = 0
         . not (f = 0)

end
%% Axioms of a pointed abelian group
spec PUNGROUP=
     sort ElemPlus
     op   zero: ElemPlus;
     sort Elem3={ x: ElemPlus. not x = zero }
     ops 1: Elem3;
         __ * __: ElemPlus * ElemPlus -> ElemPlus

         forall x : Elem3; y : Elem3 . x * y = y * x
         forall x : Elem3; y : Elem3; z :Elem3
. (x * y) * z = x * (y * z)
         forall x : Elem3 . x * 1 = x /\ 1 * x = x
         forall x : Elem3 . exists x' : Elem3 . x' * x = 1
         forall x : ElemPlus . x * zero = zero
/\ zero * x = zero
end

spec GENERIC =
     sort Elem
     op   elt: Elem
     sort SubElem={ x : Elem . not x = elt }
end

view I1:
     GENERIC to GROUP =
     Elem |-> Element, elt |-> 0, SubElem |-> Elem2
end

view I2:
     GENERIC to PUNGROUP =
     Elem |-> ElemPlus, elt |-> zero, SubElem |-> Elem3
end

spec Colimit = combine I1, I2
```

After computing the colimit we obtain exactly the concept of a bigroup:

```
logic CASL.SulFOAlg=

sorts Elem, SubElem
sorts SubElem < Elem
op 1 : SubElem
op __*__ : Elem * Elem -> Elem
op __+__ : Elem * Elem -> Elem
op f : Elem
op zero : Elem
forall x : Elem . x in SubElem <=> not x = zero %(Ax1)%
forall x, y : Elem . x + y = y + x %(Ax2)%
forall x, y, z : Elem . (x + y) + z = x + (y + z) %(Ax3)%
forall x : Elem . x + zero = x /\ zero + x = x %(Ax4)%
forall x : Elem . exists x' : Elem . x' + x = zero %(Ax5)%
. not f = zero %(Ax6)%
forall x, y : SubElem . x * y = y * x %(Ax2_8)%
forall x, y, z : SubElem . (x * y) * z = x * (y * z) %(Ax3_9)%
forall x : SubElem . x * 1 = x /\ 1 * x = x %(Ax4_10)%
forall x : SubElem . exists x' : SubElem . x' * x = 1 %(Ax5_11)%
forall x : Elem . x * zero = zero /\ zero * x = zero %(Ax6_12)%
```

This proves our first theorem:

Theorem 7.1 *The concept of* Bigroup *is the formal colimit of* Abelian Group *and* Pointed Abelian Group, *in the category of many-sorted first-order theories with axiom-preserving signature morphisms.*

Second, we show how to obtain the concept of field as a blend of the concepts of bigroup and a distributive space; i.e.,

$$Field = BiGroup \vee_{G_1} DistSpace = (AbGroup \vee_G PoinAbGroup) \vee_{G_1} DistSpace.$$

The specification of the concepts and morphisms is the following:

```
spec BIGROUP=
    sort Element
    ops 0,1: Element;
        __ + __ : Element * Element -> Element
        __ * __ : Element * Element -> Element
    sort Elem2= { x: Element. not x = 0 }
        forall x : Element; y : Element . x + y = y + x
        forall x : Element; y : Element; z : Element
. (x + y) + z = x + (y + z)
        forall x : Element . x + 0 = x /\ 0 + x = x
        forall x : Element . exists x' : Element . x' + x = 0
        . not (1 = 0)
        forall x : Element; y : Element . x * y = y * x
        forall x : Element; y : Element; z :Element
. (x * y) * z = x * (y * z)
        forall x : Element . x * 1 = x /\ 1 * x = x
        forall x : Elem2 . exists x' : Elem2 . x' * x = 1
    end
```

```
spec DISTRISPACE=
     sort Elemdistr
     ops __ oo __: Elemdistr * Elemdistr -> Elemdistr
         __ && __: Elemdistr * Elemdistr -> Elemdistr
         forall p : Elemdistr; q : Elemdistr; r : Elemdistr
. p && (q oo r) = (p && q) oo (p && r)

end

spec GEN=
     sort Elem
     ops __ + __: Elem * Elem -> Elem
         __ * __: Elem * Elem -> Elem

end

view I1:
     GEN to BIGROUP =
     Elem  |-> Element, __ + __ |-> __ + __,
__ * __ |-> __ * __
end

view I2:
     GEN to DISTRISPACE =
     Elem  |-> Elemdistr, __ + __ |-> __ oo __,
__ * __ |-> __ && __
end

spec Colimit = combine I1, I2
```

After doing the computation of the colimit we obtain the classic concept of a field:

```
sorts Elem, Elem2
sorts Elem2 < Elem
op 0 : Elem
op 1 : Elem
op __*__ : Elem * Elem -> Elem
op __+__ : Elem * Elem -> Elem
forall x : Elem . x in Elem2 <=> not x = 0 %(Ax1)%
forall x, y : Elem . x + y = y + x %(Ax2)%
forall x, y, z : Elem . (x + y) + z = x + (y + z) %(Ax3)%
forall x : Elem . x + 0 = x /\ 0 + x = x %(Ax4)%
forall x : Elem . exists x' : Elem . x' + x = 0 %(Ax5)%
. not 1 = 0 %(Ax6)%
forall x, y : Elem . x * y = y * x %(Ax7)%
forall x, y, z : Elem . (x * y) * z = x * (y * z) %(Ax8)%
forall x : Elem . x * 1 = x /\ 1 * x = x %(Ax9)%
forall x : Elem2 . exists x' : Elem2 . x' * x = 1 %(Ax10)%
forall p, q, r : Elem
. p * (q + r) = (p * q) + (p * r)       %(Ax1_11)%
```

Thus, the former implementations constitute a formal proof of the following theorem:

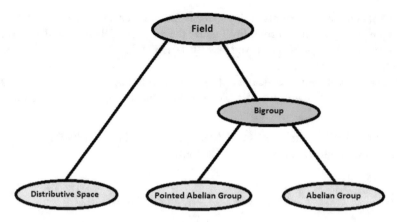

Fig. 7.1 Diagrammatic representation of the generation of the concept of (mathematical) fields through formal conceptual blending

Theorem 7.2 *The concept of* Field *is the formal colimit of* Bigroup *and* Distributive Space, *in the category of many-sorted first-order theories with axiom-preserving signature morphisms.*

One can find a schematic representation of the former collection of blends in Fig. 7.1. The way to read the graphic is as follows: Going upwards, the blending concept is located where two lines (coming from two concepts) meet. For example, the concepts *Punc. Abelian Group* and *Abelian Group* meet in their blend *Bigroup*.

7.3.2 Field Extensions

Let us blend two copies of the concept of an algebraic substructure, identifying a minimal amount of sorts, in order to obtain a concept which we call *algebraic bi-substructure*, necessary for getting the concept of a field extension in the next step. The explicit theorem is the following:

Theorem 7.3 *The concept of* Alg. Bi-Substructure *is the formal colimit of two copies of the notion of* Alg. Substructure, *in the category of many-sorted first-order theories with axiom-preserving signature morphisms, i.e.,*

$$AlgBiSubstruc = AlgSubStruc \vee_{G_2} AlgSubStruc.$$

Now, the blending space is basically a subset (GSUBALGSTRUC) of the base space (GALGSTRUC) having simultaneously two algebraic substructures given by two binary operations.

The specific list of axioms of this space is found in the next implementation (see the repository given in a former link) under the name ALGBISUBSTRUC. In fact, the notation "$Ax *\_*$" is preserved exactly, as HETS displays it when one asks for the theory of the colimit node.

So, we now obtain a particular list of axioms defining an equivalent version of the concept of field extension as a blend of the concepts of field and algebraic bi-substructure:

Theorem 7.4 *The concept of* Field Extension *is the formal colimit of* Field *and* Alg. Bi-Substructure*, in the category* $\mathbb{C}$*oncepts*.

$$FieldExt = Field \vee_{G_3} AlgBiSubStruc.$$

After computing the colimit, we obtain an equivalent axiomatization for describing the concept of field extension.[10]

7.3.3 Group of Automorphisms of a Field

We want to specify spaces of functions from a set X into itself in order to be able to manage concepts like the group of automorphisms of a field extension E/F, fixing the base field F; i.e., $\text{Aut}_F E$. For this purpose, we need to define a new sort $Func$, implicitly carrying such a space of functions. This is because HETS works with a language of many-sorted first-order logic. We are able to specify notions like the image of a function f on a point $x \in X$ by defining an operation $eval : Func \times X \to X$.

In this way, we are going to express these spaces of functions. Of course, when we talk about the collection of models for these spaces, then spaces of functions (in Zermelo–Fraenkel set theory) would be a particular prototype of models, but, in principle, there could be additional kinds of models as well.

However, for any kind of model fulfilling a concept C involving an evaluation function $eval : Func \times X \to X$, we can always recover an interpretation of any element $f \in Func$ as a function $f : X \to X$ in the classical sense, by means of defining $f(x) := eval(f, x)$, for any $x \in X$. In this sense, we are going to interpret this kind of concept. More generally, one equivalent form of describing a fixed collection A of functions from a set X into itself is given by a pairing $e : A \times X \to X$. In fact, such pairing is defined as the "evaluation" meta-function, which generates for any element $g \in A$, the function $g_e : X \to X$ defined by $g_e(a) := e(g, a)$ for any $a \in X$.

---

[10]The specific implementation can be seen in the repository under the name "EquivalentAxomatizationFields."

This is a technical trick that we use to express formally the structure AutL in terms of a many-sorted first-order specification, namely, in terms of the evaluation function $eval : \text{Aut}L \times L \to L$. So, this evaluation function allows us to express all the algebraic properties of AutL.

Our first goal is to generate the space AutL of automorphisms of a field L as a blending of structural concepts. The set of automorphisms of L has the structure of a group with the operation of composition of functions $\circ$ and with the identity on L as neutral element.

It is straightforward to verify that in the case of non-trivial fields (i.e., $0 \neq 1$), an automorphism f, so defined, fulfills additionally $f(1) = 1$ (that means in conclusion that f is an homomorphism of commutative rings with unity) and its inverse function is also an automorphism.[11]

One combines the notions of action and distributive space to obtain what we call here a *distributive action*,[12] i.e.:

Theorem 7.5 *The concept of* Distributive Action *is the formal colimit of* Action *and* Distributive Space, *in the category* $\mathbb{C}oncepts$.

$$DistrivAct = Action \vee_{G_4} DisSpace.$$

By combining this concept with one of a field we get the concept of bijections of a field, which are group homomorphisms with the sum:

Theorem 7.6 *The concept of* Bijections of a Field *which are Additive Group Homomorphisms is the formal colimit of* Distributive Action *and* Field, *in the category* $\mathbb{C}oncepts$.

$$BiyGHomField = DistrivAct \vee_{G_5} Field.$$

Summarizing, the colimit of the last implementation corresponds to the concept of bijections of a field, which are at the same time group homomorphisms regarding to the addition of the corresponding field.

So, we combine again the concept of distributive space with the concept obtained before in order to obtain the notion of group of automorphisms of a field;[13] i.e.,

Theorem 7.7 *The concept of* Automorphisms of a Field *is the formal colimit of* Bijections of a Field *which are Additive Group Homomorphisms and* Distributive Space, *in the category* $\mathbb{C}oncepts$.

---

[11] One can see a description of this space in CASL under the name of AUTOMORPHSPACE in the file "GroupAutFieldExt" within the repository.

[12] The resulting blending space is shown in the next implementation under the name DISTRIBU-TIVEACTION located in the folder GroupofBiyectionAdditiveHomField.

[13] See the file GroupAutField.

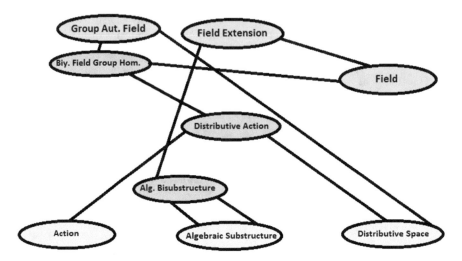

Fig. 7.2 Diagrammatic representation of the recursive generation of the concepts of field extension and the group of automorphisms of a field through formal conceptual blending

$$Aut\,Field = BiyGHomField \vee_{G_6} Dist\,Space.$$

The colimit obtained here is exactly equivalent to the concept of $AutL$ given before.

We see an additional schematic summary of the former results in the Fig. 7.2.

7.3.4 Aut(E/F)

We generate the concept $Aut_F E$ by simply blending two more concepts to the notion of $AutE$ as follows: Firstly, we generate the concept of group of homomorphisms of a field, which is the extension of another field as a blend of the concepts of group of automorphisms of a field and the concept of algebraic bi-substructure, namely,

Theorem 7.8 *The concept of* Automorphisms of a Field Extension *is the formal colimit of* Automorphisms of a Field *and* Alg. Bi-Substructure, *in the category* Concepts.

$$Aut\,FieldExt = Aut\,Field \vee_{G7} Alg\,Bi\,SubStruc.$$

Lastly, we blend the last colimit with the concept of a space of fixed points in order to obtain the notion of $Aut_F E$.

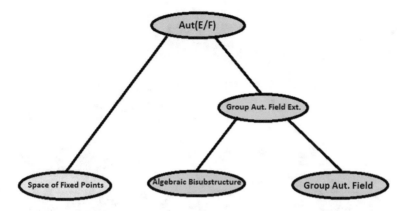

Fig. 7.3 Diagrammatic representation of the recursive generation of the concept of Meta-Galois group of a field extension through formal conceptual blending

Theorem 7.9 *The concept of* Group of Automorphisms of a Field Extension fixing the Base Field *is the formal colimit of* Automorphisms of a Field Extension *and* Space of Fixed Points, *in the category* Concepts.

$$\text{Aut}(F/E) = Aut\,Field\,Ext \vee_{G_8} Fix\,Point\,Spc.$$

We see the last two blends in the Fig. 7.3.[14]

7.4   Additional Evidence: The Theory of Lie Groups

In order to support the general claim that our specific formalization of formal conceptual blending, in terms of colimits of many-sorted first-order theories, it can be considered as a meta-generator of mathematical theories whose applicability goes beyond abstract algebra and elementary arithmetic. We will indicate in this section how to obtain the seminal notion of Lie Group in terms of more elementary notions by using the same methodology.

Explicitly, in order to be able to describe the notion of Lie Group in a many-sorted first-order logic setting, it is enough to characterize the concept of smooth manifold in terms of the following:

A sort X for the grounding set of the manifold $Tr(X)$ denotes the (sort of) trivializations (or charts) of X whose domains are elements of a fixed open covering of X; $Tr^{-1}(X)$ denotes the inverse charts whose domains are $\mathbb{R}^n$; where n is

---

[14]For the sake of completeness one can see the explicit specification of $Aut_F E$ obtained in the file ExplicitMetaGaloisGroup.

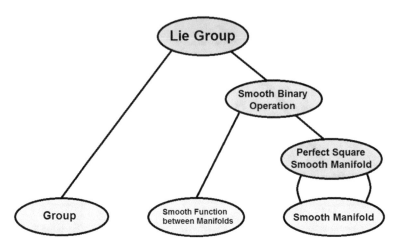

Fig. 7.4 Diagrammatic representation of the recursive generation of the concept of lie group through formal conceptual blending

the dimension of X;[15] $P\,Smooth(\mathbb{R}^n)$ denotes sort for the collection of partial smooth functions from $\mathbb{R}^n$ to $\mathbb{R}^n$ with open domain; and $P\,Functions(\mathbb{R}^n)$ denote the (super-)sort[16] of partial functions from $\mathbb{R}^n$ to $\mathbb{R}^n$. One should also consider functional symbols I_1 and I_2 for denoting the canonical isomorphisms between $Tr(X)$ and $Tr^{-1}(X)$[17] and a functional symbol

$$o : Tr^{-1}(X) \times Tr(X) \to P\,Functions(\mathbb{R}^n)$$

for the compositions that generate the transition maps.

So, one can express in this context that X is a smooth manifold of dimension n if and only if for all $\alpha : Tr^{-1}(X)$ and for all $\omega : Tr(X)$, it holds $o(\alpha, \omega) : P\,Smooth(\mathbb{R}^n)$.

In the following graphic (Fig. 7.4), we show how to obtain the notion of Lie Group as an iterative collection of blends (colimits) starting with the three basic notions of Smooth Manifold, Smooth Function between Manifolds, and Abelian Group. In addition, it is an elementary (but considerably long) process to write all the corresponding specifications in HETS/CASL, so for space considerations, we allow this to be an exercise for the interested reader.

---

[15]Here, we assume manifolds with the same global dimension.

[16]This sort contains the elements of the last one.

[17]This map sends a function ϕ to its inverse ϕ^{-1}.

7.5 Generating Genuine Concepts from Formal Weakening of Inconsistent Ones

In this section, we describe formal examples of inconsistent mathematical concepts emerging from conceptual blending which can be made consistent by using the cognitive ability of syntactic weakening (of axioms). Let us keep in mind the formalization of blending developed at the beginning of this chapter.

7.5.1 Non-trivial Space with a Transitive Divisibility Relation

Let us describe explicitly the two initial concepts that we need for doing our initial blend.

7.5.1.1 First Concept: Non-trivial Space with Additive Divisibility

A set D with a binary operation $*$, a binary relation $|$ and a distinguished (neutral) element e is called an *Non-Trivial Space with Additive divisibility* if the following axioms holds:

1. $(\forall a \in D)(a * e = e * a = a)$
2. $(\forall a, b \in D)(a | b \leftrightarrow [\exists c \in b = a * c])$
3. $(\forall a \in D)(a \neq e \rightarrow \neg(e | a))$
4. $(\forall a, b, c \in D)((a | b \wedge a | c) \rightarrow a | b * c)$
5. $(\exists a \in D)(a \neq e)$

7.5.1.2 Second Concept: A Partially Ordered Space with Minimum

A set E with a binary relation $\prec$ and a distinguished element n is called a *Partially Ordered Space with Minimum* if the following conditions holds:

1. $(\forall a \in E)(a \prec a)$
2. $(\forall a, b \in E)((a \prec b) \wedge (b \prec a) \rightarrow a = b)$
3. $\forall a, b, c \in E)((a \prec b) \wedge (b \prec c) \rightarrow a \prec c)$
4. $(\forall a \in E)(n \prec a)$

7.5.1.3 Generic Space

The generic space consists of a set G with a distinguished element g and a binary relation $||$.

7.5.1.4   Doing the blend

Let us do the natural conceptual morphisms from the generic spaces into the input spaces, namely, the first one sends G to D, g to e, and $||$ to $|$, and the second one sends G to E, g to n, and $||$ to $\prec$.

So, after computing the blend, we obtain an inconsistent space, because the translations of the axioms (3) and (5) of the first space and the axiom (4) of the second space impliy the following contradiction:

$$(\exists a \in G)(a \neq g \wedge \neg(g||a) \wedge (g||a)).$$

However, if we remove the axiom (3) from the first space and we blend the remaining weakened space with the second one, we obtain a consistent mathematical concept which can be called a *Non-trivial Space with a Transitive, Reflexive and Anti-symmetric Divisibility Relation*, since the resulting notion is the formal union of both input axiomatizations with the corresponding sort's identifications. A model of this blended theory is the set of Natural numbers with the product operation and the 1 as neutral minimal element.

7.5.2   Goldbach's Rings

7.5.2.1   First Concept: (Non-trivial) Goldbach's Structures

One of the simplest (and somehow quite mysterious) questions in number theory emerged from a letter that the mathematician Christian Goldbach wrote to Leonhard Euler. In modern terms, Goldbach's question (currently known as Goldbach's conjecture) asks if any even number larger than 2 can be written as the sum of two prime numbers [25]. Inspired by Goldbach's conjecture, we define a *Non-trivial Goldbach's Structure* by the following data: a set A, binary operations $+$ and $*$, elements 0 and 1 acting as the neutral elements with respect to each of the operations, a divisibility relation $|$ (defined in terms of the operation $*$), and an unary relation P in A defining the prime numbers of this structure, i.e.

$$[\forall d \in A(P(a) \leftrightarrow [\forall a, b \in A(d|a * b \rightarrow (d|a \vee d|b))])];$$

a distributivity axiom

$$(\forall a, b, c \in A)(a * (b + c) = a * b + a * c \wedge (b + c) * a = b * a + c * a),$$

the statement guaranteeing the non-triviality of the structure

$$(\exists a, b \in A)(a + a \neq b + b),$$

and the "Goldbach's condition"

$$[(\forall a \in A)[\exists p_1, p_2 \in A(P(p_1) \wedge P(p_2) \wedge P(p_2) \wedge a + 1 + a + 1 = p_1 + p_2)]].$$

The last condition guaranteed that each sufficiently bigger even number can be expressed as the sum of two prime numbers.

7.5.2.2   Second Concept: Groups of Torsion Two

A Group of Torsion two is a set B together with a binary operation $\oslash$ and neutral element e is Group of Torsion two if $(B, \oslash, e)$ is a group, i.e.

1. $(\forall a \in B)(a = e = a)$.
2. $(\forall a \in B)(\exists b \in B)(a = b = e)$.
3. $(\forall a, b, c \in B)((a) = a \oslash (b)))$.

Each element has order (torsion) at most two, namely,

$$(\forall a \in B)(a = e).$$

7.5.2.3   Generic Space

We define the generic space as a set G with a binary operation $\oplus$ and a neutral element n.

7.5.2.4   Doing the Blend

Let us send the sort G to A and B, respectively. Furthermore, we send $\oplus$ to $+$ in the first space, and to $\oslash$ in the second space. Finally, n is sent to 0 and e, respectively.

By doing the blend in HETS as a formal colimit, we obtain an inconsistent space. In fact, the non-triviality axiom guarantees that there exists at least one non-trivial (non-neutral) even number, and, at the same time, the torsion-two condition states that each "even" number is the neutral element.

If we do syntactic weakening with the non-triviality condition from the first space (i.e., we remove it), we get an enriched Goldbach's Ring, since the binary operation $+$ fulfills the axioms of a group and the Goldbach's space describes the distributive condition for $*$ with respect to $+$, and neutral condition for the element 1. Even more, the commutative ring with unity $(\mathbb{Z}/9\mathbb{Z})$ together with the (natural) operations and constants is a model for this conceptual space.

7.6   (Co-)Inventing Experiences of Students Using Formal Conceptual Blending

The author gave an intensive 1-week seminar titled "Creating new and classical mathematical concepts through a categorical formalization of conceptual blending" to bachelor's and master's students of the program of cognitive sciences in the university of Osnabrück.

During the seminar, the cognitive ability of conceptual blending and some seminal notions of (many-sorted) first-order logic and category theory were introduced. This was done in order to prepare the students for understanding a formalization of conceptual blending in terms of colimits of "mathematical concepts" expressed by finitely axiomatizable theories in a language of many-sorted first-order logic. In addition, it was shown how to generate specific concepts like the pseudo-integers, prime ideals of commutative rings of unity, and (mathematical) fields as formal blends by means of the former formalization. Finally, the students were asked to produce their own blends based on well-known classical concepts which could be freely chosen.

The most outstanding results by the participants are the following:

7.6.1   Normed Groups

One of the students of the master program in cognitive sciences, who has a bachelor's degree in communication sciences and had no prior post-high school formation in formal mathematics, was able to essentially (co-)discover the well-known notion of normed group. Specifically, he wished to blend formally the concepts of (abelian) group and "normed space." After some work during the seminar trying to find a sufficiently general notion of normed space, he was able to describe explicitly a V-diagram in HETS consisting of the input concepts of (abelian) group, normed space, a simple generic space, and the corresponding conceptual morphisms; such that the formal blending of V (i.e., its colimit) gives an equivalent formalization of the notion of normed group.

This student, with a high-school level in mathematical reasoning, was able to (co-)discover by means of formal conceptual blending a quite sophisticated mathematical notion (e.g., normed group) in a non-standard way; i.e., without working with specific collections of groups which can be enriched with a norm in a classical sense. Therefore, conceptual blending seems to inspire, in this case, a fast re-invention of a notion that, by standard means, would require a longer mathematical training in order to be discovered. Thus, this (co-)discovery was driven in part through the explicit description of this cognitive process as a fundamental ability for conceptual creation.

7.6.2 Quotient Groups

A student with a more sophisticated formation in pure mathematics, namely, holding a bachelor's degree in mathematical economics, was able to generate the notion of group of cosets G/N of an initial group G divided by a normal sub-group N, as a formal blend of the concepts of normal sub-group and the collection of left cosets of a group derived of a sub-group.

In this case, in contrast to the former example, the student already knew the concept of group of cosets in advance, and he was trying to find suitable input and generic concepts in order to be able to express this notion as a formal blend. Subsequently, and after some discussion during the seminar, the student was able to obtain the desired V-diagram, and so, he was able to re-express this classical notion of group theory as a colimit of two simpler ones.

7.6.3 Additional Remarks

This particular formalization of conceptual blending in terms of colimits of many-sorted first-order specifications of mathematical concepts seems to be quite fruitful when it is explicitly described as a core creative ability of the human cognition. Specifically, students without a professional formation in mathematics start to show a natural ability to (co-)discover sophisticated mathematical notions as one of normed groups, and simultaneously, to re-express concepts as one of quotient groups in a new formal way. From a pedagogical point of view, the students showed a stronger auto confidence when encountering mathematical problems, when the teacher introduced conceptual blending as a universal cognitive ability being present in the human mind.

7.7 General Conclusions

The implementations developed here start to offer initial basic features and meta-theorems playing a central role within the AMI program in the formation of mathematical concepts based on formal conceptual blending and on (a smaller part on) analogical reasoning. Specifically, in all the generated mathematical definitions, we obtained the blended concepts as combinations of two input conceptual spaces by doing basic syntactic identifications between some sort, function, and/or predicate symbols. These identifications can be seen not only as formal blends but also as being constituted through elementary instances of formal analogy making (see, for example, [21, 29]).

The collection of former examples supports the metamathematical soundness of the formalization of blending through a colimit categorical approach, since it suggests a concrete way of modeling our minds' formal ability to create new mathematical definitions. In addition, from a methodological perspective, the metamathematical statements described and proved here have all the formal rigor and clarity desired by researchers in basic sciences like mathematics, and at the same time, the essential formalizations are inspired by results from cognitive sciences, AI, and theoretical computer sciences. In addition, the results showed in this chapter also support the fact that general theories in (cognitive) sciences arise not only through abstract and general notions but also by means of collections of simple and clear examples which validate the soundness of a particular model. This heuristic (mathematical) meta-fact can be more widely seen in Chap. 11.

Furthermore, from a computational creativity perspective, these explicit implementations also suggest that formal conceptual blending (as colimits) routines could be subsequently implemented in automatic theorem provers and automated (mathematical) reasoning systems in order to make such systems more efficient and to improve the quality of their models of creative reasoning.

Acknowledgments The author wishes to thank all other members of the project COINVENT, specially Joseph Corneli, Felix Bou, Ewen Maclean, and Kai-Uwe Kühnberger for valuable and enlightening discussions. More explicitly, he thanks Alan Smaill for all the valuable inspiration and support on the meta-generation of the notion of bigroup. The author was funded firstly by the European Commission (SP7-ICT-2013-10), FET-Open Grant number: 611553, and subsequently he was also supported by the Vienna Science and Technology Fund (WWTF) as part of the Vienna Research Group 12-004.

References

1. Alexander, J.: Blending in mathematics. Semiotica **187**, 1–48 (2011)
2. Bidoit, M., Mosses, P.D.: CASL User Manual. Lecture Note in Computer Science 2900, Springer-Verlag, Berlin, Heidelberg (2004)
3. Bou, F., Corneli, J., Gomez-Ramirez, D., Maclean, E., Peace, A., Schorlemmer, M., Smaill, A.: The role of blending in mathematical invention. Proceedings of the Sixth International Conference on Computational Creativity (ICCC). S. Colton et. al., eds. Park City, Utah, June 29-July 2, 2015. Publisher: Brigham Young University, Provo, Utah. pp. 55–62 (2015)
4. Codescu, M., Neuhaus, F., Mossakowski, T., Kutz, O., Gomez-Ramirez, D.: Conceptual blending in dol: Evaluating consistency and conflict resolution. In: R. Confalonieri, A. Pease, M. Schorlemmer, T. Besold, O. Kutz, E. Maclean, M. Kaliakatsos-Papakostas (eds.) Concept Invention: Foundations, Implementation, Social Aspects and Applications, pp. 69–96. Springer International Publishing, Cham (2018)
5. Diaconescu, R.: Institution-Independent Model Theory. Studies in Universal Logic. Birkhäuser Basel (2008)
6. Edwards, H.M.: Galois theory., *GTM*, vol. 101. Springer-Verlag New York (1984)
7. Fauconnier, G., Turner, M.: The Way We Think. Basic Books (2003)
8. Fleuriot, J., Maclean, E., Smaill, A., Winterstein, D.: Reinventing the complex numbers (2014)
9. Ganter, B., Wille, R.: Formal concept analysis: mathematical foundations. Springer Science & Business Media (2012)
10. Goguen, J.: An introduction to algebraic semiotic with application to user interface design. *In* Computation for methaphors, analogy and agents. C. L. Nehaniv, Ed. Vol. 1562 pp. 242–291 (1999)
11. Goguen, J.: Towards a design theory for virtual worlds: algebraic semiotics and scientific visualization as a case study. *In* Proceedings Conference on Virtual Worlds and Simulation

(Phoenix AZ, 7–11 January 2001). C. Landauer and K. Bellman, Eds. Society for Modelling and Simulation pp. 298–303 (2001)

12. Goguen, J.: Steps towards a design theory for virtual worlds. *In* Developing future interactive systems. M. Sanchez-Segura, ed. Idea Group publishing, 116–152. pp. 116–128 (2003)

13. Goguen, J.A., Burstall, R.M.: Introducing institutions. In: E. Clarke, D. Kozen (eds.) Logics of Programs: Workshop, Carnegie Mellon University Pittsburgh, PA, June 6–8, 1983, pp. 221–256. Springer, Berlin, Heidelberg (1984)

14. Goguen, J.A., Burstall, R.M.: Institutions: Abstract model theory for specification and programming. Journal of the ACM (JACM) **39**(1), 95–146 (1992)

15. Gomez-Ramirez, D.: Conceptual blending as a creative meta-generator of mathematical concepts: Prime ideals and Dedekind domains as a blend. C3GI at UNILOG 2015, Workshop on Computational Creativity, Concept Invention, and General Intelligence. Tarek B. Besold et. al., eds. Publications of the Institute of Cognitive Sciences, PICS series, University of Osnabrück Vol. 2 (2015)

16. Gomez-Ramirez, D., Fulla M. Rivera, I., Velez, J., Gallego, E.: Category-based co-generation of seminal concepts and results in algebra and number theory: Containment division and Goldbach rings. JP Journal of Algebra, Number Theory and Applications **40**(5), to appear (2018)

17. Gomez-Ramirez, D., Smaill, A.: Formal conceptual blending in the (co-)invention of (pure) mathematics. In: R. Confalonieri, A. Pease, M. Schorlemmer, T. Besold, O. Kutz, E. Maclean, M. Kaliakatsos-Papakostas (eds.) Concept Invention: Foundations, Implementation, Social Aspects and Applications, pp. 221–239. Springer International Publishing, Cham (2018)

18. Kaliakatsos-Papakostas, M., Cambouropoulos, E., Kühnberger, K.U., Kutz, O., Smaill, A.: Concept invention and music: Creating novel harmonies via conceptual blending. In: In Proceedings of the 9th Conference on Interdisciplinary Musicology (CIM2014), CIM2014. Citeseer (2014)

19. Lakoff, G., nez, R.N.: Where Mathematics Comes From. Basic Books (2000)

20. Lang, S.: Algebra (revised third edition). Graduate Texts in Mathematics 211, Springer-Verlag, New York (2002)

21. Martinez, M., Abdel-Fattah, A., Krumnack, U., Gómez-Ramírez, D., Smail, A., Besold, T., Pease, A., Schmidt, M., Guhe, M., Kühnberger, K.U.: Theory blending: Extended algorithmic aspects and examples. Annals of Mathematics and Artificial Intelligence pp. 1–25 (2016)

22. Meinke, K., Tucker, J.V.: Many-sorted logic and its applications. John Wiley & Sons, Inc., New York, NY, USA (1993)

23. Mossakowski, T.: Colimits of order-sorted specifications. In: F.P. Presicce (ed.) Recent trends in algebraic development techniques. Proc. 12th International Workshop, Vol 1376, pp. 316–332, Lecture Notes in Computer Sciences, Springer-Verlag, London (1998)

24. Mossakowski, T., Maeder, C., Codescu, M.: Hets user guide (version 0,99) (2014). URL www.informatik.uni-bremen.de/agbkb/forschung/formal_methods/CoFI/hets/UserGuide.pdf

25. Richstein, J.: Verifying the Goldbach conjecture up to 4×10^{14}. Mathematics of computation **70**(236), 1745–1749 (2001)

26. Rotman, J.: An Introduction to algebraic topology. Springer (2004)

27. Schmidt, M., Krumnack, U., Gust, H., Kühnberger, K.U.: Heuristic-driven theory projection: An overview. In: H. Prade, G. Richard (eds.) Computational Approaches to Analogical Reasoning: Current Trends, pp. 163–194. Springer, Berlin, Heidelberg (2014)

28. Schorlemmer, M., Smaill, A., Kuehnberger, K.U., Kutz, O., Colton, S., Cambouropoulos, E., Pease, A.: COINVENT: Towards a computational concept invention theory. In: 5th International Conference on Computational Creativity (ICCC)

29. Schwering, A., Krumnack, U., Kuehnberger, K.U., Gust, H.: Syntactic principles of heuristic driven theory projection. Cognitive Systems Research **10**(3), 251–269 (2009)

Chapter 8
Formal Analogical Reasoning in Concrete Mathematical Research

8.1 Introduction

The ability to find commonalities between two objects living in multiple conceptual environments seems to be an omnipresent cognitive process within all scientific disciplines. In particular, the invention of well-known results in physics (e.g., the Rutherford atom model) and in pure mathematics (e.g., the discovery of the complex numbers) can be seen as emerging from an analogical process [5, 12]. In general, outstanding mathematicians of the twentieth century, such as André Weil, pointed out the prominent role that analogical reasoning plays in the development of modern mathematics [15].

Usually, analogy making is modeled more from a perspective of transferring logical information from a source domain to a target domain in order to obtain new qualitative insights regarding the proof of a particular logical or mathematical statement described in the target [14]. Other approaches aim to find analogical matches for sentences in monadic second-order logic by computing the least general generalization of pairs of terms defined through a fixed collection of "basic (first- and higher-order) substitutions" [12].

Some of these abstract frames have the limitation of not being able to define in a precise formal way and in a sufficient general spectrum what the "analogy" of two formulas could be. For instance, if two (first-order logic) formulas A and B have different orders in the way the quantifiers appear in each of them, then what should be a well-defined and cognitively-sound notion of "formal analogy" between them? However, there are other kinds of valuable (and informal) works concerning the importance of analogy in mathematics, which are based on a more solid collection of working examples and, simultaneously, tend to be grounded in more (general) cognitive and philosophical considerations, like [9, 10] and [7]. So, in this chapter, we start to develop a more formal and specialized notion of analogy, in part inspired by the former works, which is more suitable for constructing

© Springer Nature Switzerland AG 2020
D. A. J. Gómez Ramírez, *Artificial Mathematical Intelligence*,
https://doi.org/10.1007/978-3-030-50273-7_8

sound artificial implementations toward the main goals of the artificial mathematical intelligence meta-program (see Chaps. 1 and 3).

In particular, we start to answer the former question in the initial setting of propositional and first-order logic, and we show initial evidence for the claim that our approach is cognitively-sound and computationally-feasible for modeling formal reasoning related to proving constructively the validity of arguments. Furthermore, we take as an initial deductive framework a classic version of a Hilbert style propositional calculus[1] because of its simplicity and, at the same time, enough complexity for getting stronger intuitions about the cognitive features that our minds use for solving problems. Nonetheless, from a purely algorithmic perspective the problem of generating proofs of valid arguments in propositional logic is a classic issue solved in several ways during the past decades.[2] Although, as far as the author knows such solutions have, in general, as a main goal efficient solubility instead of cognitive plausibility, which is one of our main aims here.

Finally, the fundamental focus of this chapter is based on the fact that we want to formalize the (potential) analogical information between arbitrary formulas, instead of distinguishing explicitly between formulas that can (or cannot) be analogous. We aim to materialize the cognitive-syntactic information behind such analogical matches independently of the fact that the resulting formalization belongs (or does not) to the same collection of objects where the input formulas are taken.

8.2   Basic Notions

Since we want to meta-model the cognitive processes used to prove the (non-)validity of abstract arguments in systems like propositional (and first-order) logic, we adopt here a syntactic-structural approach for analogical reasoning. This means that our notion of analogy of two propositions would capture the purely syntactic similarities in the way in which both propositions are built. For example, let us consider the propositions $W_1 = P \rightarrow Q$ and $W_2 = R \rightarrow S$. Then, a straightforward observation shows that what W_1 and W_2 have in common is that both of them are of the form $X_1 \rightarrow X_2$, where X_1 and X_2 are propositional variables taking values over propositional letters only.

As second example, let us consider $W_1 = A \leftrightarrow B$ and $W_2 = A \leftrightarrow C$. In this case, we can infer that the commonalities (or analogical matches) between them have the form $A \leftrightarrow X_1$, where X_1 is again a (letter) variable. Let us now increase the complexity of the propositions. For instance, let $W_1 = (H \rightarrow I) \wedge J$ and $W_2 = (A \rightarrow B) \wedge C$. So, the syntactic commonalities of both can be represented by the (generic) proposition $(X_1 \rightarrow X_2) \wedge X_3$, for (letter) variables X_1, X_2, and X_3. Now, if we consider $W_1 = (A \rightarrow B) \leftrightarrow C$ and $W_2 = E \leftrightarrow (F \wedge G)$, then by

---

[1] See, for example, [8, Ch.1].
[2] See, for example, [1] and [11] and the references therein.

trying to codify the maximal amount of common (syntactic) information between these propositions as before, we find as first candidate the (generic) proposition $Y_1 \leftrightarrow Y_2$, where Y_1 and Y_2 are variables that should have values over more complex propositions than just propositional letters. For example, Y_1 stands for $A \to B$ on the one hand and for E on the other hand. Thus, this symbolic convention allows us to save more data regarding the syntactic commonalities between W_1 and W_2. One can also continue recursively this meta-process for more complex propositions.

8.2.1 Syntactic and Generic Depth

Inspired by the former examples, we will define the notion of *atomic analogy* of two propositions. However, in order to do this, we need to first define a "complexity measure" of a proposition.

Definition 8.1 Let W be a proposition (resp., generic proposition). We define the (syntactic) depth of W, denoted by $D(W)$, recursively as follows:

1. If W is a letter (resp., propositional variable), $D(W) = 1$.
2. $D(\neg W) = D(W) + 1$.
3. If $W = W_1 \rightleftharpoons W_2$, where W_1 and W_2 are propositions and $\rightleftharpoons$ stands for a binary logic connective, then $D(W) = max \{D(W_1), D(W_2)\} + 1$.

This notion of syntatic depth coincides with the number of levels that the corresponding tree of W has. Recall that the notion of the tree of a proposition is informally reconstructed by putting the main logical connective on the top and then writing downward at the end of the branches the two (resp., one) sub-propositions and repeating this process until one gets as sub-proposition in any branch a propositional letter [13].

Let Y be a propositional variable that can take only the propositional values $V_1, \cdots, V_n$, where $n \in \mathbb{N}$ and $D(V_i) = d_i$ for $i = \{1, \cdots, n\}$ are the syntactic depths of $V_1, \cdots, V_n$, respectively. Then, we define the *generic depth* of Y (sometimes denoted by "g-depth" or $D_G(Y)$) to be

$$max\{D(V_1), \cdots, D(V_n).\}$$

In other words, the generic depth denotes the maximal syntactic depth of all the potential concrete values of Y. Note that the one can also consider the syntactic depth of Y as atomic syntactic unit (i.e., letter), which is exactly one.[3]

We define a second "dissimilarity measure," namely, the *index* of a generic proposition Z; that is, a proposition involving letters, variables X_r of g-depth one, and variables Y_s of higher g-depth $m_s \geq 2$. Specifically, we define the index of a

---

[3]In some cases, we will simply talk about the "depth" of a proposition (resp., propositional letter) when it is clear from the context what is the implicit choice, i.e. syntactic of generic depth.

propositional letter as zero, the index of a variable X_r as one, and the index of a variable Y_s as m_s.[4] Moreover, if $V(Z)$ denotes the collection of propositional letters and variables appearing in Z, then we define the index of Z by the formula:

$$I(Z) = \sum_{\alpha \in V(Z)} I(\alpha),$$

where the sum is taken counting repetitions.

8.2.2 Atomic Analogy

We define the *atomic analogy*, of two propositions W_1 and W_2, recursively by induction on $S := D(W_1) + D(W_2)$.

If $S = 2$, then W_1 and W_2 are letters. When both of them are equal (to the letter A, for example), then

$$atoman(W_1, W_2) = A.$$

If they represent different letters, then $atoman(W_1, W_2) = X_1$, where X_1 is a propositional variable of g-depth one, since it generates the (trivial) *analogical replacement* (a. r.)[5] $W_1 = X = W_2$.

For the case $S = 3$, let us assume that $D(W_i) = 2$ and $D(W_j) = 1$. Therefore, $atoman(W_i, W_j) = P$, if $W_i = P \rightleftharpoons Q$ (or $Q \rightleftharpoons P$) and $W_j = P$; or $W_i = P$; and $W_j = \neg P$. On contrary, if $W_i = P \rightleftharpoons Q$, $W_j = R$, $P \neq R$, and $Q \neq R$, $atoman(W_i, W_j) = Y$, where Y is a variable of g-depth two, since it produces the a. r. $W_1 = Y = W_2$, where one can define the g-depth of this replacement as the maximum of the depths of the involved propositions.

Finally, for the general case, assume that $W_1 = V_1 \rightleftharpoons_1 V_2$, and $W_2 = U_1 \rightleftharpoons_2 U_2$, where V_1, V_2, U_1, and U_2 are sub-propositions and $\rightleftharpoons_1$ and $\rightleftharpoons_2$ are connective variables. Now, we have the following cases: If $\rightleftharpoons_1 \neq \rightleftharpoons_2$, then $atoman(W_1, W_2) = Y$, where Y is a variable of g-depth

$$max \{D(W_1), D(W_2)\},$$

namely, the g-depth of the corresponding replacement.

---

[4]Sometimes, it is useful to declare artificially propositional variables of the form $Y_{a,b}$ with the convention $D_G(Y_{a,b}) = a$. For the purposes of this chapter, this will not be necessary.

[5]In this context, we use the expression "pairing" as well. Strictly speaking, such a replacement measures, in a syntactic sense, where the analogical information ends, so, the name disanalogical correspondence could also be used. However, we prefer the adjective "analogical" correspondence, because it emerges directly by using analogy making.

When $\rightleftharpoons_1=\rightleftharpoons_2=\rightarrow$, $atoman(W_1, W_2) = atoman(V_1, U_1) \rightarrow atoman(V_2, U_2)$, it is well-defined since the corresponding sub-propositions have a strictly smaller depth. Suppose that $\rightleftharpoons_1=\rightleftharpoons_2\neq\rightarrow$. In this case, the corresponding connective is symmetric; i.e., one obtains an equivalent proposition when one commutes the arguments, e.g. $U_1 \wedge U_2 = U_2 \wedge U_1$. Here, we should choose the best atomic analogies between the two possible natural options given by comparing the sub-propositions of W_1 and W_2. So, a simple way to compare them is using as a parameter the sum of the depths of the atomic analogies of the corresponding matches between sub-propositions. That is, choose a pair of indexes i_1, i_2, j_1, and j_2 in $\{1, 2\}$ such that $\{i_1, i_2\} = \{j_1, j_2\} = \{1, 2\}$ and $D(atoman(U_{i_1}, V_{i_2})) + D(atoman(U_{j_1}, V_{j_2}))$ is maximal. In the case that at least two pairs of indexes produce the same maximal value for the addition of the corresponding depths, we use as second criterion, the minimal value of the sum of the indexes of the respective sub-propositions.[6] We define

$$atoman(W_1, W_2) = atoman(U_{i_1}, V_{i_2}) \rightleftharpoons atoman(U_{j_1}, V_{j_2}).^7$$

In the case that $W_1 = \neg U_1$ and $W_2 = V_1 \rightleftharpoons V_2$ (or vice versa), we define again $atoman(W_1, W_2) = Y$, since the principal logical connectives are different. Lastly, if $W_1 = \neg(U_1)$ and $W_2 = \neg(V_1)$, we define

$$atoman(W_1, W_2) = \neg(atoman(U_1, U_2)).$$

Any of the propositional variables of an atomic analogy induce a natural (syntactic) pairing between sub-propositions of W_1 and W_2.

Let us illustrate more explicitly with a collection of examples how this notion captures the intuition of syntactic similarities among propositions and generates explicit pairings among sub-propositions, which reflect the syntactic "bounds" of the analogical similarities between the corresponding propositions.

Effectively, it is straightforward to check the following equalities:

Let P, Q, R, and S be different letters

$$atoman(P \rightarrow Q, P \rightarrow R) = P \rightarrow X,$$

with pairing $Q = X = R$;

$$atoman(P \leftrightarrow Q, Q \leftrightarrow P) = P \leftrightarrow Q,$$

---

[6]Informally, two propositions are more similar if their index is smaller and the depth of their atomic analogy is bigger. For example, two propositional letters A_1 and A_2 are identical if (and only if) their index is zero (here the depth of the atomic analogy is always one).

[7]In the case of multiple optimal candidates with maximal depth-sums and minimal index-sums, we choose any of them.

with no pairing;

$$atoman(P \wedge (Q \rightarrow R), Q \wedge R) = X \wedge Y,$$

with pairings $P = X = Q$ and $Q \rightarrow R = Y = R$;

$$atoman((P \leftrightarrow Q) \wedge R, (P \leftrightarrow Q) \wedge S) = (P \leftrightarrow Q) \wedge X,$$

with pairing $R = X = S$;

$$atoman(\neg(P \wedge Q), \neg(S \wedge Q)) = \neg(X \wedge Q),$$

with pairing $P = X = S$;

$$atoman(P \wedge Q, R \wedge S) = X_1 \wedge X_2.$$

with pairing $P = X_1 = R$ and $Q = X_2 = S$.

The last example is a case where we obtain the same (maximal) values for the sum of the depths and the same (minimal) values for the sum of the indexes. So, we need to choose authentically between any of the (analogical) matches. The choice is decoded by the pairing, so a possible option was shown before and the other coherent matching generates the pairings $P = X_1 = S$ and $Q = X_2 = R$.

Let us define the tree of a proposition by (complete) induction on the depth. Effectively, let W be a proposition. If $D(W) = 1$, then $T(W) := \{W\}$. Assume that $D(W) > 1$. So, when $W = \neg U$, then $T(W) = (\neg, T(U))$. In the case that $W = U_1 \leftrightharpoons U_2$, then $T(W) = \{\leftrightharpoons, W, T(U_1), T(U_2)\}$. Informally, the tree of W contains in hierarchical order all the sub-propositions of W together with the respective logical connectors. In fact, we say formally that a proposition V is a sub-proposition of W if there exists a finite chain of propositions $W_1 = U, W_2, \cdots, W_n = W$, such that $W_i \in T(W_{i+1})$, for all $i = 1, \cdots, n - 1$. It is easy to check that this notion corresponds with the standard one. Let us denote the collection of all sub-propositions of W by $SP(W)$.

8.2.3 Analogical Space

In some cases, it is useful to be able to compare two propositions not only in a global way, as before, but also at a local level. Namely, it is worth verifying if one proposition is very similar to some sub-proposition of a second one and vice versa. This idea can also be seen in some sub-routines of system hdtp, for example, re-representation [12]. Therefore, the next step is to extend the former notion to sub-propositions and to collect all these "formal analogies" in an *analogical space*. This motivates the following definition:

Let W_1 and W_2 be propositions. Then, their analogical space is defined as

$$AS(W_1, W_2) := \{atoman(U_1, U_2)/U_1 \in SP(W_1), U_2 \in SP(W_2)\}.$$

The elements of this space are called *analogical propositions*, since they arise as atomic analogies of two standard propositions.

Among all the possible (local) atomic analogies, we want to choose the ones showing the maximal (local) similarities between W_1 and W_2. Thus, as argued before, we want to focus our attention on the atomic analogies with the biggest depth and, within this sub-group, we choose one with the smallest index.

So, let us define the space of the *best analogies* as follows:

$$BA(W_1, W_2) = \{A \in I = AS(W_1, W_2)/\forall(B \in I)D(B) \le D(A)$$

$$\wedge(\forall C \in I)(D(C) = D(A) \Rightarrow I(C) \ge I(A))\}.$$

Thus, the adjective "best" is justified by the simple fact that we prefer the syntactically most complex similarities (in the formal way of atomic analogies) between sub-propositions of W_1 and W_2. And, again, among these highly complex generic analogies we prefer the ones with the biggest number of specific identifications of propositional letters and the smallest number of analogical mismatches. Effectively, the last facts are quantitatively measured through maximal depths and minimal indexes.

8.2.4 Extension for Predicate Logic

For the case of first-order logic, we can generalize the former collection of notions in the following way: first, we add two additional kinds of meta-variables, namely, one type of variable that generates analogical correspondences to terms or to a special symbol (e.g., θ) denoting the absence of terms, and the other type that generates as purely symbolic analogical replacements the quantifiers (i.e., $\forall$ and $\exists$) or a special symbol (e.g., Θ) denoting the absence of quantifiers. Moreover, we extend the notion of tree for a first-order formula by adding new vertices for each quantified syntactic configuration for the form $\forall\eta$ or $\exists\iota$. We also extend the notions of (syntactic and generic) depth and index accordingly. Finally, we use the same heuristic principles as before for defining recursively the corresponding analogical notions, but taking into consideration the presence of terms and quantifiers bounding variables, e.g. if there is a syntactic match in position and form between both formulas, we include it (in the atomic analogy for example); if not, we include the corresponding type of meta-variable, that implicitly generates an analogical replacement. Let us illustrate this with some examples:

Let x be a variable, and let R be an unary relation in a FOL language L. Then

$$atoman((\forall x)(R(x)), (\exists x)(R(x))) = (\Xi x)(R(x)),$$

where $\varXi$ is a syntactic variable for quantifiers. Lastly, a more complex instance is the following:

$$analog((\exists x)(\forall y)(R_1(x, y)), (\exists v)(\forall w)(\exists u)(R_2(v, w, u))$$

$$= (\varXi_1 v_1)(\varXi_2 v_2)(\varXi_3 v_3)(M_R(v_1, v_2, v_3)),$$

where v_1, v_2, v_3 are meta-variables for terms; $\varXi_1, \varXi_2$ are variables for quantifiers; and M_R is a relational meta-variable, which plays the role of a propositional meta-variable in the former case. The analogical replacements are $\theta = v_1 = v, x = v_2 = w, y = v_3 = u; \Theta = \varXi_1 = \exists, \exists = \varXi_2 = \forall, \forall = \varXi_3 = \exists$ and $R_1 = M_R = R_2$.

This case of formalizing analogical reasoning in a first-order setting is important from a purely logic point of view. However, from a pragmatic and cognitive perspective, it is slightly more fundamental to find formalizations of analogical reasoning within the specific heuristic framework where the mathematician's mind does pragmatic inferences, sketches, and uses more compact and dense symbolic configurations for generating new conceptual information (see, for example, the cognitive generations presented in Chap. 11).

8.2.5 An Additional Approach to Conceptual Blending Between Propositions

Since we have a concrete way of describing global analogical matches between two propositions W_1 and W_2, namely, by means of their atomic analogy, we can use now this notion to define an alternative formalization of a conceptual blending between them. Effectively, $atoman(W_1, W_2)$ codifies the initial global syntactic commonalities of W_1 and W_2, and it also codifies where these similarities ends; i.e., where the propositional variables X_i and Y_j appear. So, let us assume that

$$atoman(W_1, W_2) = G(X_1, \cdots, X_n, Y_1, \cdots, Y_m),$$

where $X_1, \cdots, X_n, Y_1, \cdots, Y_m$ are the variables appearing in the definition, with pairings given by $V_{1,i} = X_i = V_{2,i}$, and $U_{1,j} = Y_j = U_{2,j}$, such that $V_{1,i}$ and $V_{2,i}$ are propositional letters, and $U_{1,j}$ and $U_{2,j}$ are sub-propositions of higher depth of W_1 and W_2, respectively. Here, $G(\_, \_)$ denotes the corresponding propositional function sending $(n + m)$−tuples of propositions into single propositions.

Thus, we can interpret these pairings in a natural way as the positional "analogical connections" between them [4, Pag. 47], since exactly at these positions in the corresponding trees the syntactic similarities end. Thus, as typical from a conceptual blending construction, we want to integrate these dissimilarities in a unique manner into the blended proposition. Therefore, we want to codify each sub-proposition $V_{r,i}$ and $U_{s,j}$ into the blend in order to be able to reconstruct the whole network of propositions from the blended proposition alone. This is sometimes called the "unpacking principle" [3]. Thus, we define the formal blend between W_1 and W_2 as follows:

$$B(W_1, W_2) = G(V_{1,1} \wedge V_{2,1}, \cdots, V_{1,n} \wedge V_{2,n}, U_{1,1} \wedge U_{2,1} \cdots, U_{1,m} \wedge U_{2,m}).$$

With this notion, we are codifying the external basic similarities between the input propositions in an internal manner into the blend, since the blend B has the same structural global form as $atoman(W_1, W_2)$. This fact guarantees the "topology" postulate required on a blending [3]. In addition, it is straightforward that B was constructed from a genuine integrative process using symmetrically syntactic aspects of both propositions. Finally, if we manipulate B as a syntactic unit in our framework, it will preserve the same kind of formal connections with the input spaces, so B fulfills the "web" precept as well [3].

In conclusion, we have an initial formalization of formal conceptual blending, which can be integrating when constructing cognitively-based algorithms for finding human-style proofs of valid arguments based on the former formalizations of analogical reasoning.

8.3 Our Formal Framework

Please see Sect. 2.2.1 of Chap. 2 for the description of the exact deductive propositional system that we will use throughout the rest of this chapter.

There are several approaches to find efficient algorithms for verifying that a wf is (or is not) a theorem in propositional logic. Among them, some of the most outstanding methods are the Davis–Putnam Procedure, the DPLL algorithm, the Stålmarck's method, and the binary decision diagrams, among others [6, Ch. 2]. There is also a valuable and classical work of S. A. Cook and R. A. Reckhow on the estimation of lower bounds for the lengths of proofs of a tautology, in particular, proof systems in terms of the length of the tautology [2]. There is also at least one online assistant, which helps in verifying that a specific proof is formally correct.[8]

Any of the former formal/practical frameworks give, at most, indirect procedures in order to decide whether a particular wf is a tautology (resp., satisfiable or contradictory). In fact, some of these methods used implicitly the completeness theorem in order to guarantee that the corresponding wf is a theorem, and they employ demonstrations by the sake of contraction (e.g., the DPLL algorithm). So, none of them offer explicit step-by-step constructive proofs for arbitrary tautologies.

Moreover, from a cognitive perspective, it is more valuable and challenging to model how such deduction steps are generated rather than abstractly verifying that a wf is a tautology using true table methods and similar approaches, which are

---

[8]http://philosophy.lander.edu/~jsaetti/ToolBox/dojoProof2/proof1_7.html.

"computationally" more efficient but offer little information about how our mind is able to find an explicit formal proofs following a fixed collection of logic rules.

8.4 Toward an Analogy-Based Deduction Algorithm for Propositional Logic

We will show in this section how the core and, in principle, ingenious ideas of some concrete proofs of basic tautologies can be modeled naturally from our formalism.

We need to define a couple of sub-routines that we will use often in our initial cognitively-inspired deduction method.

Assume that we should find a formal proof of $\mathscr{H}_1, \cdots, \mathscr{H}_n \vdash \mathscr{T}$. The first fact that we need to check is that $\mathscr{T}$ is either some $\mathscr{H}_j$ or it could be directly obtained as an exemplification of an axiom scheme (Ai). If this is true, the proof of the former facts consisting of indicating either the specific sub-index j such that $\mathscr{H}_j = \mathscr{T}$ or the concrete replacements that should be done in order to obtain $\mathscr{T}$ from some axiom scheme (Ai), for some $i \in \{1, 2, 3\}$. Besides, it is elementary that the last fact is equivalent to the following implementable conditions:

1. $D(atoman(\mathscr{T}, Ai)) = D(Ai)$
2. The sub-propositions of (Ai) involved in each of the replacements generated by the former atomic analogy are exactly all the propositional variables $\#_r$, for $r \in \{1, 2\}$ (i=1,3) or $r \in \{1, 2, 3\}$ (i=2).
3. The replacements induced for each $\#_r$ by the former atomic analogy are all coherent, and the particular replacements induced by each instance of $\#_r$ correspond exactly to identical sub-propositions of $\mathscr{T}$.

Let us denote these sub-routines by $Rhyp(\Gamma, \mathscr{T})$ and $Rax(A, \mathscr{T})$, where $A = \{A1, A2, A3\}$.

Additionally, we want to make a general classification of *analogical dissimilarity* among all the elements of the analogical space of two (generic and/or variable) propositions Ω_1 and Ω_2,[9] in terms of the former numerical and conceptual notions defined before. Concretely, we define for $r = 0, 1, \cdots, max(D(\Omega_1), D(\Omega_2))$:

$$AS_r(\Omega_1, \Omega_2) = \{V = Atoman(A_1, A_2) : A_i \in SP(\Omega_i) \wedge (max(D(A_1), D(A_2))$$

$$-D(atoman(A_1, A_2)) \leq r) \wedge (I(atoman(A_1, A_2)) \leq r)\} \setminus AS_{r-1}(\Omega_1, \Omega_2).$$

Here, by convention $AS_{-1}(\Omega_1, \Omega_2) := \emptyset$. We say that the analogical propositions of $AS_r(\Omega_1, \Omega_2)$ have *degree of analogical dissimilarity* (d.a.d.). Informally, the analogical propositions with d.a.d. zero are exactly the ones emerging from

---

[9]This means that Ω_1 and Ω_2 can have as propositional "pseudo-letters" either fixed propositional variables (X_i, Y_j), arbitrary variables ($\#_k$), or standard propositions ($\mathscr{A}$).

identical sub-propositions of Ω_1 and Ω_2. Furthermore, there is natural preference for analogical propositions with high depth and small d.a.d., since these facts reflect on the one hand wide global structural similarities on the configuration of the logical connectives, and, on the other hand, high specific matches between the particular propositional letters involved.

We will also need to define a more specialized modification of the notion of best analogy between a (generic) proposition W and an axiom scheme (Ai). Effectively, we define the *best thesis analogy* $BA_T(W, Ai)$ as the generic proposition V of $BA(W, Ai)$ such that it is induced as the atomic analogy of a sub-proposition A' of A_i being at the rightest position in comparison with the other sub-propositions of Ai generating $BA(W, Ai)$. Informally, we want to find analogical matches placing (sub-propositions of) W as a thesis within Ai. Typically, W is the thesis $\mathscr{T}$, so we need to find an algorithmic way to estimate syntactically how near are our axiom schemes to producing $\mathscr{T}$ as a thesis.

Similarly, we define the *best hypothesis analogy* $BA_H(W, Ai)$ replacing "rightest" by "leftest" in the former definition. In the next section we will see a direct application of these notions.

Another sub-routine is what we call *thesis reduction*, namely, if we are considering the proposition $T = T_1$ as actual thesis, and we obtain as current hypothesis a proposition of the form $S = T_2 \to T_1$ (or of the form $(\cdots \to (T_2 \to T_1)))$, then we update temporarily T with the new thesis T_2, since by MP it is enough to prove T_2 in order to obtain T_1 (or we can continue this process until we update the leftest hypothesis). If after further steps one finds that the new hypothesis T_2 cannot be deduced, then one updates T again with the former value T_1 and continues the next sub-routine. We denote this by $Redth(T = T_1, S)$. Clearly, if S does not have the suitable form $\# \to T_1$, then T remains the same. This step can be called abductive sub-routine, since we are trying to find adequate propositions implying our thesis.

8.5   Analogical Meta-Modeling of Specific Proofs

In this section, we will meta-model formal proofs of some basic theorems using our analogical schematic framework. One of the main purposes is to be able to simulate in a natural way the "ingenious" ("out of the blue") parts of the proofs.

8.5.1   First Theorem

Let us start with the cognitively-inspired generic description of the basic theorem $\vdash \mathscr{A} \to \mathscr{A}$ [8, Ch.1, Lemma 1.8]. First, we save the thesis in a variable T, i.e. $T \twoheadrightarrow (\mathscr{A} \to \mathscr{A})$. We check that $Rax(A, T)$ does not hold, since $D_G(T) = 2$ is smaller than the depths of the axiom schemes. We continue by computing $BA_T(T, A1) = X_1 \to X_2$. It generates the replacements $\#_2 = X_1 = \mathscr{A}$ and $\#_1 = X_2 = \mathscr{A}$, and, therefore, we can integrate as a new hypothesis the corresponding exemplification of (A1): $H_1 = \mathscr{A} \to (\mathscr{A} \to \mathscr{A})$. Next, we apply $Redth(T, H_1)$ and obtain as a new thesis $T = \mathscr{A}$. It is a trivial fact to see that the new thesis is not in general a theorem, since $\mathscr{A}$ denotes an arbitrary wf. So, we update T again to the original value $\mathscr{A} \to \mathscr{A}$. Furthermore, we continue with (A2) and compute $BA_T(T, A2) = X_1 \to X_2$. It generates the replacements $\#_3 = X_1 = \mathscr{A}$ and $\#_1 = X_2 = \mathscr{A}$. So, we obtain the following generic proposition as a potentially new hypothesis:

$$H(\#_2) = ((\mathscr{A} \to (\#_2 \to \mathscr{A})) \to ((\mathscr{A} \to \#_2) \to (\mathscr{A} \to \mathscr{A}))).$$

Thus, we need to find a suitable propositional value for $\#_2$ such that from the corresponding new hypothesis $H(\#_2)$, we could potentially obtain our desired thesis more directly. Note that $\mathscr{A} \to \mathscr{A}$ is the thesis of the thesis of $H(\#_2)$.

So, a simple abductive analysis of $H(\#_1)$ looking for a reduction of thesis leads us to the proposition:

$$TH(\#_2) = (\mathscr{A} \to \#_2) \to (\mathscr{A} \to \mathscr{A}).$$

The hypothesis of this proposition is $U = \mathscr{A} \to \#_2$. Thus, the next step is to obtain U from an exemplification an axiom scheme. Therefore, the binary operation of atomic analogy is a direct way of generating the corresponding replacements (i.e., pairings). In fact, if we compute $atoman(A1(\#'_i), U)$ renaming the propositional variables of $A1$ in order to avoid confusion with the symbol $\#_2$ in U, we obtain

$$atoman(A1(\#'_i), U) = X_1 \to Y_1,$$

with pairings $\#'_1 = X_1 = \mathscr{A}$ and $\#'_2 \to \#'_1 = Y_1 = \#_2$. In this way we obtain a concrete replacement for U, namely,

$$U' = \mathscr{A} \to (\#'_2 \to \#'_1) = \mathscr{A} \to (\#'_2 \to \mathscr{A}),$$

where $\#'_2$ is again a free propositional variable.

We apply the sub-routine $Redth$ to the current thesis and $TH(\#'_2 \to \mathscr{A})$ for obtaining as temporary new thesis U'.

Now, if we apply the former replacement to the hypothesis of $H(\#_1)$, we get the generic proposition:

$$HH(\#'_2 \to \mathscr{A}) = \mathscr{A} \to ((\#'_2 \to \mathscr{A}) \to \mathscr{A}).$$

We employ again $Redth$ with the new thesis and $H(\#'_2 \to \mathscr{A})$ in order to get as new thesis $HH(\#'_2 \to \mathscr{A})$.

Finally, we verify $Rax(A, HH(\#2' \to \mathscr{A}))$, since the new thesis can be obtained from $A1$ through the replacements $\#_1 = \mathscr{A}$ and $\#_2 = \#2' \to \mathscr{A}$. So, if we replace $\#'_2$ by a particular (arbitrary) proposition, let us say $\mathscr{A}$, then we can simply reproduce a constructive proof of our original thesis $\mathscr{A} \to \mathscr{A}$, by going backward in the presented order of deduction. In fact, the proof presented in [8, Ch. 1,Lemma 1.8] seems to have at first sight an innovative sequence of replacements. Additionally, regarding this proof, E. Mendelson justifies the particular exemplifications of the axiom schemes used there as a result of abductive arguments (e.g., $Redth(-, -)$) coming from the MP and "a mixture of ingenuity and experimentation" [8, fn. pag. 37]. This type of "ingenuity" is what we meta-model as an essential kind of analogical reasoning ability.

8.5.2 Second Theorem

Let us apply the former procedure to the theorem $\vdash (\neg\mathscr{A} \to \mathscr{A}) \to \mathscr{A}$. Effectively, it is straightforward to verify that $Rax(T, A)$ does not hold, where $T = (\neg\mathscr{A} \to \mathscr{A}) \to \mathscr{A}$. So, we use the sub-routine $BA_T(T, Ai)$, for $i = 1, 2, 3$. After dismissing the first case (i=1) due to an unsolvable thesis' update $T = \mathscr{A}$, we obtain for the second axiom scheme the following generic proposition as generic new hypothesis:

$$(T \to (\#_2 \to \mathscr{A})) \to ((T \to \#_2) \to (T \to \mathscr{A})).$$

By proceeding exactly as in the former theorem, we obtain the replacement $\#_2 = \neg\mathscr{A}$, since this is generated through the pairings from $Atoman(A1, T \to \mathscr{A})$. So, by doing two times thesis' reduction we get as new thesis the proposition:

$$(\neg\mathscr{A} \to \mathscr{A}) \to (\neg\mathscr{A} \to \mathscr{A}).$$

But, this is exactly an instance of the former theorem. In conclusion, we proceed as in the first theorem to find a formal proof of this new thesis in order to complete a constructive proof of our result.

It is worth noting that one can obtain an additional proof by applying $BA_T(T, A3)$, since it generates as a new hypothesis the proposition:

$$(((\neg\mathscr{A}) \to (\neg\mathscr{A})) \to (((\neg\mathscr{A} \to \mathscr{A}) \to \mathscr{A})).$$

Thus, one obtains as new thesis again $\neg\mathscr{A} \to \neg\mathscr{A}$ and the final procedure is the same as before.

8.6 Conclusions

The formal framework developed so far starts to show that it is possible to model the creative aspects involved in the generation of constructive proofs for the validity of arguments in propositional logic, in terms of concrete types of analogical matches between the propositions involved. Due to their particular formal and recursive constructions, it can be proved that all the concepts and methods involved in this analogical model can be implemented. Finally, this chapter is an initial step toward getting a more solid, pragmatic, and formal understanding of the seminal cognitive–metamathematical features of this fundamental cognitive ability.

Acknowledgments The author would like to sincerely thank his friend Edisson Gallego Gonzales for all his support during all these years and for all the inspiring conversations (regarding analogical reasoning in mathematics, among many others). Besides, many thanks to his important suggestions regarding initial sketches of this chapter.

References

1. Biere, A., Heule, M., van Maaren, H.: Handbook of satisfiability, vol. 185. IOS press (2009)
2. Cook, S.A., Reckhow, R.A.: The relative efficiency of propositional proof systems. The Journal of Symbolic Logic **44**(01), 36–50 (1979)
3. Fauconnier, G., Turner, M.: Conceptual blending, form and meaning.   Recherches en communication **19**(19), 57–86 (2003)
4. Fauconnier, G., Turner, M.: The Way We Think. Basic Books (2003)
5. Fleuriot, J., Maclean, E., Smaill, A., Winterstein, D.: Reinventing the complex numbers (2014)
6. Harrison, J.: Handbook of practical logic and automated reasoning. Cambridge University Press (2009)
7. Leatherdale, W.H.: The role of analogy, model, and metaphor in science. North Holland Pub. Co. (1974)
8. Mendelson, E.: Introduction to Mathematical Logic (Fifth Edition). Chapman & Hall/CRC (2010)
9. Pólya, G.: Mathematics and plausible reasoning: Induction and analogy in mathematics, vol. 1. Princeton University Press (1990)
10. Pólya, G.: Mathematics and plausible reasoning: Patterns of plausible inference, vol. 2. Princeton University Press (1990)
11. Robinson, A.J., Voronkov, A.: Handbook of automated reasoning, vol. 1. Elsevier (2001)
12. Schwering, A., Krumnack, U., Kuehnberger, K.U., Gust, H.: Syntactic principles of heuristic driven theory projection. Cognitive Systems Research **10**(3), 251–269 (2009)
13. Smullyan, R.M.: First-order logic. Courier Corporation (1995)
14. Boy de la Tour, T., Peltier, N.: Computational Approaches to Analogical Reasoning: Current Trends, chap. Analogy in Automated Deduction: A Survey, pp. 103–130. Springer-Verlag, Berlin, Heidelberg (2014)
15. Weil, A.: De la métaphysique aux mathématiques (in french). Science **60**, 52–56 (1960). Reprinted in [André Weil Collected Papers II, p. 406–412]

Chapter 9
Conceptual Substratum

9.1 Introduction

One of the most powerful and useful abilities of the mind consists of abstracting the essential features of a concept, an object, an idea (or collection of them), to understand and subsequently manipulate the corresponding cognitive seminal configurations for solving conceptual and concrete tasks.[1] Let us illustrate the scope and generality of such cognitive ability within the context of mathematical research with an enlightening example. Suppose that we need to prove the following arithmetical statement: "Prove that there are arbitrarily large sequences of consecutive composite natural numbers."

Firstly, let us rewrite the task in a symbolic way. We need to prove that given an arbitrary natural number $b \in \mathbb{N}$, there exists a $m \in \mathbb{N}$ such that the sequence $m, m+1, \cdots, m+b$ consisting of $b+1$ consecutive natural numbers is entirely made of composite numbers. Now, the typical form of a number in this sequence is $m+k$, where $k \in \{0, 1, \cdots, b\}$. From the former representation, we know that m is a fixed number and that k varies. So, it would be useful if we could factor k from $m+k$ for all k big enough (e.g., $k \geq 2$). Such a condition forces k to divide m for all $k \in \{2, \cdots, b\}$. The former collection of requirements suggests that a suitable number m should have the form $2 \times \cdots \times b = b!$. So, we make a kind of "symbolic check" with a generic expression of the form

$$b! + k : k \in \{2, \cdots, b\},$$

and we simply do a factorization of the form $b! + k = k((b!/k) + 1)$, to verify that such a number is composite (because $k, ((b!/k) + 1 \geq 2)$. In conclusion, we

---

[1]For a more general overview of the most fundamental cognitive metamathematical mechanisms used in mathematical research, see Chap. 10.

© Springer Nature Switzerland AG 2020
D. A. J. Gómez Ramírez, *Artificial Mathematical Intelligence*,
https://doi.org/10.1007/978-3-030-50273-7_9

generate a sequence of $b - 1$ consecutive composite natural numbers, which solves the problem due to the fact that $b - 1$ can be arbitrarily large.

In the former proof, we employ a useful and fundamental trick for generating the desired sequence, namely we were able to abstract the conceptual substratum of a generic element of such a sequence; i.e., $b! + k : k \in \{2, \cdots, b\}$.

This creative "trick" is an elementary instance of a much more general mechanism that allows us to find morpho-syntactic configurations capturing the essential information of arbitrary (mathematical) concepts needed for doing general mental inferences.

Assuming a Lockean perspective for a moment, we can see that taking the conceptual substratum of a specific concept C is very similar to considering the (m.-s. version of the) nominal essence of C (see, for instance, [13]).

In this chapter, we will assume a pragmatic approach, i.e., we will center on the identification and characterization of conceptual substrata in terms of (domain-specific) symbolic configuration explicitly used (mostly in sketches) by the mathematician's mind during the creative process of generating mathematical information. In particular, issues involving the ontological nature of this kind of substrata are viewed here from a mental, linguistic, and nominal dimension.

We call this cognitive ability *(formal) conceptual substratum*, and we use the notation $CS(\mathscr{C})$, when it is applied to a (mathematical) concept $\mathscr{C}$. The approach developed here has some similarities with the notion of (proto-)typicality developed in [12], although ours has emerged more strongly from a mathematical setting.

An additional form of understanding conceptual substratum from a cognitive perspective is being seeing it as a (metamathematical) way of generating (abstract) "line drawings" of a specific concept, where the starting input corresponds to a formal explicit characterization of the (mathematical) structure in consideration (given, for example, in a first-order logic conceptual environment) [8]. Note that since doing line drawing (or informally, doing sketches) involves processes like memory, perception, and judgment, there can be multiple conceptual substrata of a single mathematical notion, varying according to the concrete conceptual environment and to the particular goals involved [8, §2].

In the next sections, we study a wider variety of examples coming from different mathematical fields to get a wider intuition about what conceptual substrata look like, we present initial meta-formalizations in a (many-sorted) first-order and in a general setting, and we present basic arithmetic and proof-theoretic results related with a functional approach for conceptual substratum. Finally, we describe the dual cognitive mechanism of conceptual substratum; i.e., conceptual lining together with an initial meta-formalization of it.

9.2   Additional Conceptual Support from Several Mathematical Domains

Let us move to a more abstract mathematical field; i.e., modern algebraic geometry [4, 6]. In order to analyze some sophisticated notions of this quite elegant mathematical area, let us start with some of the algebraic and topological concepts involved.

First, let us consider the concept of abelian group [2]. The essential information about this basic notion lies in the equations describing the algebraic properties of all the elements in the underlying set. In particular, the existence of "inverses" for arbitrary elements can be represented by using the auxiliary "minus" symbol, denoting the inverse element. Thus,

$$CS(\text{Abelian Groups}) = [(x_1 + x_2 = x_2 + x_1), (x_1 + e = x_1),$$

$$(x_1 + (x_2 + x_3) = (x_1 + x_2) + x_3), (x_1 + (-x_1) = e)].$$

Note that in the former representation we avoid explicitly the quantifiers because when we consider those expressions involving letters usually named like variables (i.e., x_1, x_2 and x_3), we assume that the relations shown through them hold for every element in the corresponding set. Thus, the concrete form of the names used here starts to matter from the semantic viewpoint. In addition, we also mention explicitly the inverse of a generic element x_1 by using the syntactic symbol $-x_1$, instead of adding an extra variable. Finally, we obtain a compact expression (e.g., sentence) with a minimal amount of syntactic complexity, but being able to compress all the fundamental information defining an abelian group. The symbol "$\wedge$" is replaced by a comma because we usually save in our minds conceptual units without giving special attention to adding artificial syntactic symbols to joint them (for example, $\wedge$). Very oft we will write these conceptual units between parenthesis in order to guarantee more clarity in the exposition, although we do not frequently include "mental parenthesis" in our (cognitive) representations.

For the notion of group we can write

$$CS(\text{Groups}) = [(x_1 + e = e + x_1 = x_1),$$

$$(x_1 + (x_2 + x_3) = (x_1 + x_2) + x_3), (x_1 + (-x_1) = (-x_1) + x_1 = e)].$$

The next concepts that we can analyze based on the former ones are the notions of (commutative) ring (with unity). In this case, it is enough to show just explicitly the substratum of the most complex one; i.e., commutative rings with unity:

$$CS(\text{Commutative Rings with unity}) = [(x_1 + x_2 = x_2 + x_1), (x_1 + e = x_1),$$

$$(x_1 + (x_2 + x_3) = (x_1 + x_2) + x_3), (x_1 + (-x_1) = e), (x_1 \cdot x_2 = x_2 \cdot x_1),$$

$$(1 \cdot x_1 = x_1), (x_1 \cdot (x_2 \cdot x_3) = (x_1 \cdot x_2) \cdot x_3),$$

$$(x_1 \cdot (x_2 + x_3) = x_1 \cdot x_2 + x_1 \cdot x_3)].$$

Now, let us move to the notion of ideals of a commutative ring with unity. Since, this notion emerges from the a grounding concept (e.g., commutative ring with unity), the corresponding conceptual substratum in this case contains in some sense the conceptual substratum of the grounding notion.

$$CS(\text{Ideals of a Commutative Ring with Unity}) = [CS(C.R.U.),$$

$$x_1 \cdot i_1 + (-i_2) \in I, x_1 \in R; i_1, i_2 \in I].$$

Here, when certain variables are bounded by universal quantifiers (resp. existential) we include all the corresponding cognitive codifications in more sophisticated ways, where the degree of m.-s. complexity increases gradually.[2]

For the concept of a prime ideal P (of a commutative ring with unity), the first option for describing the substratum is

$$CS(\text{Prime Ideals}) = [CS(\text{Ideals}), a \cdot b \in P \to (a \in P \vee b \in P)].$$

In fact, we could omit the parenthesis, since in this case we (typically) save in our minds the structural properties involving the membership in P some of the two (generic) elements, assuming the membership of their product. In this case, we omit the quantifiers because they are implicitly assumed from this particular kind of expression.

An equivalent way of describing this concept requires that we assume the concept consisting of all the ideals of a commutative ring (with unity). So, if R denotes a commutative ring with unity, let us denote by $Id(R)$ the collection of all ideals contained in R.

In this way, we can find an alternative form of the conceptual substratum of a prime ideal as follows

$$CS(\text{Prime Ideals}) = [CS(\text{Ideals}), I, J \in Id(R), I \cdot J \subseteq P \to (I \subseteq P \vee J \subseteq P)].$$

Here, it is worth clarifying that when we think about the conceptual substratum of a concept which is partially based on other concepts (as in the former cases), then in the entire conceptual substratum the expression having a bigger preponderance is the newest one; i.e., the one belonging only to the new notion in consideration (e.g., the specific prime ideal condition), but not to the former notions (e.g., ideals, commutative rings with unity). For example, for the notion of prime ideal, the two former equivalent conditions defined in terms of the membership and containment

---

[2]We will study such cases concretely later on.

relations, respectively, are the most immediate ("new") for the concept. Therefore, they appear explicitly in the substratum. However, although the conditions defining a commutative ring with unity are also necessary in the description of (a purely formal axiomatization of) the notion of prime ideal, they were already assumed and "cognitively stored" when the notion of commutative ring with unity was described ("learned"). This is the reason why we do not write again the substratum of (the concept of) ideals in any of the conceptual substrata of prime ideals.

Let us now focus on topological notions for a while. Specifically, we can describe the substratum of a topological space [11] as follows:

$$CS(\text{Topological Spaces}) = [\emptyset, X \in \mathcal{T}(X), U_j \in \mathcal{T}(X) \to \bigcup_{j \in J} U_j \in \mathcal{T}(X),$$

$$\bigcap_{i=1}^{n} U_i \in \mathcal{T}(X)].$$

Again, informally we do not care so much about the universal quantifier for the open sets U_j or the index set J when we think about topological spaces. We also assume that n is a natural number without mentioning it explicitly.

For the notion of a compact topological space, we get

$$CS(\text{Compact Top. Spaces}) = [U_j \in \mathcal{T}, \bigcup_{j \in J} U_j = X \to \bigcup_{r=1, j_r \in J}^{n} U_{j_r} = X].$$

Here, we write the seminal fact describing compactness and we replace the symbol for conjunction by a comma because we have a stronger tendency to juxtapose several facts, one after the next one, and thus it is more natural to separate them (cognitively) with a comma (or with a blank space). The additional properties of j_r and n are indirectly inferred from the expression.

For the concept of Hausdorff space we obtain the following expression:

$$CS(\text{Hausdorff Top. Spaces})$$

$$= [a, b \in X, a \neq b \to U_a, U_b \in \mathcal{T}(X), U_a \cap U_b = \emptyset, a \in U_a, b \in U_b].$$

We continue analyzing a basic notion which emerges from the cornerstone concept of affine scheme, namely "a section of an affine scheme"; i.e., a function f on the structure sheaf O_X, where $X = \text{Spec} R$, and R denotes a commutative ring with unity. Let us call this concept $S.A.S.$

So, one compact way of describing the conceptual substratum of this notion would be the following:

$$CS(\text{S.A.S.}) = \left[f(w) = r_j/q_j \in R_w : w \in \mathcal{U}_j; \mathcal{U} = \bigcup_{j \in J} \mathcal{U}_j \right].$$

In fact, the most essential property of such a section is that they are locally given as a quotients of elements of global ring R. In the syntactic information given before one can infer the existence of the corresponding open covering of $\mathcal{U}$, from the relation $\mathcal{U} = \bigcup_{j \in J} \mathcal{U}_j$. This is a consequence of a more general "cognitive" phenomenon, namely when we explicitly write down some formal syntactic relations describing mathematical entities, we implicitly assume that they exist, at least at a formal "mental" level.[3] So, in the former description it is not completely necessary to describe explicitly existential quantifiers for the objects appearing there.[4]

9.3   Introducing a Suitable Notation

Within our global taxonomy containing, among others, morphological units, abstract (mathematical) objects, and cognitive representations, conceptual substrata of mathematical notions and proofs belong to a special new class of entities combining cognitive, linguistic, and mathematical features. Therefore, during the formalization process we will "jump" semantically to new kinds of formal beings going beyond the standard mathematical realm.

So, a conceptual substratum of a (meta-)mathematical object (e.g., classic mathematical concepts, formal proofs, theories) consists of a finite and ordered sequence of morphological packages (units) (e.g., $A \cup B, x + y, 2 \cdot n$), which will be separated by commas for the sake of the clarity in this presentation.[5]

Each of these packages corresponds to a term/atomic relation which, in general, does not contain symbols representing logical connectives. For example, when we think about the disjunction of two mathematical expressions Φ and Θ, we save in our minds each of them independently and then we choose the most adequate depending on the goal that we want to achieve (e.g., to prove, to assume, to generalize, etc.)

When we think about a concept C defined by an implication of the form $\Phi \to \Theta$, we used to save the sequence Φ, Θ in exactly this order; i.e., during the process of consciously "refreshing" in our minds the meaning of C, we tend to think first

---

[3]For example, when we want to verify if a hypothetical mathematical structure exists, or when we aim to prove an statement by contradiction.

[4]In Sect. 9.4, we will present cognitive codifications taking into account the type of quantifiers in a more precise way.

[5]When we create mental representations of mathematical phenomena, in (almost) all cases we do not imagine commas separating the objects.

about Φ and immediately after this we think about Θ, without thinking between these two stages about an implication symbol " $\rightarrow$". In conclusion, the implication symbol is, pragmatically speaking, only a meta-artifact used for writing proofs and for describing metamathematical and logic properties.

Another important fact is that our minds use "positive" sentences in order to understand even formal negations of concepts (see Chap. 3). For instance, if we need to perform a mental inference (e.g., to use cognitively) a natural number which is not an even number, typically what we do is transform the former concept into another one, which is logically equivalent to the first one, but with the additional property that it does not contain any formal negation. In our current example, we will understand the negated concept of even natural number, pragmatically speaking, as the equivalent transformed concept describing natural numbers which can be written as the sum of an even number plus one; i.e., the conceptual substratum of the resulting working concept would be of the form $[2 \cdot n + 1, n \in \mathbb{N}]$, which does not contain a negation symbol.

Additionally, if we need to use the negation of an atomic formula $R(t_1, \cdots, t_n)$ within a proof, often we transform unconsciously such a "negative" expression into a "positive" one, by considering the complementary relation $\neg R$. For example, if we need to use the fact that x is not equal or bigger than 2, what we concretely use is the equivalent fact stating that $x < 2$. In conclusion, our minds transform (unconsciously) "negative statements" into "positive" (and more concrete) ones in order to be able to perform "successful" inferences involving them.

Another enlightening example is the concept of continuous real functions. Effectively, if we say "Let us consider a non-continuous function $f : \mathbb{R} \rightarrow \mathbb{R}$," generally, we save this fact in our minds in the form of an equivalent statement describing a function $f : \mathbb{R} \rightarrow \mathbb{R}$ such that exists a point x_1 where f has (either a removable or an essential) discontinuity. If we go deeper and try to be more specific, then we imagine a fixed $\epsilon_1 > 0$ and a sequence $\{a_i\}_{i \in \mathbb{N}} \rightarrow x_1$, such that $|f(x_1) - f(a_i)| > \epsilon_1$. Again, we obtain a "positive" description of a "negative" statement, which is more suitable to be concretely manipulated in subsequent reasoning steps within formal (mathematical) research.

9.4 A Formalization of Conceptual Substratum in a (Many-Sorted) First-Order Framework

Based on the former facts, we propose an initial formalization of conceptual substratum for a mathematical notion C expressed by a (many-sorted) first-order formula $\Psi(x_1, \cdot, x_n)$, where $x_1, \cdots, x_n$ are exactly the collection of free variables of Ψ parametrizing C.

First of all, the remarks of the former section imply that it is more convenient to use a completely new notation for the way in which conceptual substrata are described, since here we are denoting cognitive representations of mathematical concepts and proofs. In addition, the collection of former examples suggests that conceptual substrata should consist of families of atomic formulas arranged in a very specific order. Thus, we will simulate morphologically the way in which our minds "understand" each logical connective in a very unique fashion. So, let us start by updating the notation used in the description of conceptual substrata, since the symbols "[" and "]" are used very frequently in mathematics for denoting, for instance, intervals of real numbers and (more generally) totally ordered sets. Therefore, we use instead, the delimiters "⌜" and "⌝" only in the case that a conceptual substratum consists of a single atomic formula.

Let us present this meta-notion recursively starting with atomic formulas. If Ψ has the form $R(t_1, \cdots, t_n)$, then we define $CS(\Psi) = \ulcorner R(t_1, \cdots, t_n) \urcorner$. Now, in the case that Ψ is the negation of an atomic formula, i.e., $\neg R(t_1, \cdots, t_n)$, we will introduce a new relational symbol $R_\neg(t_1, \cdots, t_n)$, with the expected meaning; i.e., the relation defined as the formal negation of R. Thus, $CS(\Psi) = \ulcorner R_\neg(t_1, \cdots, t_n) \urcorner$. In the cases that Ψ has the form $\Psi_1 \wedge \Psi_2$, $\Psi_1 \vee \Psi_2$, $\Psi_1 \leftrightarrow \Psi_2$ and $\Psi_1 \rightarrow \Psi_2$, we define recursively the corresponding conceptual substrata as $\|CS(\Psi_1), CS(\Psi_2)\|$; $\langle CS(\Psi_1), CS(\Psi_2) \rangle$; $\langle\langle CS(\Psi_1), CS(\Psi_2) \rangle\rangle$; and $[\![CS(\Psi_1), CS(\Psi_2)]\!]$; respectively.

Let us assume we are considering a formula of the form $\neg\Phi$. In this case, it is an elementary exercise to see that $\neg\Phi$ is logically equivalent to a formula Ξ in *affirmative normal form*, namely Ξ has no negation symbols acting over non-atomic sub-formulas; i.e., all the negations where gradually transformed (preserving logical equivalence) until one obtains a formula involving only quantifiers, the logical connectives $\wedge, \vee, \rightarrow, \leftrightarrow$, and the occurrences of $\neg$ act only over atomic formulas $R_j(t_1, \cdots, t_n)$. So, these occurrences are replaced (similarly as before) by new atomic formulas of the form $R_j^{[n]}(t_1, \cdots, t_n)$, which should be interpreted in the natural way. This re-representation is similar to the negation normal form, but strictly speaking, it is not the same construction because we are not interested in eliminating the implications in the formula. Here, one of the main principles is to preserve the "essential" and (cognitively speaking) "natural" aspects of the concepts. Therefore, we also need to be careful with the way in which the morphological aspects of the formulas evolve when we perform several kinds of transformations, even if those transformations preserve logical equivalence. So, we will adopt the following (standard) conventions when we negate formulas:

$$\neg(A \rightarrow B) \cong A \wedge \neg B,$$

$$\neg(A \leftrightarrow B) \cong (A \wedge \neg B) \vee (B \wedge \neg A),$$

$$\neg(A \wedge B) \cong (\neg A \vee \neg B),$$

$$\neg(A \vee B) \cong (\neg A \wedge \neg B),$$

$$\neg\forall x \, P(x) \cong \exists x \neg P(x),$$

and

$$\neg\exists x \, P(x) \cong \forall x \neg P(x).$$

In our formalization, we will fix symbols for the constants and variables in the following way: $\{c_i\}_{i\in I}$ denotes the collection of (potential) constants to be used, where I is a finite collection of parameters.[6]

$\{x_i\}_{i\in I}$ denotes the collection of free variables, $\{y_i\}_{i\in I}$ denotes the collection of universally quantified variables, and $\{z_i\}_{i\in I}$ denotes the collection of existentially quantified variables.

In order to eliminate quantifiers from the expressions in the conceptual substrata, we will perform a procedure very similar to Skolemization in nature, but include extra information regarding the scope of the existential quantifiers that are codified.

So, let us continue with the recursive definition of conceptual substratum. We will distinguish the cases of classic and many-sorted first-order logic. The reason is that with many-sorted first-order logic we can explicitly represent highly abstract concepts in a simple fashion. Meanwhile, with a classic first-order framework we should add a lot of extra information which can be codified in the choice of the sorts.

Taking inspiration from Skolemization [14, Ch. 6] to some degree, we would like to define formal expressions for conceptual substratum of the following form

$$CS((\forall x_1)(\exists x_2)(x_1 + x_2 = 0)) = \ulcorner\ulcorner x_{(1)} + x_{[2]}^{(1)} = 0 \urcorner_{[2]}\urcorner^{(1)},$$

where the sub-index in parenthesis written below a variable (e.g., (1)) (resp. in brackets (e.g., [2])) indicates that such a variable is universally (resp. existentially) quantified and the identical corresponding symbol written below $\urcorner$ (resp. above $\ulcorner$corners$\urcorner$) indicates the scope of the universal quantifier of the corresponding variable. Moreover, the super-index in parenthesis on a variable (e.g., $x_{[2]}^{(1)}$) indicates which universal quantified variables are outside it syntactically (i.e., such variable is inside the scope of the other universal quantifiers (e.g., x_1) that quantified the variables with the super-indices in parenthesis). In this way, we can codify an implicit functional dependence among variables, without introducing new artificial functional/constant symbols and a second-order logic as in the case of Skolemization. In this case, we take direct inspiration from a method used in automated reasoning, because it reflects a seminal aspect related to the way in which our minds understand the interaction between universal and existential quantifiers in a pragmatic way. For instance, the former sentence is usually comprehended in an algebraic framework (e.g., Group theory) in a functional way, namely the variable x_2 is perceived and modeled as the formal inverse of x_1, and implicitly denoted (in a lot of cases) in a functional way, i.e., $x_2 = -(x_1)$.

In general, let us assume that any formula Φ is written in such a way that there are no repetitions of the same symbols for (bounded or free) variables among syntactically disjoint parts of it (e.g., there are no two bound variables in disjoint sub-formulas of Φ with the same name).

---

[6]This is enough, since the specific morpho-syntactic representations of mathematical entities are finite because they are bounded by the number of discrete symbols that a human mind can generate.

Firstly, assume that A is a quantifier-free formula and Ψ has the form $\exists z_r A(z_r)$ (resp. $\forall z_r A(z_r)$), then $CS(\Psi)$ is $\ulcorner A(z_{[r]}) \urcorner_{[r]}$ (resp. $CS(\Psi)$ is $\ulcorner A(z_{(r)}) \urcorner^{(r)}$), where the ("slightly new") meta-variable $z_{[r]}$ (resp. $z_{(r)}$) represents the way in which we "mentally" understand the fact that z_r is an existentially (resp. universally) bounded variable in Ψ. And, the sub-index $\ulcorner - \urcorner_{[r]}$ (resp. super-index $\ulcorner - \urcorner^{(r)}$ denotes the fact that our minds understand implicitly the scope of the existential quantifier bounding z_r.[7]

Secondly, suppose that A denotes an arbitrary FOL formula, and let Ψ be a formula of the form $\forall x_r A(x_r)$ (resp. $\exists x_r A(x_r)$), for some $r \in I$. Assume that

$$CS(A) = \mathscr{C}on Sub(x_{[n_1]}^{(m_{1,1},\cdots,m_{1,a_{n_1}})}, \cdots, x_{[n_k]}^{(m_{n_k,1},\cdots,m_{n_k,a_{n_k}})}, x_r),$$

where a meta-variable $x_{[n_*]}^{(m_{n_*,1},\cdots,m_{n_*,a_{n_*}})}$ codifies in its representation the fact that the (existentially quantified) variable x_{n_*} is under the scope of the universally quantified variables $x_{m_{*,1}}, \cdots, x_{m_{*,a_{n_*}}}$. So, since implicitly we assume that we are not repeating (bounded and free) variables into the representation of a single formula, we can define the conceptual substratum of Ψ as

$$CS(\Psi) = \ulcorner \mathscr{C}on Sub(x_{[n_1]}^{(m_{1,1},\cdots,m_{1,a_{n_1}},r)}, \cdots, x_{[n_k]}^{(m_{n_k,1},\cdots,m_{n_k,a_{n_k}},r)}, x_{[r]}) \urcorner_{[r]},$$

$$(\text{resp. } CS(A) = \ulcorner \mathscr{C}on Sub(x_{[n_1]}^{(m_{1,1},\cdots,m_{1,a_{n_1}})}, \cdots, x_{[n_k]}^{(m_{n_k,1},\cdots,m_{n_k,a_{n_k}})}, x_{(r)}) \urcorner^{(r)}).$$

Note that in the last representation the free variables are left intact. In fact, this is a way of codifying cognitively with a simple notation that such variables are not quantified, which is for our minds a clear difference with the implicit assumption of taking the classical universal closure of it.

From the whole construction, it is a computation to verify that from conceptual substrata of the above specific form we can reconstruct the original formula in a unique fashion. Intuitively, this is a natural fact in the sense that the conceptual substratum should allow us to reconstruct completely the corresponding source concept.

In the setting of a many-sorted first-order logic, we the construction is basically the same as before, but with the addition that one adds to the right side of the substratum an expression of the form

---

[7]Note here that the new kind of symbols introduced before have mainly a referential purpose; i.e., we aim to find a more precise morpho-syntactic approximation of the way in which our minds save and process mathematical contents of any sort. However, it is clear that the specific kinds of symbolic units that we introduce (e.g., $\ulcorner - \urcorner^{(-)}$) play an auxiliary role within our formalization and are used for the sake of developing a clear exposition of some of the (cognitively-inspired) parts of our model.

$$\ulcorner x_{p_1} : S_{h_{p_1}}, \cdots, x_{p_m} : S_{h_{p_m}}, y_{r_1} : S_{q_{r_1}}, \cdots, y_{r_n} : S_{q_{r_n}}, z_{i_1} : S_{j_{i_1}}, \cdots, z_{i_k} : S_{j_{i_k}},$$

$$f_{h_1} : S_{h_{1,1}} \times \cdots \times S_{h_{1,w_1}} \rightarrow S_{h_{1,w+1}}, \cdots, f_{h_v} : S_{h_{v,1}} \times \cdots \times S_{h_{v,w_v}} \rightarrow S_{h_{w_v+1}},$$

$$R_{g_1} : S_{g_{1,1}} \times \cdots \times S_{g_{1,u_1}}, \cdots, R_{g_t} : S_{g_{t,1}} \times \cdots \times S_{g_{t,u_t}} \cdot \urcorner$$

Nonetheless, this additional information is quite important, it takes a secondary role in the conceptual substratum of a concept.[8]

Remark 9.1 In this kind of multidisciplinary approach, a "theoretical soundness" is as important as a "pragmatic" coherence for our cognitively-inspired model. Therefore, although with the former notion of conceptual substratum we can (virtually) cover any kind of mathematical definition due to the fact that the first-order framework of Zermelo–Fraenkel set theory with Choice (ZFC) serves as a global logic foundation for modern (standard) mathematics [10]; from a practical point of view, the working mathematician typically does not understand sophisticated mathematical notions by means of long explicit first-order ZFC sentences, but instead, (s)he builds several "conceptual layers" materialized through the specific way in which we recursively write (mathematical) morphemes at several levels of complexity in order to describe more structured concepts. For example, this occurs frequently in highly sophisticated mathematical disciplines like algebraic geometry and algebraic topology. For instance, if we want to express the concept of a section of an affine scheme strictly by a first-order formula in ZCF, it would take several lines of symbols. However, as we showed before, and due to this gradual way of building concepts, we just need one line of symbols for describing the conceptual substratum of such a mathematical notion.

Finally, for the sake of simplicity during the subsequent applications regarding the generation of proofs and concepts in several mathematical domains in further chapters, we will write conceptual substrata often in a way slightly simpler than the one in which the former formalization was described. The main reason is to maximize clarity in the exposition and the fact that for most of the applications developed here, there is no need to write too many super- and sub-indexes. In fact, the previous formalization is meant more for further goals regarding the AMI program (see Chap. 12) and for pedagogical reasons in this exposition.

---

[8]Despite the fact that larger instantiations of the former m.-s. descriptions of conceptual substrata could be challenging to grasp pragmatically, they should not represent major challenges for sufficiently robust intelligent agents.

9.5 Functional Conceptual Substratum

In [3], an initial meta-formalization of (functional) conceptual substratum for notions given by $n-$ary relations in a first-order setting is presented. For the sake of completeness, we recall it here together with the most outstanding results proved there:

Definition 9.1 Let us fixed a first-order language L. An $n-$ary property Θ in (a set-theoretical structure) M (i.e., $\Theta \subseteq M^n$) possesses a *functional conceptual substratum* (over L and M), if there exist terms t_i (for $i = 1, \ldots, r$) and atomic formulas $F_1, \ldots, F_m$ whose variables are contained in $\{x_1, \ldots, x_s\}$, such that for all $b_1, \ldots, b_r \in M$, $(b_1, \ldots, b_n) \in \Theta$ if and only if

$$M \models (\exists x_1) \cdots (\exists x_s)(y_1 = t_1 \wedge \cdots \wedge y_n = t_n \wedge F_1 \wedge \cdots \wedge F_m)[y_1 \mapsto b_1, \cdots, y_r \mapsto b_r],$$

where $t_1, \ldots, t_r$ are L-terms whose variables are among $x_1, \ldots, x_s$.

Functional conceptual substratum can be used for generating more cognitively generated versions of the sequential calculus. In fact, let LK_e be the sequent calculus for first-order predicate logic with equality over a language L, with the standard inference rules (see [1, 15]). Let us add to our language L a new $n-$ary predicate symbol D defined in terms of functional conceptual substratum, i.e., by the expression

$$D(b_1, \ldots, b_n) \Leftrightarrow (\exists y_1 \cdots y_s)(b_1 = t_1 \wedge \cdots \wedge b_n = t_n \wedge F_1 \wedge \cdots \wedge F_m),$$

where $t_1, \ldots, t_n$ are $L-$terms and $F_1, \ldots, F_m$ are L-atoms whose variables are among $y_1, \ldots, y_s$.

Now, one can define two extensions of LK_e, denoted by $LK_e(D)$ and $LK_e^{fcs}(D)$ by adding the two pairs of inference rules, respectively:

Firstly,

$$\frac{\phi(b_1, \ldots, b_n), \Gamma \to \Delta}{D(b_1, \ldots, b_n), \Gamma \to \Delta} D_l \qquad \text{and} \qquad \frac{\Gamma \to \Delta, \phi(b_1, \ldots, b_n)}{\Gamma \to \Delta, D(b_1, \ldots, b_n)} D_r$$

where $\phi(b_1, \ldots, b_n)$ abbreviates the formula defining $D(b_1, \ldots, b_n)$ as above.

And secondly,

$$\frac{b_1 = t_1[\underline{y}\backslash\underline{\zeta}], \ldots, b_n = t_n[\underline{y}\backslash\underline{\zeta}], F_1[\underline{y}\backslash\underline{\zeta}], \ldots, F_m[\underline{y}\backslash\underline{\zeta}], \Gamma \to \Delta}{D(b_1, \ldots, b_n), \Gamma \to \Delta} D_l^{fcs}$$

and

$$\frac{\Gamma \to \Delta, F_1[\underline{y}\backslash\underline{u}] \quad \cdots \quad \Gamma \to \Delta, F_m[\underline{y}\backslash\underline{u}]}{\Gamma \to \Delta, D(t_1[\underline{y}\backslash\underline{u}], \ldots, t_n[\underline{y}\backslash\underline{u}])} D_r^{fcs}.$$

The last two rules are generalized ways of formalizing typical statements declaring that a (generic) representative of a concept has the suitable (functional) conceptual substratum; e.g., Due to the fact that t is an odd prime, there exists $s \in \mathbb{N}$ such that $t = 2s + 1$, "2(u+v+w) is an even number," respectively.

Let ψ be a formula, then we denote by $\psi[D \backslash \phi]$ the formula obtained after replacing D by ϕ in ψ. Then, one can prove that $\mathrm{LK}_e(D)$ is a conservative extension of LK_e, i.e., for any formula ψ, $\mathrm{LK}_e(D) \vdash \psi$ if and only if $\mathrm{LK}_e \vdash \psi[D \backslash \phi]$. Furthermore, $\mathrm{LK}_e^{\mathrm{fcs}}(D)$ is sound and complete with respect to $\mathrm{LK}_e(D)$, i.e., for any formula ψ, $\mathrm{LK}_e^{\mathrm{fcs}}(D) \vdash \psi$ if and only if $\mathrm{LK}_e \vdash \psi[D \backslash \phi]$. This implies, in particular, that for any formula ψ, $\mathrm{LK}_e^{\mathrm{fcs}}(D) \vdash \psi$ if and only if $\mathrm{LK}_e(D) \vdash \psi$.

With the former systems, we keep at least the same deductive power as with classic ones and, simultaneously, we are closer to developing meta-formalizations with a stronger cognitively-inspired nature.

9.6 Metamathematical and Cognitive Interpretation of the Church–Turing Thesis

In this section, we will give an additional metamathematical and cognitive interpretation of the Church–Turing thesis in terms of an ideal UMAA (see Chap. 1, Sect. 1.1). In other words, how should the recursively enumerable (resp. decidable) sets be syntactically represented by an ideal UMAA, after doing all the formal constructions that we have developed so far? By simplicity, we will consider computability over the integers. More specifically, let us consider the structure given by $\mathbb{Z}$ with the language $L = \{0, 1, +, -, *, =, <\}$.

In the next proposition, we will find a characterization of the recursively enumerable relations in $\mathbb{Z}$ exactly in terms of the expressions involving functional conceptual substratum.

Proposition 9.1 *Let n be a positive natural number, and let $\mathscr{C}$ be a concept defining a n−ary relation Ω in $\mathbb{Z}$. Then $\mathscr{C}$ (resp. Ω) has a functional conceptual substratum if and only if Ω is a recursively enumerable set.*

Proof Let us assume that $\mathscr{C}$ has a functional conceptual substratum as described above.

So, every F_j can be written on the form of either $w_1(x_1, \ldots, x_n) = w_2(x_1, \ldots, x_n)$ or $w_1(x_1, \ldots, x_n) < w_2(x_1, \ldots, x_n)$, where w_1 and u_2 are the corresponding polynomials in $\mathbb{Z}[x_1, \ldots, x_n]$ representing the terms described in F_j.

In the first case, F_j can be rewritten in the form $Q(x_1, \ldots, x_n) = 0$, where $Q = w_1 - w_2$. For the remaining case, we can use Lagrange's theorem (i.e., any natural number can be written as the sum of four perfect squares [5]) for expressing the condition described by F_j in a Diophantine manner, i.e.,

$$(\exists a_1, \cdots, a_4)(w_1 - w_2 = z_1^2 + a_2^2 + a_3^2 + a_4^2 + 1).$$

Furthermore, one can describe a finite number of conjunctions of polynomial equations through a single equation. Effectively, this is clearly shown by using the fact that over the integers $\sum_{i=1}^{m} b_i^2 = 0$ if and only if each $b_i = 0$. Thus, combining the former steps one can generate an explicit polynomial $E(y_1, \ldots, y_r, x_1, \ldots, x_n)$ such that for all $a_1 \ldots, a_r \in \mathbb{Z}$, $a_1, \ldots, a_r \in \Omega$ if and only if

$$\mathbb{Z} \models (\exists x_1, \cdots, x_n)(E(y_1, \ldots, y_r, x_1, \ldots, x_n) = 0))[y_1 \mapsto a_1, \cdots, y_r \mapsto a_r].$$

In other words, Ω defines a Diophantine set [9, Ch. 1].

Now, Ω defines a recursively enumerable set by the MRDP theorem [9, Ch. 2].

Conversely, let C be a concept defining a recursively enumerable property Ω, then by the MRDP theorem Ω is Diophantine. Thus, for all $b_1 \ldots, b_r \in \mathbb{Z}$, $b_1, \ldots, b_r \in \Omega$ if and only if

$$\mathbb{Z} \models (\exists x_1, \cdots, x_m)(F(y_1, \ldots, y_r, x_1, \ldots, x_m) = 0))[y_1 \mapsto b_1, \cdots, y_r \mapsto b_r].$$

We can rewrite the former expression as follows:

$$\mathbb{Z} \models (\exists x_1, \cdots, x_m, x_1', \cdots, x_r')(b_1 = x_1' \wedge \cdots \wedge b_r = x_r' \wedge F_1))$$

$$\times [y_1 \mapsto b_1, \cdots, y_r \mapsto b_r].$$

where F_1 denotes the atom $F(x_1', \ldots, x_r', x_1, \ldots, x_m) = 0)$. This means that $\mathscr{C}$ has functional conceptual substratum. Note that in this case, the essential information of the concept can be, at least formally, codified more in the atom F_1 rather than in the initial polynomial expressions.

Remark 9.2 Due to the former equivalence we see that one of the most outstanding and important arithmetic concepts, namely the prime numbers, has a functional conceptual substratum (over L' and $\mathbb{Z}$). Even more, based on a classic result presented in [7], we can explicitly describe a polynomial inequality characterizing the positive prime numbers:

$$CS(\text{Positive Prime Numbers}) = [p, (p+2)(1 - (wz + h + j - q)^2$$

$$-((gk + 2g + p + 1)(h + j) + h - z)^2 - (2n + k + q + z - e)^2 -$$

$$(16(p + 1)^3(p + 2)(n + 1)^2 + 1 - f^2)^2 - (e^3(e + 2)(a + 1)^2 + 1 - o^2)^2$$

$$-((a^2 + 1)y^2 + 1 - x^2)^2 - (16r^2 y^4(a^2 - 1) + 1 - u^2)^2$$

$$(((a + u^2(u^2 + a))^2 - 1)(n + 4dy)^2 + 1 - (x + cu)^2)^2 - (n + l + v - y)^2$$

$$-((a^2 - 1)l^2 + 1 - m^2)^2 - (ai + p + 1 - l - i)^2$$

$$-(k + l(a - n - 1) + b(2an + 2a - n^2 - 2n - 2) - m)^2$$

$$-(q + y(a - k - 1) + s(2ak + 2a - k^2 - 2k - 2) - x)^2$$

$$-(z + kl(a - k) + t(2ak - k^2 - 1) - mk)^2) > 0,$$

$$(a, b, c, d, e, f, g, h, i, h, j, k, l, m, n, q, r, s, t, u, v, w, x, y, z) \geq (\bar{0}),$$

$$a, b, c, d, e, f, g, h, i, h, j, k, l, m, n, q, r, s, t, u, v, w, x, y, z \in \mathbb{Z}].$$

Another way of describing a conceptual substratum with only one equation is obtained with the help of Lagrange's Theorem; i.e., every natural number can be expressed as the sum of at most four squares [5]. Due to this classic theorem, we can add 4 new existentially quantified variables and replace the former (26) ones by the corresponding quadratic forms, in order to obtain a polynomial $Q(y_1, \ldots, y_{104})$, with integer coefficients, fulfilling

$$CS(\text{Positive Prime Numbers}) = [y_1, Q(y_1, \ldots, y_{108}) > 0, y_2, \ldots, y_{108} \in \mathbb{Z}].$$

On the other hand, if we restrict the atoms A_i to expressions of the form $x_{r_i} < c_i$ or $c_i < x_{c_i}$, where c_j denotes a constant symbol, then one can prove that the prime numbers do not have a functional conceptual substratum.

Remark 9.3 Another important fact is the arithmetical invariance of functional conceptual substratum when we extend the conceptual framework of notions from the natural numbers to the integers. Effectively, let $L_1 = \{0, 1, +, *, =, <\}$ be the (arithmetical) language avoiding the subtraction symbol. Thus, assume that a concept $\mathscr{C}$ is given by a $n-$ary relation R in $\mathbb{N}$. Then, $\mathscr{C}$ has a functional conceptual substratum in L_1, if and only if $\mathscr{C}$ seen as a concept described by the corresponding $r-$ary relation $D \subseteq \mathbb{Z}^r$ has a functional conceptual substratum in L.

Now, in virtue of the former proposition we can state the following cognitive interpretation of the Church–Turing Thesis:

Thesis 9.1 (Cognitive Characterization of the Church–Turing Thesis) *Let $\mathscr{C}$ be a concept defined by a relation $\Omega \subseteq \mathbb{Z}^n$ ($n \in \mathbb{N}_{>0}$). Then, $\mathscr{C}$ is decidable in the informal sense (i.e., Ω is decidable in the informal sense) if and only if both $\mathscr{C}$ and it complementary concept $\mathscr{C}^+$ (defined by the relation $\mathbb{Z}^n \setminus \Omega$) have functional conceptual substratum.*

Although from the purely logic perspective the proof of Proposition 9.1 seems to be an elementary fact, based on a more sophisticated one (i.e., the MRDP theorem), from the cognitive and metamathematical point of view this result possesses a stronger and deeper importance. Effectively, if we remember the way in which we came up with the definition of functional conceptual substratum, we see that it emerges as the natural way of writing explicitly each of the components of the generic form of the elements of the relation describing a concept $\mathscr{C}$ defined over a fixed structure M. In other words, this special kind of concepts has a generic and transparent syntactic description which can be directly grasped and understood by an ideal UMAA from a m.-s. perspective. For example, in the case of the prime numbers an ideal UMAA would have no deep trouble in understanding them in terms of the former long polynomial description, since (informally speaking) it implicitly possesses a quite bigger ram and rom memories than an average mathematician. Furthermore, the former cognitive and metamathematical characterization of the Church–Turing Thesis shows a direct link between what can be decidable in the classic sense of recursion theory and what can be explicitly described in terms of an ideal UMAA for both the original concept (i.e., $\mathscr{C}$) and its corresponding conceptual complement (i.e., C^+ (see Chap. 10)). Moreover, as we mentioned in the first part of the book, the complement or, in some sense, the formal negation of a concept is understood by our mind, pragmatically speaking, using m.-s. characterizations which create conceptual equivalences with new notions which do not use negations, and are described in a more atomic and affirmative way. So, this is exactly what this Thesis says about a decidable concept, namely that not only it, but also its conceptual complement can be explicitly described in a wide sense, i.e., in terms of explicit defining terms and defining atomic formulas.

9.7 Conceptual Lining

There is a natural conceptual operation which can be seen as a "dual form" of conceptual substratum, namely one starts with seminal and specific conceptual information and then our mind generates a more robust and contextualized concept/object around it. In other words, this cognitive ability is one of the mental engines responsible for the "materialization" of abstract ideas in the real world through language (and subsequently with the creation of new (visible) concepts, entities, or objects). For instance, through the repetitive usage of conceptual lining, among many other mental processes, we are able to transform a simple mental idea into a book, a company, an invention, a painting, or a successful result in sports, among others.

From the mathematical point of view, conceptual lining consists of recovering from a syntactic expression describing a (mathematical) concept or a m.-s. configuration (e.g., substratum) $\mathfrak{C}$, the corresponding (uniquely defined) (mathematical) concept "behind" it. We will call this cognitive ability *conceptual lining* and it will be denoted by $CL(\mathfrak{C})$.

Let $\varXi$ be a (mathematical) concept, then it holds the rule $CL(CS(\varXi)) = \varXi$. And, if $\Theta = CS(\Psi)$, for some concept Ψ, then $CS(CL(\Theta)) = \Theta$.

Acknowledgments The author thanks William Restrepo and family for all the support, transparency and sincere friendship.

References

1. Buss, S.R.: An introduction to proof theory. Handbook of proof theory, in Studies in Logic and the Foundations of Mathematics **137**, 1–78 (1998)
2. Fraleigh, J.B.: A first course in abstract algebra. Pearson Education India (2003)
3. Gomez-Ramirez, D.A.J., Hetzl, S.: Functional conceptual substratum as a new cognitive mechanism for mathematical creation. arXiv preprint arXiv:1710.04022 URL https://arxiv.org/pdf/1710.04022.pdf
4. Görtz, U., Wedhorn, T.: Algebraic Geometry: Part I: Schemes. With Examples and Exercises. Springer (2010)
5. Hardy, G.H., Wright, E.M.: An introduction to the theory of numbers, (Sixth Edition). Oxford University Press (2008)
6. Hartshorne, R.: Algebraic Geometry. Springer-Verlag, New York (1977)
7. Jones, J.P., Sato, D., Wada, H., Wiens, D.: Diophantine representation of the set of prime numbers. The American Mathematical Monthly **83**(6), 449–464 (1976)
8. Liu, Y., Yu, M., Fu, Q., Chen, W., Liu, Y., Xie, L.: Cognitive mechanism related to line drawings and its applications in intelligent process of visual media: a survey. Frontiers of Computer Science **10**(2), 216–232 (2016)
9. Matiyasevich, Y.V.: Hilbert's tenth problem, *Foundations of Computing*, vol. 105. MIT Press Cambridge (1993)
10. Mendelson, E.: Introduction to Mathematical Logic (Fifth Edition). Chapman & Hall/CRC (2010)
11. Munkres, J.: Topology. Second Edition. Prentice Hall, Inc (2000)
12. Osherson, D., Smith, E.E.: On typicality and vagueness. Cognition **64**(2), 189–206 (1997)
13. Owen, D.: Locke on real essence. History of Philosophy Quarterly **8**(2), 105–118 (1991)
14. Robinson, A.J., Voronkov, A.: Handbook of automated reasoning, vol. 1. Elsevier (2001)
15. Takeuti, G.: Proof theory (Second Edition). Dover Publications (2013)

Chapter 10
(Initial) Global Taxonomy of the Most Fundamental Cognitive (Metamathematical) Mechanisms Used in Mathematical Creation/Invention

10.1 Introduction

What are the most fundamental cognitive abilities used by our minds during the process of mathematical creation/invention? Is there a general way to formalize them beyond a specific mathematical discipline? Could we do this in a unified way?

The aim of the present chapter, among others, is to give general (positive) answers to these questions. We present an initial global taxonomy of all the cognitive (metamathematical) mechanisms (or abilities) used by the mind in the particular intellectual activity of solving abstract mathematical inquiries at arbitrary levels of abstraction.

We present them in a unified way and without doing any kind of "internal prioritization," mainly based on the (metamathematical) heuristic principle that all of them play a constituent role in mathematical generation, sometimes with different "frequencies of application" depending on the specific sub-disciplines, but always at a constituent level.

Our classification is made on two thematic pillars: the first one consists of strong support and inspiration for a wide spectrum of (modern and classic) result in cognitive science, psychology, linguistics, and related areas; the second pillar is based on the implicit guidelines that a global metamathematical program should fulfill; e.g., meta-explanatory soundness and pragmatic coherence (of the corresponding mechanisms).

Frequently, when we talk about cognitive abilities (or mechanisms), we are implicitly looking for cognitive metamathematical abilities. On the other hand, in this chapter we do not deal with the (purely cognitive) question of deciding if any of the cognitive metamathematical abilities is, in fact, a seminal cognitive ability or if it can be grounded in terms of other (more fundamental) ones. This is an important aspect to be clarified to avoid confusion with terminology used (locally) in modern research belonging to cognitive science and related scientific disciplines.

© Springer Nature Switzerland AG 2020
D. A. J. Gómez Ramírez, *Artificial Mathematical Intelligence*,
https://doi.org/10.1007/978-3-030-50273-7_10

The name "cognitive metamathematical ability" (or mechanism) was chosen mainly due to (1) the cognitive basis grounding it and (2) its general presence as a natural and practical tool during (abstract) mathematical reasoning.

The pragmatic support of our taxonomy can be appreciated more clearly in the next chapter, where we use all its deductive and cognitive power during the meta-generation of dozens of fundamental mathematical concepts and proofs belonging to several sub-disciplines.

Finally, we claim that from the point of view of modern cognitive science, the present taxonomy can serve as a valuable case of study toward the general quest of identifying the general processes used by the mind in all the intellectual and concrete tasks of daily life going beyond mathematical creation.

10.2   General Mathematical Concepts

Inspired by the former meta-definition presented in Chap. 7 and, in some sense, by the general considerations of J. Goguen developed in [27], we will present here a general notion of mathematical concept with a strong pragmatic component. Namely, in our approach, not only is the semantic dimension fundamental, but so too is the morphological-syntactic one (i.e., the specific symbols and the explicit way in which the defined axioms are described). This aspect is foundational from a cognitive perspective because the specific form in which we "visualize" concepts (even in mathematics) influences considerably the way in which we subsequently understand them [22].

Furthermore, our formalization of (mathematical) concepts exploits to a large extent, the m.-s. dimension. The semantic component is formalized in a more compact way by means of considering semantic closures of the corresponding conceptual entities. This is presented in this specific way due, on the one hand, to the pragmatic advantages that the m.-s. descriptions of concepts possess, and on the other hand, due to the challenge that the semantic dimension of mathematical structures represents within the new cognitive foundations program (see Chaps. 3, 5, and 6).

Definition 10.1 A *mathematical concept* $\mathscr{C}$ consists of the following data:

$$\mathscr{C} = \langle L_{og}, \Sigma, A, \mathbb{M} \rangle,$$

where L_{og} is a logic,[1] Σ denotes the totality of the symbols to be used,[2] A is an explicit collection of "sentences" written only with symbols from Σ and under the

---

[1] Here, we mean implicitly a logic rich enough for having well-defined notions of syntax, models, satisfiability, and truth.

[2] We include also non-logical symbols like variables, constants, functional symbols, etc.; logical symbols; auxiliary symbols like parenthesis, comas, $\prod, \int, \bigcap$ or $\otimes$ that must be used explicitly. In addition, these collections of symbols will be separated in groups defined according to their particular function, e.g., logic, auxiliary, variables, constants, functional and relational symbols.

syntactic formation rules determined by L_{og}, and $\mathbb{M}$ is the corresponding semantic closure of A in L_{og}, i.e., the collection of all possible models L_{og}−satisfying each sentence in A.[3] Furthermore, a concept $\mathscr{C}' = \langle L_{og}, \Sigma, A', \mathbb{M}' \rangle$ is called *a sub-concept of* $\mathscr{C}'$ if all the axioms of A' are contained in A. In this case, $\mathscr{C}$ is called a super-concept of $\mathscr{C}'$.

For most of the applications presented here, (many-sorted) first-order logic will be enough. In fact, an implicit heuristic principle that the reader can observe throughout the whole book is that, for the sake of a cognitive meta-generation of theorems and notions in several mathematical fields, (many-sorted) first-order logic is virtually enough for simulating even concepts that classically were described in a higher-order logic setting. The main reason for this is that one can add enough sorts for replacing the need of quantifying over subsets, relations, and functions, without affecting the essence of cognitive characterization of the concepts done by the mind (see Chap. 11 for illustrating examples). This can be done only locally (i.e., when one fixes a finite number of concepts to generate), and this is enough for our purposes due to the local nature of the conscious mind (Sect. 3.4 Chap. 3).

On the other hand, our syntactic-semantic scope is big enough to allow the description of mathematical concepts given in natural language and using more intuitive and natural "logics" like the ones used for expressing concepts in classic Euclidean geometry, graph and knot theory, and (naive) set theory. In other words, the methods presented here are also able to offer formal explanations for the cognitive reconstructions of mathematical results in more informal contexts involving persons who are not working mathematicians, e.g., high school, college and university students, professionals and researchers in math-related areas like physics, biology, chemistry, informatics, economics, engineering, among many others.

10.3 Formal (Meta-)Exemplification and Generic Exemplification

10.3.1 Formal Examples as One of the Most Important Cognitive Sources for Doing Research in Mathematics

Since the very beginning of ancient forms of mathematical research thousands of years ago, concrete instances of mathematical structures like the natural, prime, rational, real and complex numbers; points, lines, triangles, polygons, circles

---

[3]One can assume, for the sake of cognitive completeness of our definition, that A and Σ are robust enough to contain a particular conceptual substratum of $\mathscr{C}$, and the corresponding symbols describing its m.-s. configuration. However, this additional requirement is not mandatory because from the former data one can reconstruct subsequently desired forms of conceptual substratum for $\mathscr{C}$ as wished. Moreover, in the cognitive meta-generations presented in Chap. 11, this additional requirement is not needed.

(among many others) have been one of the most powerful inspirations for the development of more sophisticated forms of mathematics. In fact, a closer look into the history of mathematics as a whole suggests that thousands of notions and results used in mathematical research today were discovered/created starting with collections of (basic) examples and then generalizing gradually specific properties of them until the "right" and "compact" conditions were found, which, at the same time, could fulfill the desired conjecture/conditions [29].

Within the development of general and fundamental theories, the role that small, simple, and basic examples play is as important as the one played by sophisticated and quite general notions. Effectively, sometimes the "right" examples are able to modify the direction of development of an entire theory.

For instance, an outstanding result in this direction is the counterexample to the localization problem obtained by H. Brenner and P. Monsky [8], which answered a fundamental question in Tight Closure, one of the most important theories of commutative algebra [35]. Since the late 1980s, when this theory firstly appeared, the question of determining if this particular closure operation commutes with localization was subsequently on the focus of intensive research. Nonetheless, neither the creators nor the researchers in commutative algebra came to the solution to this problem. In fact, the main intuitions for constructing a counterexample came from algebraic geometry. Concretely, from the work of H. Brenner relating vector bundles and torsors [5]. His work is rich of examples and geometrical and homological intuitions of the corresponding algebraic phenomena.

In this particular case, a specific (algebraic-geometric) construction turned out to be as important as a whole theory. In fact, after the acceptance of the counterexample by the mathematical community, the horizon in the theory of Tight Closure changed substantially. In particular, new trends on this field look at the creation of new theories repairing what Tight Closure could not fulfill, for example, theories based on closure operations that "commute with localization" [19].

More generally, similar phenomena occur when central conjectures, whose positive answer would play a fundamental role within the development of a specific theory, are answered in a negative manner. In this case, concrete (counter-)examples of mathematical structures serve as "turn signals" for either abandoning whole theories or modifying them. In addition, a more moderate version of the former cases occurs when instead of just one outstanding example, an entire collection of "formally milder" constructions are also able to conduct fundamental aspects of whole theories (for specific constructions (meta-)exemplifying the former phenomena see [25, 65, 66] and [40]).

Classic examples of how elementary mathematical structures have inspired quite abstract and general theories can be found in Galois theory, number theory and (real) analysis, among many others.

For instance, E. Galois got inspiration for the development of the seminal cornerstones of the theory carrying his name after studying, in detail, groups of permutations of the roots of polynomials [54]. Currently such structures can be seen as extremely elementary in comparison with the levels of generality that (extensions of) this theory have been able to achieve within the contemporary mathematical panorama, see for instance [16, 32, 39, 59] and [38].

In number theory, one of the most influential examples is the one described by the famous (elementary) exponential Diophantine equation $x^n + y^n = z^n$, for $x, y, z, n \in \mathbb{Z}$ and $n \geq 3$. This basic and simple statement originates one of the most outstanding and intradisciplinary results that modern arithmetic has seen on the last 40 years, which originated the whole proof of Fermat's Last Theorem. Explicitly, it involves new connections between the theory of elliptic curves, the theory of modular forms, Iwasawa Theory, and Galois Cohomology, among others [13, 67]. In this context, one qualitative question appears naturally; i.e., from a metamathematical point of view, what is more valuable and important, the single and elementary Fermat's challenge or the whole technical apparatus constructed by Wiles et al. for settling it?

The answer to this question emerges clearly by means of a deeper study of the entire history of Fermat's Last theorem and the subsequent amount of work that it inspired by hundreds of professional and amateur mathematicians throughout more than three centuries [17, 60]. Effectively, this (in some sense) "tiny" Diophantine sentence is at least as valuable and important as the corresponding formal constructs obtained afterwards, since it played the role of an attractive mathematical goal, or in other words, an outstanding cognitive "token," whose explicit statement was so "simple" and easy of comprehend, that the implicit belief of finding a formal proof in accordance with the sophistication of the original statement seems to be inserted in the minds of a considerable number of people interested in its solution throughout the centuries [63].[4]

In the case of (real) analysis, there have been thousands of pages about the way in which Newton and Leibniz obtained the seminal intuitions for constructing the foundations of what we call today classic analysis (over the real line), based implicitly on elementary mathematical entities and results obtained previously and going back to the Greeks and the Arabians. Additionally, there is also a huge literature about the subsequent formalizations of the so-called calculus until we were able to arrive at the modern version of (real and complex) analysis that we have today. Specifically, the working entities in this context were mainly (the graphics of) curves, given (in most of the cases) by polynomials (with rational coefficients), (elementary versions of) trigonometric, exponential and logarithmic functions (see, for instance, [4, 62] and [30]). Again, from a qualitative perspective these (nowadays quite elementary) structures were as fundamental as the whole formal apparatus that was constructed subsequently, because they were an "ontological source" of inspiration for most of the intuitions and frameworks used.

Another important aspect lying behind suitable mathematical structures is that, sometimes, a single example serves as the "meta-intersection point" where several

---

[4]The most curious reader can find in the literature a lot of similar examples of solved problems in number theory with similar "qualitative" properties as Fermat's Last Theorem, but perhaps not so famous.

propositions and theorems (within a fixed theory) are genuinely instantiated. For example, in the context of forcing algebras over polynomial rings in several variables with coefficients over a perfect field, and related matters, the forcing algebra given as $R[T_1, T_2]/H$, where $R = k[x, y]$, k is a perfect field and

$$H = (h_1, h_2) = (xT_1 + yT_2, yT_1 + xT_2) = \left(\begin{pmatrix} x & y \\ y & x \end{pmatrix} \cdot \begin{pmatrix} T_1 \\ T_2 \end{pmatrix} \right);$$

serves as an "enlightening" example, where not only properties involving normality can be well instantiated but also a wide variety of topological phenomena of the corresponding prime spectrum [7, Pag. 4787], [6].

Thus, within the meta-program of Artificial Mathematical Intelligence concrete examples (at several degrees of complexity) play a fundamental role. In fact, we will show in Chap. 11 that several fundamental mathematical notions can be reconstructed from initial elementary mathematical structures after applying only finitely many instances of the (meta-formalizations of the) cognitive mechanisms described in this part of the book.

In conclusion, not only can general and abstract theorems and principles construct whole theories but also "tiny," simple, and elementary examples of mathematical structures.

10.3.2   Formal (Meta-)Exemplification

Based on all the considerations given in the last section, we offer here a definition of the cognitive mechanism behind the explicit (and subsequently conscious) consideration of concrete mathematical structures as a natural way for gaining more "intuition" (or "feeling") with respect to the general solution of a specific conjecture.

Definition 10.2   A *Formal exemplification* is a tuple $\mathbb{E} = (\mathscr{L}_{an}, \mathscr{L}_o, \mathscr{A}, \mathscr{E})$, where $\mathscr{L}_{an}$ is a fixed language, $\mathscr{L}_o$ is a fixed logic described with $\mathscr{L}_{an}$, $\mathscr{A}$ is a collection of statements (e.g., sentences) describing the explicit properties of the structure $\mathbb{E}$, and $\mathscr{E}$ is the collection of (main) mathematical objects (e.g. sets, classes) in consideration with fixed labels defining $\mathbb{E}$, which satisfy the axioms $\mathscr{A}$ within the logic $\mathscr{L}_o$. Furthermore, a *Formal Meta-exemplification* $\langle \mathbb{E}_i \rangle_{i \in I}$ is a collection of formal exemplifications $\mathbb{E}_i$ described in the language $\mathscr{L}_{an}$ by using the logic $\mathscr{L}_o$, where I is an index set. In addition, an exemplification $\mathbb{E}' = (\mathscr{L}_{an}, \mathscr{L}_o, \mathscr{A}', \mathscr{E}')$ is called *a sub-exemplification of $\mathscr{E}$* if all the axioms of A' are axioms of A, and $\mathscr{E}'$ is a subset of $\mathscr{E}$. Dually, $\mathbb{E}$ is called *a super-exemplification* of $\mathbb{E}'$.

We assume a wide understanding of the notions of logic, language, axiom and label, preponderantly from the point of view of cognitive science; i.e., we are more interested in the identification of the seminal cognitive heuristics behind the actual reasoning processes involving concrete (meta-)examples. Specifically, "logic" could mean (but is not restricted to) first-order (many-sorted), higher-order, paraconsistent or/and modal logic, among many others, together with the cognitive (grounding) inference rules and principles structure them.[5] In addition, $\mathbb{E}$ can contain specific sets (describing main structures), $n-$ary relations, partial and total functions, and constants. So, we also allow partial functions in our formalism because they better represent the actual way in which new mathematical results are cognitively found (during the creative process of the researcher/student).

Moreover, by the term "language" we also include the morpho-syntactic atoms needed for doing sketches and reasoning within the (meta-)example in consideration. For instance, if we are talking about a particular graph, then generic symbols like "•" (for denoting vertices) and "−" (for denoting edges) would also be included in the language. The former definition is strongly related and coincides in some special cases with the more informal notion of "specialization" described in [56, Ch.1].

By axioms we mean any explicit syntactic description of the sufficient and necessary condition characterizing the (meta-)example.

Lastly, by "labels" we mean either a temporal or a canonical token (resp. collection of tokens) used as the explicit referential symbols (resp. explicit referential symbols) of the (meta-)example in consideration. Instances of canonical tokens are $\mathbb{N}$, $\mathbb{Q}$, and $\mathbb{R}$ (denoting what we call "the natural, rational, and real numbers"); instances of temporal tokens are $\mathbb{P}$ (for denoting "the prime numbers"), X (for denoting a fixed topological space), F (for denoting a fixed field), $C(\mathbb{R}, \mathbb{R})$ (for denoting "the collection of real valued functions"), and $2 \cdot \mathbb{Z}$ (for denoting the even numbers). In addition, we assume implicitly that the token (resp. collection of tokens) is included in the language. Typical examples of chains of sub-exemplifications are the natural numbers as sub-exemplification of the integers; the integers as sub-exemplification of the rational numbers; and the rational numbers as sub-exemplification of the real numbers.

By analyzing conceptual generation in mathematics from an historical perspective, we see how powerful and, at the same time, necessary is the cognitive mechanism of formal (meta-)exemplification and the foundational role it plays in mathematical creation. Nonetheless, this preponderant role is not necessarily exclusive, because as we showed in the chapter devoted to conceptual blending (see Chap. 9), one can generate entire collections of mathematical concepts starting from basic ones and without considering only examples as the main cognitive source of inspiration.

---

[5]From an additional perspective of the notion of logic from a point of view of computer science and within an institution-independent model theoretical approach see [14].

10.3.3   Syntactic Restriction to Exemplification(s)

One of the main principles of research in model theory is a more precise understanding of the general relation between concrete mathematical structures (semantics) and the purely symbolic (minimal and maximal) axiomatizations of them (syntax) [46]. Daily mathematical research is full of similar phenomena, but from a more pragmatic point of view. In general, if we are looking for the solution of a conjecture C involving the sentences $H_1, \ldots, H_r$ as hypothesis and T as thesis (described in some specific logic), then, as mentioned before, one of the typical methods for getting initial evidence for the veracity of C (in the generality that it is described) is to look for a particular mathematical structure E and to test if (all the sentences described in) C are fulfilled (or satisfied) by E.

Even if the testing process E does not satisfy all $H_1, \ldots, H_r$, and T, we do not (always) immediately discard E (or C), but try to figure out with more accuracy exactly which sentences are satisfied by E in order to "polish" the general statement of C (improve the form of the sentences), or even to disprove it (in the case that E satisfied $H_1, \ldots, H_r$ but not T). Often, the manner in which a mathematical conjecture is formed is by abstracting syntactic commonalities essentially based on the study of several specific mathematical structures. In this process, the $H_1, \ldots, H_r$ are carefully selected by minimizing the quantity of conditions required; and this is done with the aim of preserving the highest level of generality.

Now, we will give the explicit definition of this additional technical realization attached to formal (meta-)exemplification.

Definition 10.3 Let $\mathbb{E} = (\mathscr{L}_{an}, \mathscr{L}_o, \mathscr{A}, \mathscr{E})$ be a formal exemplification and Δ a collection of $\mathscr{L}_{an}$−sentences described using the logic $\mathscr{L}_o$ (resp. a (formal) mathematical concept). The *syntactic restriction* of Δ to the exemplification $\mathbb{E}$, denoted by $\Delta_{|\mathbb{E}}$, consists of all the $\mathscr{L}_{an}$−sentences ψ of Δ which are satisfied by $\mathscr{E}$ in $\mathscr{L}_o{}^6$ (resp. in the case of a mathematical concept one keeps only the function's symbols corresponding to total functions in $\mathbb{E}$.).

10.3.4   Generic Exemplification and Generic Generalization

Another directly related metamathematical cognitive mechanism that we use not only in mathematical research but also in daily life is *(abstract) generic exemplification*. This means that, for the purposes of facilitating the deductive process required for solving a particular (abstract or concrete) problem, our mind fixes a concrete, but at the same time generic, instance of a concept. For example, imagine that someone says, "let us consider a prime number p and construct with it the number $2^{2^p} - 1$. Could this number be composite?" or "Assume that you have bought a novel written

---

[6]Implicitly, we assume here that there is a formal notion of satisfaction in the logic $\mathscr{L}_o$.

in Spanish, then count the number of words in the book containing the vocal a and divide this number by the total number of words. Do you believe that there is general lower bound for this ratio?"

To understand the beginning of these statements, our minds should be able to imagine a fixed (single) prime number (resp. a fixed (single)novel written in Spanish), and at the same time our mind should stay at a generic (abstract) level in order to be able to "understand" the former statements in their full generality.[7]

In other words, this cognitive ability allows the mind to generate a purely symbolic and generic instance of a concept and to apply similar inference rules also used when one is reasoning with concrete examples.

Let us introduce the corresponding formal notion within our AMI framework:

Definition 10.4 Let $\mathscr{C} = \langle L_{og}, \Sigma, A, \mathbb{M} \rangle$ be a mathematical concept or its morpho-syntactic part, then the *generic exemplification* of $\mathscr{C}$, denoted by $GE(\mathscr{C})$ (resp. $GE(MorSyn(\mathscr{C}))$) is defined as

$$GE(\mathscr{C}) = GE(MorSyn(\mathscr{C})) = \langle L_{og}, \Sigma, A, M_G \rangle,$$

where M_G is an abstract generic model for $\mathscr{C}$; i.e., M_G contains fixed (morpho-syntactic) copies of all the symbolic units (e.g., sorts) denoting (exactly) the mathematical structures needed for describing a concrete instance of $\mathscr{C}$. Dually, $\mathscr{C}$ is called the *generic generalization of* $GE(\mathscr{C})$, and is typically denoted with the symbols $GG(-)$. Finally, let $\mathscr{C}'$ be a sub-concept of $\mathscr{C}$. Then, one can also define the generic exemplification (resp. generalization) of $\mathscr{C}'$ within the morpho-syntactic part of $\mathscr{C}$ in a similar way by fixing (resp. generalizing) only the abstract generic model for (the sorts corresponding to) $\mathscr{C}'$. This is denoted by $GE_{MorSyn(\mathscr{C})}(\mathscr{C}')$ (resp. $GG_{MorSyn(-)}(*)$).

For example, let $\mathscr{G} = \langle FOL, \Sigma, A', \mathbb{M}' \rangle$ be the concept of (mathematical) group, and let us assume that the sorts denoting structures involving this notion are H, $\oplus$, and e. Then the generic exemplification of $\mathscr{G}$ is

$$\mathscr{C} = \langle FOL, \Sigma, A', M_G' \rangle,$$

where M_G' consists of the generic symbols H_{ge}, $\oplus_{ge}$, and e_{ge}.

It is clear from the prototypical nature of a generic exemplification that one can recover the original concept through the application of conceptual lining; i.e., for any mathematical concept $\mathscr{C}$, it holds

$$CL(GE(\mathscr{C})) = \mathscr{C}.$$

---

[7]Otherwise, one would "under-understand" such clauses. For example, if someone understands the beginning of the second statement as taking only one novel written in Spanish and computing the ratio, he/she will not be able to comprehend the whole semantic content of the subsequent question. The reason is that the question implies that the hypothesis was understood in a generic way, i.e., where potentially a plurality of novels can be checked by computing their corresponding ratios.

Remark 10.1 Generic generalization and conceptual lining are closely related cognitive mechanisms, because both capture essential information of (mathematical) concepts. However, conceptual lining produces the explicit m.-s. configurations characterizing the working heuristic aspects of a concept, whereas generic generalization implicitly summarizes a similar information in a "hidden" fashion (for the conscious mind) by means of the corresponding generic symbols. So, from a morphological perspective, generic generalization acts on a higher symbolic (cognitive) level, which uses and (mentally) re-labels the lower and, at the same time, more sophisticated m.-s. configurations used when conceptual substratum is applied.

10.4 Syntactic and Conceptual Strengthening and Weakening

During daily mathematical research it turns out that in a lot of cases, when one wishes to prove a conjecture C of the form $H_1 \wedge \cdots \wedge H_m \rightarrow T$, it is not clear how much of the hypotheses are really needed. On the one hand, one could deduce T only from a small sub-collection of $\{H_1, \cdots, H_m\}$. On the other hand, one could need additional hypotheses for deducing T. So, the researcher should do several "formal (cognitive) trials" by adding new hypotheses (resp. erasing old ones) often in sequential fashion, until C can be finally solved.

Additionally, the local nature of the (conscious) mind (see Sect. 3.4 Chap. 3) implies in a natural way the pragmatic principle that our daily (conscious) reasoning consists hugely in selecting (resp. adding) exactly those facts that we consider (consciously) important at a given moment and doing (conscious) inferences based (temporally) only on them; although, we potentially could have several additional facts to use at the same spatio-temporal event.

This selective nature of the conscious (mathematical) mind, is, in more general terms, supported and specified by the sophisticated way in which cognitive control is structurally influenced by the corresponding attentional functions through the prioritization of semantic content, among others [45]. Additionally, selective attention is strongly influenced by a dynamic interplay where the (conscious) suppression of some irrelevant information simultaneously interacts with the processing of a second type of perceptual data receiving most of the attentional focus. This perceptual network is globally influenced by goal-relevant stimuli, which, in our case, would consist partially of the cognitive information provided by the thesis T [43].

All these cognitive phenomena occur at different levels of semantic depth, for instance, sometimes the (conscious) mind performs deductive processes with a strong focus on the semantic structures behind.[8] In other cases, most of the cognitive attention is given to the morpho-syntactic configurations of symbols, where there is only a secondary focus on the semantic content of the underlying mathematical notions.[9]

Let us present an initial formalization of the notions syntactic and conceptual strengthening and weakening of collections of axioms and concepts:

Definition 10.5 Let $\mathscr{C} = (L_{og}, \Sigma, A, \mathbb{M})$ be a mathematical concept. Then a sub-concept $\mathscr{C}'_1 = (L_{og}, \Sigma', A', \mathbb{M}')$ of $\mathscr{C}$ is called a *conceptual weakening* of $\mathscr{C}$; and its morpho-syntactic part $MorSyn(\mathscr{C}') = (\Sigma', A')$ is called a *syntactic weakening* of $\mathscr{C}$. On the other hand, a super-concept of $\mathscr{C}$ (i.e., a concept $\mathscr{C}^* = (\Sigma^*, A^*, \mathbb{M}^*)$ having $\mathscr{C}$ as a sub-concept) is called a *conceptual strengthening* of $\mathscr{C}$; and its morpho-syntactic part $MorSyn(\mathscr{C}^*) = (L_{og}, \Sigma^*, A^*)$ is called a *syntactic strengthening* of $\mathscr{C}$.[10]

10.5 Formal Metaphorical Reasoning

Metaphorical reasoning, or the mechanism allowing us to understand a concept in terms of another, is one of the most well-known and most important cognitive abilities of our minds. It has been abundantly studied in cognitive science, psychology, education, and philosophy [26, 44, 50, 53] and [42].

In particular, pure mathematics has seen a tremendous explosion of creative results based on metaphorical reasoning being more visible during the last decades. For instance, new entire areas of mathematical research have appeared based on the effort of trying to understand one specific mathematical sub-domain in terms of others. As illustrative examples, we can mention the development of modern algebraic geometry mainly within Grothendieck's school by means of introducing categorical and homological methods to the classic framework described in terms of classic varieties and their rings of coordinates [18]; the outstanding solution of a classic number theoretical problem as Fermat's Last Theorem by means of the metaphorical usage of new conceptual frameworks coming from Iwasawa Theory, the theory of modular forms and the theory of elliptic curves [13, 67]; and the integration of seminal methods coming from algebraic topology to fundamental

---

[8]In this case, there is a strong perceptual focus on the corresponding mathematical models underlining the morpho-syntactic descriptions. In such a case, we often use the adjective *conceptual* in our formalizations.

[9]In such a case, we tend to use the adjective *syntactic* or *morpho-syntactic*, where the exact meaning and usage slightly differ depending on the specific context of exposition. In fact, this is a general linguistic principle of words in natural language, whose meaning depends strongly on the corresponding context-dependent and context-independent information defining them [2].

[10]In most cases, we do not mention explicitly the logic underlining the sub-concept since it should be clear from the context.

notions of modern algebraic geometry as the development of a homotopy theory for schemes [49], among others.

In addition, one of the most outstanding works concerning the central importance of metaphorical thinking in mathematical research is due to Lakoff and Núñes [42]. Specifically, they argue that lots of fundamental notions grounding, for instance, the most important numerical systems used in mathematics, are concepts emerging from several kinds of metaphors.

An important aspect of a metaphor, which is implicitly associated with it, is its "scope," namely, the conceptual range of effectiveness of the source concept over the target concept. For example, let us consider the classic metaphor "the real numbers are points in a specific (physical) line."[11] Implicitly, the range of this metaphor is limited to finite constructions, because if such a scope were total then one could find a corresponding conceptual translation of the sentence "consider the set of natural numbers and its multiplicative inverses embedded into the reals," as the physical realization of a statement of the form "paint into the line all the physical places (points) corresponding to each of the natural numbers and its multiplicative inverses." Nonetheless, due to physical principles based mainly on the quanta of space and time and the finiteness of the quantity of energy in our universe the last statement lacks a physical fulfillment [1, 3, 24, 52].

Let us keep in mind the notion of mathematical concept developed in the first section of Chap. 7 in terms of many-sorted first-order theories with specific signatures. In that context, we will define the following notion of local, global, and total metaphor.

Definition 10.6 Let $\mathscr{C}_1 = \{Log1, \Sigma_1, A_1, \mathbb{M}_1\}$ and $\mathscr{C}_2 = \{Log2, \Sigma_2, A_2, \mathbb{M}_2\}$ be mathematical concepts (described, for example, as many-sorted first-order theories in the category $\mathbb{C}oncepts^{12}$). A *total metaphor* from $\mathscr{C}_1$ to $\mathscr{C}_2$ is simply a morphism $\eta : \mathscr{C}_1 \rightarrow \mathscr{C}_2$ (denoted also as $\eta : \mathscr{C}_1 \rightrightarrows_T \mathscr{C}_2$), which is given through a correspondence $\lambda : \Sigma_1 \rightarrow \Sigma_2$, respecting the internal taxonomy of the symbols.[13] In addition, there is a canonical way (given by λ) of translating each of the sentences in A_1 into a collection of well-formed sentences $\eta(A_1)$ within the conceptual environment given by $\mathscr{C}_2$. Moreover, each sentence in $\eta(A_1)$ can be deduced syntactically or semantically from the axioms of $\mathscr{C}_2$.[14] A *global metaphor* is a weaker notion which involves all the requirements of a total morphism $\mu : \mathscr{C}_1 \rightarrow \mathscr{C}_2$ (denoted by $\mu : \mathscr{C}_1 \rightrightarrows_G \mathscr{C}_2$) excepting the axiom-preserving

---

[11]Let us imagine that this metaphor involves the drawing of a *real* (painted) line in the physical world.

[12]Clearly, we do not restrict our deductive and ontological scope to (many-sorted) first-order concepts. In fact, the definition presented here can be applied to any sufficiently robust logic possessing its own type of signatures, sentences, satisfaction, and models. For example, suitable kinds of institutions can be used (see, for example, [14]).

[13]Logic symbols are sent to logic symbols, etc.

[14]These requirements are a generalization of the conditions imposed in the case of the category $\mathbb{C}oncepts$ (see Sect. 7.2 Chap. 7).

condition (i.e., the translation of the axioms in $\mathscr{C}_1$ do not necessary are consequences of the axioms of $\mathscr{C}_2$). If $\mathscr{C}_1' = (\Sigma_1', A_1', \mathbb{M}_1')$ and $\mathscr{C}_2' = (\Sigma_2', A_2', \mathbb{M}_2')$ are sub-concepts of $\mathscr{C}_1$ and $\mathscr{C}_2$, respectively (i.e., Σ_i' is contained in Σ_i and A_i' is contained in A_i, for $i = 1, 2$), a *local metaphor* from $\mathscr{C}_1$ to $\mathscr{C}_2$ (or simply a *metaphor*) is a global metaphor $\kappa : \mathscr{C}_1' \rightrightarrows_G \mathscr{C}_2'$. The *(conceptual) scope* of κ is the concept $(\Sigma_2', A_2' \cup' \kappa(A_1'), \mathbb{M}_3)$, where $A_2' \cup' \kappa(A_1')$ is the collection of all the sentences in A_2' and $\kappa(A_1')$, and $\mathbb{M}_3$ is the corresponding semantic closure.[15] Finally, the former definitions are extended in a natural way in the cases that $\mathscr{C}_1$ and $\mathscr{C}_2$ are also exemplifications and/or generic exemplifications as follows:[16] in the case of a total metaphor where $\mathscr{C}_2$ is an exemplification, the axiom-preserving condition should be only verified on the fixed model described within $\mathscr{C}_2$; the semantic scope includes only such a model as well. In such a case, a total metaphor will be defined only if the fixed model of $\mathscr{C}_2$ satisfies all the axioms of $\kappa(A_1')$. In all other cases, the axiom-preserving condition of a total metaphor is verified from the perspective of the morpho-syntactic part of the mathematical structure $\mathscr{C}_2$.

The main idea behind this formalization is that from a pragmatic point of view the metaphors usually occur between local parts of concepts instead of the whole corresponding mathematical structures. For example, one identifies just one operation in an algebraic structure A with another operation in B and one makes useful formal inferences on the solution of a conjecture without caring too much of the other defining components of A. This is related with a (cognitive) specialization's ability of the mind.

As a matter of notation, we can also denote a metaphor κ more explicitly by describing the specific images of the (finite) sorts. For instance, let us assume that κ send the sorts X and Y to A and B, respectively; the function f to h and the relation R to S, then we can use the notation:

$$\kappa = \lceil X \rightrightarrows A, Y \rightrightarrows B, f \rightrightarrows h, R \rightrightarrows S \rfloor .$$

From a practical point of view, the most common kind of metaphors used in mathematical research are local, followed by global, where total metaphors tend to be used less due to the strong syntactic and semantic requirements (see Chap. 11).

Due to the local nature of the conscious mind, metaphors are described in general between single (meta)mathematical structures. However, depending on the level of technical accuracy that one wishes to have, one can *conceptually extend* such metaphors by means of the explicit descriptions of deeper equivalences of "initially atomic" relational and functional symbols. For instance, if we want to be completely "purist" from a set-theoretical point of view, then we should extend any formal metaphor between sophisticated mathematical structures (e.g., topological spaces,

---

[15] In most cases, we mention the signature and the collection of axioms of the conceptual scope assuming implicitly its semantic closure. In such cases, we can talk about the "syntactic scope" of a metaphor.

[16] All the possible combinations are included.

manifolds, Riemann surfaces) to the main axioms of ZFC and consider the validity of their translations to the target structure. Nonetheless, from a cognitive perspective this is not relevant, since mathematicians' minds can make productive inferences about such structures without knowing all the ZFC axioms by heart. So, throughout the meta-generation of mathematical structures in Chap. 11, we will indicate the essential aspects of the formal metaphors that we apply, based on the desired purposes. One could potentially extend such metaphors conceptually (to further sorts) if there is a "deeper" or "more technical" need.

10.6   Conceptual and Morpho-Syntactic Generalization and Particularization

The cognitive ability of expanding the inferential scope of phenomena observed initially among some entities sharing minimal conceptual commonalities (i.e., generalization) is one of the more omnipresent abilities of our minds. Similarly, the dual ability of applying in more specific instances, the conclusions/constructions obtained at a wider conceptual level (i.e., particularization, or specialization) is of, at least, the same importance during mental creation/invention [56] and [55, 57, 69].

Specifically, mathematics is one of the scientific disciplines employing a highly sophisticated usage of these seminal cognitive abilities, e.g., from the natural until the complex numbers, from commutative rings with unity and (classic) varieties until affine schemes; from rings of continuous real functions until sheaves of rings; from the complex plane until Riemann surfaces, from even number until fractional ideals; from circles until smooth n-dimensional real manifolds with boundary, and from the set of the prime (natural) numbers until Zariski's topology of the prime spectrum of a commutative ring with unity, just to mention a few examples.

The way in which our minds generalize and particularizes is, on its own, quite sophisticated and wide. This is essentially due to the fact that these cognitive abilities are extremely sensitive to the morpho-syntactic representations that we use to describe mathematical structures; i.e., a lot of generalizations done in mathematical research depend directly on the explicit way in which the mathematical concepts involved are described.[17]

Moreover, other cognitive abilities like (meta-)exemplification, generic exemplification, weakening, and strengthening are closely involved within the generation of conceptual particularization and generalization, because all of them directly affect the specialization's degree of the (mathematical) concepts, sentences, and conjectures involved.

---

[17]For an explicit example, see the analysis done later in this section with polynomials in one variable over a commutative ring with unity.

Definition 10.7 Let $\mathscr{C} = \mathscr{C}(\mathscr{C}_1, \cdots, \mathscr{C}_n)$ be a mathematical structure (i.e., a mathematical concept, (generic) (meta-)exemplification, a collection of sentences, a formal conjecture) depending explicitly and separately on the (sub-)structures $\mathscr{C}_1, \cdots, \mathscr{C}_n$.[18] Assume that for each index $i \in \{1, \cdots, n\}$ we choose either a (meta-)exemplification, a generic exemplification, a super-exemplification or a super-concept of $\mathscr{C}_i$, denoted by $\mathscr{D}_i$.[19]

Then, $CP_{\mathscr{C}_1 \to \mathscr{D}_1, \cdots, \mathscr{C}_n \to \mathscr{D}_n}(\mathscr{C}(\mathscr{C}_1, \cdots, \mathscr{C}_n)) := \mathscr{C}(\mathscr{D}_1, \cdots, \mathscr{D}_n)$ is called a *conceptual particularization* of $\mathscr{C}(\mathscr{C}_1, \cdots, \mathscr{C}_n)$, where the former expression represents the replacement of each occurrence of $\mathscr{C}_i$ (in $\mathscr{C}$) by $\mathscr{D}_i$.

Dually, $CG_{\mathscr{D}_1 \to \mathscr{C}_1, \cdots, \mathscr{D}_n \to \mathscr{C}_n}(\mathscr{C}(\mathscr{D}_1, \cdots, \mathscr{D}_n)) := \mathscr{C}(\mathscr{C}_1, \cdots, \mathscr{C}_n)$ is called a *conceptual generalization* of $\mathscr{C}(\mathscr{D}_1, \cdots, \mathscr{D}_n)$.

Finally, in the case that there are no (meta-)exemplifications among the $\mathscr{D}_1, \cdots, \mathscr{D}_n$, then $MorSynP_{\mathscr{C}_1 \to \mathscr{D}_1, \cdots, \mathscr{C}_n \to \mathscr{D}_n}(\mathscr{C}(\mathscr{C}_1, \cdots, \mathscr{C}_n)) := MorSyn(\mathscr{C}(\mathscr{D}_1, \cdots, \mathscr{D}_n))$ is a *morpho-syntactic particularization* of $\mathscr{C}$ (and of $MorSyn(\mathscr{C})$).

Dually, $MorSynG_{\mathscr{D}_1 \to \mathscr{C}_1, \cdots, \mathscr{D}_n \to \mathscr{C}_n}(\mathscr{C}(\mathscr{D}_1, \cdots, \mathscr{D}_n)) := MorSyn(\mathscr{C}(\mathscr{C}_1, \cdots, \mathscr{C}_n))$ is a *morpho-syntactic generalization* of $\mathscr{C}(\mathscr{D}_1, \cdots, \mathscr{D}_n)$ (and of $MorSyn(\mathscr{C}(\mathscr{D}_1, \cdots, \mathscr{D}_n))$).[20]

Note that the condition of separation between $\mathscr{C}_1, \cdots, \mathscr{C}_n$ is imposed simply for guaranteeing the well-definition of the expression $\mathscr{C}(\mathscr{D}_1, \cdots, \mathscr{D}_n)$. Otherwise, the formal syntactic-semantic replacement could not be defined in general. Explicitly, assume that, for example, there are only two (sub-)concepts $\mathscr{C}_1$ and $\mathscr{C}_2$, and one occurrence of $\mathscr{C}_1$ in $\mathscr{C}$ is a sub-concept of one occurrence of $\mathscr{C}_2$ in $\mathscr{C}$ as well. In addition, suppose that the languages describing $\mathscr{D}_1$ and $\mathscr{D}_2$ are disjoint. Then, an expression $\mathscr{C}(\mathscr{D}_1, \mathscr{D}_2)$ is not well-defined, because if we firstly replace the corresponding occurrence of $\mathscr{C}_2$ by $\mathscr{D}_2$ in $\mathscr{C}$, then this replacement would necessarily differ with the replacement where we first replace the same occurrence of $\mathscr{C}_1$ in $\mathscr{C}$ by $\mathscr{D}_1$ (which necessarily is an occurrence implicitly in the same corresponding occurrence of $\mathscr{C}_2$ within $\mathscr{C}$). The reason is that in the first case the replacement of this specific occurrence involves only symbols of the language of $\mathscr{D}_2$ and in the second case it involves at least one symbol of the language of $\mathscr{D}_1$.

Furthermore, in our meta-formalism the notions of conceptual and morpho-syntactic strengthening and weakening, and generic exemplification are special cases of the former ones. However, due to the seminal role that each of them play from a cognitive perspective as completely unique and, in some sense, "independent" mental mechanisms, we still will make an explicit differentiation of the exact mechanism used in any conceptual meta-construction done subsequently.

---

[18] In other words, all the occurrences of each of the (sub-)structures $\mathscr{C}_1, \cdots, \mathscr{C}_n$ appearing in $\mathscr{C}$ are completely non-interdependent; i.e., no occurrence of $\mathscr{C}_j$ in $\mathscr{C}$ involves any symbol or concept appearing in any other occurrence of any $\mathscr{C}_i$ for all $i, j \in \{1, \cdots, n\}$ with $i \neq j$.

[19] Here, the specific structure to be chosen depends on the type of structure given by $\mathscr{C}_n$.

[20] Due to the cognitive nature of this definition, the quantity of substructures n is implicitly bound to numbers where either a human or an artificial agent is able to perform concretely such a generalization (resp. particularization).

10.6.1   Instantiated Generalization and Exemplification

An additional form of generalization is typically given when we conceptually jump from an specific exemplification of a concept $\mathscr{C}$ (being part of a mathematical statement T, e.g., a conjecture) until the corresponding generic exemplification of $\mathscr{C}$ (which produces immediately an updated version of T with a higher generality scope)

Definition 10.8 Let $\mathscr{C} = \mathscr{C}(\mathscr{E}_1, \cdots, \mathscr{E}_n)$ be a mathematical concept (resp. a collection of sentences, a formal conjecture) depending explicitly and separately on the exemplifications $\mathscr{E}_1, \cdots, \mathscr{E}_n$, each of them being the exemplification of the concept $\mathscr{C}_i$, for $i \in \{1, \cdots, n\}$.[21] Assume that for each index $i \in \{1, \cdots, n\}$, $\mathscr{G}_i$ is the generic exemplification of $\mathscr{C}_1$. Then, $\mathscr{C}(\mathscr{G}_1, \cdots, \mathscr{G}_n)$ is called a *instantiated generalization* of $\mathscr{C}(\mathscr{E}_1, \cdots, \mathscr{E}_n)$, where the former expression represents the replacement of each occurrence of $\mathscr{E}_i$ (in $\mathscr{C}$) by $\mathscr{G}_i$. Dually, $\mathscr{C}(\mathscr{E}_1, \cdots, \mathscr{E}_n)$ is called a *instantiated exemplification* of $\mathscr{C}(\mathscr{G}_1, \cdots, \mathscr{G}_n)$.

10.6.2   Morpho-Syntactic Graphs

Let us take a closer look at the specific morpho-syntactic properties of the specific symbolic configurations that we use during mathematical research. A typical heuristic principle used in mathematical creation/invention is that the particular symbols and/or graphic configurations used when we describe mathematical ideas are quite less relevant than the corresponding semantic (mathematical) structures behind such configurations. In other words, one could choose virtually any kind of configuration of symbols for denoting specific structures and, if one maintains intact the deductive principles used, then the conclusion would be the same, but just expressed with other symbols and/or graphic configurations. This principle is true only to some extent. Nonetheless, it is also true that the form in which we describe (mathematical) structures and ideas can either improve (and accelerate) or delay (and decelerate) considerably our cognitive understanding about them and our deductive effectiveness at aiming to solve conjectures involving them. Examples of how a suitable notation can make the proof of an abstract mathematical and physical phenomenon easier are 1, the possibility of gluing affine schemes in algebraic geometry [33, Chapter 2, §2]. In this case, a careful writing process is essential in order to be able to "express symbolically" the core reasons lying behind this paste pattern. And 2, the well-known Dirac notation in quantum mechanics [15].

Furthermore, when we aim to find the solution of a mathematical conjecture, usually we base our analysis on the central formulas of conceptual substrata (see Chap. 7), which are given by m.-s. configurations described by a 2-dimensional

---

[21]This condition is essentially the same as in the case of (sub-)concepts (see former definition).

multi-directional juxtaposition of symbolic units that either belong to a particular structure (e.g., sort, set) or syntactically depend on one or more of the other units. For instance, the central expression of the conceptual substratum of the integer odd numbers has the form $2 * k + 1$, where the symbolic units are $2, *, k, +,$ and 1; 1 and 2 are specific exemplifications of elements in $\mathbb{Z}$, $+$ and $*$ are auxiliary symbols depending on two other symbols denoting integers; and k is a symbolic unit denoting a generic integer number.

Definition 10.9 We will associate to each (mathematical) m.-s. configuration Δ a *morpho-syntactic directed graph*[22] $DG_{m.-s.}(\Delta)$, constructed in m.-s. levels as follows: one writes down (horizontally) all the symbolic units as the initial vertices of (m.-s.) level one together with the corresponding "sorts" where they are (generically) separated by a colon (if this is the case) (e.g., $2 : \mathbb{Z}; k : \mathbb{Z}; +; *$ and $1 : \mathbb{Z}$[23]). Subsequently, one writes down (down the first line) only the m.-s. configurations containing two symbolic units (level 2), denoting a meaningful object of the corresponding (mathematical) structure (e.g., in this case there are no meaningful juxtapositions of level 2) with directed arrows coming from the corresponding symbolic units on the former level that generate the corresponding (level 2) vertices. Further, we do the same for the third level, generating directed edges from the former two levels (e.g., $2 * k, 2 + k, 2 * 1, 2 + 1, 1 + k,$ and $1 * k$). We continue in the same fashion until we exhaust all symbolic units and reach the maximal possible level (e.g., $2 * k + 1$ is the most complex m.-s. configuration of level 5).

The next example will illustrate in more detail this notion.

10.6.2.1 Polynomials in One Variable Over a Commutative Ring

With the former definition in mind, let us give a meta-description of the role that this cognitive ability plays in mathematical research.

Explicitly, let us assume that a researcher[24] (let us call him/her *Mather*) wishes to solve a concrete mathematical conjecture C involving explicitly the structure $T := R[x]$. In addition, suppose that Mather does not know in advance what a general argument for its solution looks like. Then, he/she typically starts trying to gain "insight" on C by solving it at the lowest meaningful (m.-s.) level, e.g., for the polynomial x, (Level 1). Gradually, when Mather solves C for this particular kind of polynomials (and subsequently gain more insight), he/she moves

---

[22] Here, the word "directed graph" takes inspiration from the mathematical definition. However, strictly speaking, the m.-s. directed graphs defined hereunder are a more concrete symbolic structure.

[23] In each one of the following levels, implicitly we will illustrate the case of the conceptual substratum of the odd numbers.

[24] Here, we mean any person that has reached the edge of reason and who is interested in solving mathematical problems, independently of his/her level of education.

(often unconsciously) to the next level and looks for a solution of C for constant polynomials a_i or monomials x^i (e.g., L. 2).[25] Later on, Mather finds a general argument which works for more complex monomials $a_i x^i$ (L. 4). And, Finally he/she is able to prove the general case $\sum_{i \to \alpha}^{\omega} a_i x^i$ (L. 9).

Here, it is worth mentioning that the special cases of levels 6 and 7 (i.e., the m. s. expressions $\sum_{i \to \alpha}^{\omega} a_i$, $\sum_{i \to \alpha}^{\omega} x$, and $\sum_{i \to \alpha}^{\omega} x^i$) could be consciously considered by Mather. However, the former options (appearing in increasing order of probability) are not so typical due to semantic reasons. Namely, the expression $\sum_{i \to \alpha}^{\omega} a_i$ is another more complex way of denoting constant polynomials, $\sum_{i \to \alpha}^{\omega} x$ is often included in the case $a_i x^i$, and $\sum_{i \to \alpha}^{\omega} x^i$ is frequently as hard as the general case, from an heuristic point of view.[26]

10.6.3   Generalization and Particularization on Symbolic Units

A heuristic employed often in mathematical research during the search of the solution of a conjecture C, which involves the conceptual substratum of a mathematical structure M (e.g., $\mathbb{Z}[x]$), consists of proving C firstly for the simplest (meta-)exemplifications of M regarding the complexity level of the corresponding conceptual substratum involved (e.g., monomials of the form x^n, where $n \in \mathbb{N}$), and then gradually increasing the complexity level until one gets a stronger intuition about the general (structural) reasons behind its solution which involves the conceptual substratum of M in its whole generality. An explicit example of this conceptual phenomenon is reported in [7], where a geometrical condition for the normality of forcing algebras over polynomial rings with coefficients in a perfect field is founded. In this case, the main intuitions were found analyzing very simple monomials involving two variables, and gradually adding a new variable and "slowly" changing (e.g., generalizing) the coefficients appearing in the corresponding monomial expressions.

Now, let us give a general meta-definition of cognitive heuristic at a general level.

Definition 10.10   Let $C(\mathscr{S})$ be a mathematical formula described in a logic L_{og} (e.g., a conjecture) and involving at least the mathematical structure $\mathscr{S}$ (i.e., $\mathscr{S}$ is either a mathematical concept or a (generic) exemplification described in C). Let us assume that the symbolic units of $DG_{m.-s.}(CS(\mathscr{S}))$, denoted by $u_1, \cdots, u_r$,

---

[25]From the case x until x^i, intermediate cases often emerge from specific syntactic generalizations on the index i, e.g., Mather solves C for x^2, for x^4, for x^{2j}, for x^{3j}, and finally for x^i.

[26]In Fig. 10.1, one can choose between using explicitly the symbol of the product into the m.-s. configurations or avoid it. This is the reason for the punctuated line coming from the symbol "$*$." In a classic mathematical setting this symbol is avoided. However, in more abstract theories it is more recommended to write it to maintain a higher syntactic "purity" in the presentation. Specially during proofs involving a lot of m.-s. atoms.

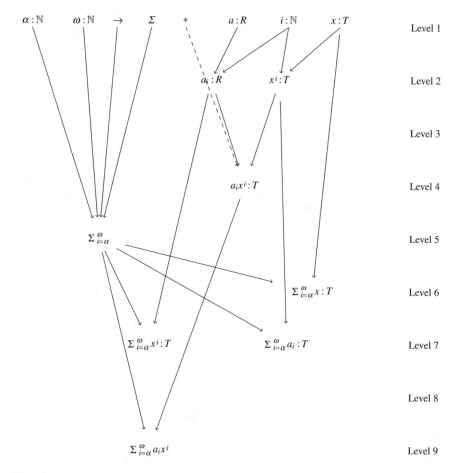

Fig. 10.1 Directed M.-S. Graphs for the ring of polynomials in one variable

involve sorts that implicitly describe mathematical structures $\mathscr{S}_1, \cdots, \mathscr{S}_r$.[27] In addition, let us assume that for some $i \in \{1, \cdots, r\}$, $\mathscr{S}'_i$ is a substructure of $\mathscr{S}_i$ (i.e., either a sub-concept or a sub-exemplification or an concrete exemplification) or that u'_i is a concrete exemplification of $\mathscr{S}_i$.[28] Then, the formula $CP_{u_i:\mathscr{S}_i \to \mathscr{S}'_i}(C(\mathscr{S}))$ obtained from $C(\mathscr{S})$ by modifying the i-th sort of $CS(\mathscr{S})$ by a sort describing the corresponding (prime) substructure (resp. the formula $CP_{u_i \to u'_i}(C(\mathscr{S}))$) obtained from $C(\mathscr{S})$ replacing u_i by u'_i) is called *a (conceptual) particularization of*

---

[27] This means that each of the corresponding sorts is understood as either a (mathematical) concept or an exemplification $\mathscr{S}_i$.

[28] For instance, in the case that u_i denotes a generic exemplification of $\mathscr{S}_i$, but is not restricted only to this possibility.

$C(\mathscr{S})$ *regarding (the symbolic unit)* u_i. *Dually,* $C(\mathscr{S})$, *is called a (conceptual) generalization of* $D := CP_{u_i:\mathscr{S} \to \mathscr{S}'}(C(\mathscr{S}))$ *(resp.* $D := CP_{u_i \to u_i'}(C(\mathscr{S})))$, *also denoted by* $CG_{u_i:\mathscr{S}' \to \mathscr{S}}(D)$ *(resp.* $CG_{u_i' \to u_i}(D))$.

10.7   Conceptual Comparison

One of the most innate and omnipresent mental processes that we use regularly during our intellectual activities consists of taking two concepts (or objects, animals, persons) and generating some kind of abstract comparison between them. The existence of such comparisons implies that our mind identifies qualitative and/or quantitative similarities among the entities involved at least at basic level. This allows us to take the same kind of conceptual attribution for generating the comparison between both concepts. Such an ability plays a considerable role in the way in which our infant mind categorizes new objects [28], and in the way in which social cognition operates [37]. In particular, this fact can provide us a deeper understanding of the way in which collective mathematical research works from an (extended) cognitive perspective [12]. On the other hand, conceptual comparison can also be seen as a required cognitive process used implicitly during the generation of conceptual blends (see Chap. 7). Effectively, in order to be able to blend (fuse) two concepts, our mind should first start an initial process of comparison between them for determining in more detail the similarities and differences which subsequently structures the basic features of the final blend [23, Ch 1].

In the context of mathematical creation/invention, conceptual comparison is materialized essentially in the form of formal relations (e.g., $A \subseteq B, C \in D$ or $E \notin F$); in the form of a common conceptual attribute (given by means of another concept) that can be tested simultaneously in both concepts (e.g., G is a compact smooth manifold and H is a non-compact smooth manifold[29]); and in combinations of the former cases (e.g., X is a topological space with more connected components than Y[30]).

Definition 10.11 Let $\mathscr{E}_1$, $\mathscr{E}_2$ be (mathematical) concepts, generic exemplifications (meta-)exemplifications, and/or symbols denoting variables, functions, relations, constants or any other kind of mathematical structures.[31] Let $\mathscr{D}$ be a (mathematical) concept (*conceptual attribute*), and let $\mathscr{C}$ be a mathematical concept (resp. the morpho-syntactic parts of it) such that both $\mathscr{E}_1$ and $\mathscr{E}_2$ are either a morpho-

---

[29] In this case, the conceptual attribute that can be tested in both manifolds is *compactness*.

[30] Here, one uses the relation $<$ in the natural numbers and the conceptual attribute of *connected component*.

[31] Mixed combinations are also allowed.

syntactic or a conceptual particularization, a generic exemplification or a (meta-)exemplification of $\mathscr{C}$. Let us denote by $\mathscr{D}[\mathscr{E}]_{is}$ the sentence "$\mathscr{E}$ is $\mathscr{D}$."[32]

Finally, assume that $\mathscr{D}[\mathscr{E}_1]_{is}$ and $\mathscr{D}[\mathscr{E}_2]_{is}$ are also sentences, i.e., both statements have a well-defined truth value. Let Φ be a symbol denoting a mathematical binary relation. Then a *conceptual comparison* between $\mathscr{E}_1$ and $\mathscr{E}_2$ is given by an expression of one of the following forms:

$$\mathscr{D}[\mathscr{E}_1]_{is} \text{ is(n)}_1 \text{true and } \mathscr{D}[\mathscr{E}_2]_{is} \text{ is(n)}_2\text{true;}$$

and

$$\mathscr{E}_1 \Phi \mathscr{E}_2;$$

where is(n)$_1$ true (resp. in(n)$_2$ true) denotes either "is true" or "is not true."

In the former definition, we assume implicitly that all the expressions written are well-defined in terms of all possible choices for the conceptual symbols involved.

A usual way of doing conceptual comparison is, firstly, forming new concepts through an common (conceptual) attribute of two concepts (e.g., the cardinality of a set) and then, secondly, comparing the new formed concepts by means of a typical (mathematical) relation between them (e.g., the cardinality of the set A is "bigger" than the cardinality of the set B).

One of the most outstanding and deepest ways of doing conceptual comparison is given by the relations which identify completely two mathematical entities, e.g., equality. In such cases the corresponding conceptual comparison takes the form of a formal identification which promotes substantially instances of our next cognitive mechanism: conceptual identification.

10.8   Conceptual Replacement, Identification, and Duplication

Suppose that we have deduced/declared that two morpho-syntactic or conceptual entities C_1 and C_2 are just two materializations of the same notion or m.-s. representation. Then, when we encounter a new conceptual framework where there are additional occurrences of C_1 or/and C_2 (for example, as a sub-concept of richer concepts), our minds tend to replace one conceptual entity with the other one with the purpose of getting an enhanced understanding of the corresponding conceptual environment (as a whole). We do this on a natural basis when we distinguish the faces of actors in new movies, when we recognize a friend within

---

[32]Here, we assume implicitly that this expression is a well-formed sentence describing one of the following cases: $\mathscr{E}$ is a (generic) exemplification (resp. particularization (on syntactic units)) of $\mathscr{D}$. For instance, $\mathbb{R}$ is a *field*, 53 is a *prime number*, and $\mathbb{R}^1$ is a *topological space*.

a new environment after a casual encounter, when we realize that a combination of letters describes our complete name written in a relevant document (e.g., test, list, contract) or when we replace the (numerical) value of a variable in a mathematical equation. In addition, conceptual replacement is an automatic process coming initially from the unconscious mind and being consciously activated during a selected collection of cognitive tasks [58]. This cognitive mechanism has been shown to be a useful cognitive tool for the way in which elementary concepts in physics are better understood through the cognitive establishment of more coherent conceptual substitutions [31].

Definition 10.12 Let $\mathscr{C}$, $\mathscr{D}$, and $\mathscr{E}$ be mathematical concepts, or collection of formulas, or m.-s. configurations.[33],[34] Suppose that $\mathscr{C} = \mathscr{C}(\mathscr{D})$ has $\mathscr{D}$ as a substructure (e.g., a sub-concept, a sub-formula or a sub-m.-s. configuration). Finally, assume that any occurrence of $\mathscr{D}$ in $\mathscr{C}$ can be replaced syntactically and semantically by $\mathscr{E}$ in a natural way such that the resulting object $\mathscr{C}(\mathscr{E})$ is a meaningful (mathematical) entity.[35] Then, $\mathscr{C}(\mathscr{E})$ is called the *conceptual replacement of* $\mathscr{D}$ *by* $\mathscr{E}$ *in* $\mathscr{C} = \mathscr{C}(\mathscr{D})$.

One of the most canonical families of conceptual replacements appears when we compare two concepts by means of the relations of equality, equivalence, isomorphism, homeomorphism, and similar identifications. In such cases, the corresponding conceptual replacements are given initially by the natural symbolic substitutions among the occurrences of the initial substructures, e.g., $\mathscr{D}$. The resulting concept $\mathscr{C}(\mathscr{E})$ is, in a lot of cases, identifiable with the original $\mathscr{C}(\mathscr{D})$ on a coherent basis, namely, in terms of equality, equivalence, isomorphism, among others.

The former cognitive ability employs, as a preliminary step, another mental process known as conceptual identification [36, 41]. In other words, the ability of identifying the core aspects of two concepts in such a way that depending on the conceptual context, both notions are simply considered as one and the same, although the m.-s. representations for them could, in principle, differ.

In mathematics, such a cognitive ability is materialized at several levels, starting from the classic notion of equality among numbers, congruence among geometrical objects, continuing with isomorphisms between objects from a fixed category (e.g., groups, rings, algebras), homeomorphisms between topological spaces, diffeomorphisms between (smooth) manifolds until homotopy between continuous functions with the same domains and codomains, among others.

The basis of a conceptual identification is, in general, a mental artificial construct, since, strictly speaking, if two concepts differ at least in one m.-s. symbol, then they

---

[33] Mixed combinations are allowed as well.

[34] Here, we assume implicitly the existence of a fixed logic L_{og}, where the involved notions are implicitly and coherently described.

[35] In this context, we do not assume that $\mathscr{D}$ and $\mathscr{E}$ are conceptually isomorphic or similar in any sense. So, our formalization would describe a wider spectrum of situations.

are not purely identifiable as concepts. In fact, in the descriptions of concepts both aspects—the morpho-syntactic and the semantic—play a quite significant role.[36]

Definition 10.13 Let $\langle \mathcal{E}_i \rangle_{i \in I}$ and $\mathcal{E}$ be (mathematical) concepts, generic exemplifications, (meta-)exemplifications and/or symbols denoting variables, functions, relations, constants or any other kind of mathematical structures. Suppose that for all $i \in I$, $\mathcal{E}_i$ is a substructure of $\mathcal{E}$ (e.g., a sub-concept, a sub-exemplification), or a (generic) exemplification of $\mathcal{E}$. Then a *conceptual identification of* $\langle \mathcal{E}_i \rangle_{i \in I}$ *regarding* $\mathcal{E}$ is a kind of *(artificial) cognitive declaration* stating that all the $\mathcal{E}_i$ will be considered as the same $\mathcal{E}$−concept.[37] Such a declaration will be globally denoted by $CI_{\mathcal{E}}(\langle \mathcal{E}_i \rangle_{i \in I})$, and locally by either $\mathcal{E}_i^{[\mathcal{E}]} \backsimeq \mathcal{E}_j^{[\mathcal{E}]}$ or $\mathcal{E}_i \backsimeq_{[\mathcal{E}]} \mathcal{E}_j$, where $i, j \in I$ and $\backsimeq$ is a symbol used to declare this special kind of 'conceptual equality'.

Typical examples are expressions of the form $x + 2 = 3$; $x \in \mathbb{Z}$. In this case, $\mathcal{E} = \mathbb{Z}$, $\mathcal{E}_1 = \langle x + 2 \rangle$ is a formal blend of a variable and the exemplification 2 and $\mathcal{E}_2 = \langle 3 \rangle$ is another exemplification.

Note that a conceptual identification can be constructed virtually from any conceptual collection and it is not necessary that the corresponding concepts possess additional commonalities apart from the one described implicitly through $\mathcal{E}$.

The dual version of conceptual identification is conceptual duplication, which consists of considering several 'mental' copies of a fixed mathematical structure within the same 'deductive' environment, where each of such copies is denotes by different m.-s. configurations. Typical instances of this kind of cognitive ability are given by statements written in natural language like, "let us take two copies of the natural numbers (resp. the reals, the projective plane, the 3-dimensional sphere)."

Definition 10.14 Let $\mathscr{S}$ be a mathematical structure, i.e., a concept, a generic exemplification, a (meta-) exemplification (resp. the m.-s. parts of the former notions), then an additional mathematical structure $\mathscr{T}$ is called *a conceptual duplication of* $\mathscr{S}$, if $\mathscr{T}$ is obtained from $\mathscr{S}$ by preserving the same logic and all the semantic structures involved and by modifying in a one-to-one way the non-logical symbols of the language and the symbolic units needed in the description of the conceptual substrata of $\mathscr{S}$, such that one can distinguish between objects involved in $\mathscr{S}$ and $\mathscr{T}$, and such that both structures are, mathematically speaking, (conceptually) isomorphic. Furthermore, $\mathscr{T}$ is *a partial conceptual duplication of* $\mathscr{T}$, the same conditions as before hold with the exception that the one-to-one

---

[36]For example, imagine that you need to work with polynomials with integer coefficients, but that you have a more primitive way of representing them using n−ary tuples codifying the coefficients positionally and, at the same time, suppose that such (integer) coefficients are described only by a finite number of vertical lines accompanied by a plus or minus sign. Under these conditions, it is clear that the deductive scope would be considerably affected when a mathematician wishes to solve a problem based on such a m.-s. representation.

[37]For example, each $\mathcal{E}_i$ will be considered as the same topological space, the same group or the same set, among others.

modification is not total but partial; i.e., one modifies only a sub-collection of symbols of the language and a sub-collection of the symbolic units.

10.9   (Un-)Conscious Conjunctive Combination

Conceptual combination is a widely studied issue in (cognitive) psychology, linguistics, and cognitive sciences. There are several models and theories aiming to ground the different materializations of it in specific contexts (see, for instance, [47, 51, 64, 68] and [9]).

From the several ways in which our minds can combine two (or more) (mathematical) concepts, there are two special instances that are relevant for us concerning the AMI program: conceptual blending (Chap. 7) and what we will call here (un-)conscious conjunctive combination, which is one of the simplest forms of conceptual combination. In other words, our minds perform (un-)conscious conjunctive combinations of two (mathematical) concepts/structures, $\mathscr{S}_1$ and $\mathscr{S}_2$, simply when both of them are brought together within the same global conceptual framework, e.g., in the common formal context aiming to solve a specific mathematical conjecture, theorem or exercise.

Definition 10.15 Let $\mathscr{S}_1$ and $\mathscr{S}_2$ be mathematical structures. Then, the *(un-)conscious conjunctive combination of $\mathscr{S}_1$ and $\mathscr{S}_2$ (or combination[38])* denoted by $\mathscr{U}_{ccc}(\mathscr{S}_1, \mathscr{S}_2)$ is simply the mathematical structure composed by both structures, i.e.,

$$\mathscr{U}_{ccc}(\mathscr{S}_1, \mathscr{S}_2) = \langle \mathscr{S}_1, \mathscr{S}_2 \rangle .$$

The motivation of the former definition is partially grounded by the remarks made in the section regarding the cognitive substratum of a mathematical proof (Sect. 3.3 Chap. 3).

Furthermore, we explicitly use the 'multi-meaningful' adjective '(un-)conscious' above because through it, we wish to emphasize the fact that the mind, as one of the most sophisticated devices in nature, brings conceptual entities together to the same deductive framework always based on strong unconscious as well as conscious reasons. In other words, we assume in our global ontology that each of the apparently little creative steps taken by the (mathematical) mind, syntactically as well as semantically, originate in a systematic and coherent way and possess no random factors.

Furthermore, as mentioned in Sect. 3.3, Chap. 3, the typical additional forms of conceptual combination generated through the logic connectives "∨" and "→" are essentially grounded in (un-)conscious conjunctive combination. This is due to the

---

[38]For the sake of simplicity in the presentation, in Chap. 11 we usually talk about "combination" or "combining" as an abbreviate form of particular applications of this ability.

fact that, in both cases, our minds save separately the two propositions connected by the corresponding symbol, and the aspect that changes is the way in which the selection process is done between them. In the first case, one can say that the choice is made in an exclusive way, and in the second case, it is made more in a sequential manner. Finally, although the logical connective "↔" is also (partially) grounded as a (un-)conscious conjunctive combination, the most immediate way in which our mind usually comprehends it is a particular form of conceptual identification (see Sect. 10.8).

10.10 Generic Conceptual Blending

In this section, we will describe a composite cognitive ability which is mainly based on the primary ones of conceptual blending and conceptual generalization and which appears often in mathematical creation. Let $\mathscr{C}$ and $\mathscr{C}_2$ be (mathematical) concepts. Then, (roughly speaking) a concept $\mathscr{G}_B$ is a *generic conceptual blend of* $\mathscr{C}$ *along* $\mathscr{C}_2$, if $\mathscr{B}_G$ is conceptually generated in a complete way by conceptual blends of (generic) exemplifications of $\mathscr{C}$, whose generic spaces can be conceptually identified on the corresponding input spaces by (generic) exemplifications of $\mathscr{C}_2$. For example, the concept of *Euclidean geometric figure* is a generic conceptual blend of the (composite) notion of *finite straight lines together with finite circle pieces*[39] along the notion of *identity functions among Euclidean points*. In addition, if we delete the finite circle pieces in the former examples we obtain the notion of 2-dimensional graph.

10.11 (General) Analogical Reasoning

Let us give a more general notion of (atomic) analogy between mathematical formulas or statements that can be expressed in natural language or even using different logical frameworks. In such cases, the analogical outputs have a stronger morphological character and they are not necessarily mathematical structures per se. However, they can be a quite fundamental starting point for using other cognitive abilities and subsequently for generating useful structures (see, for example, Chap. 11).

Definition 10.16 Let $\mathscr{A}_1 = \{Log1, \Sigma_1, A_1\}$ and $\mathscr{A}_2 = \{Log2, \Sigma_2, A_2\}$ mathematical statements (e.g., formulas), where A_1 and A_2 are single statements described within the respective logical frameworks and languages. Let us assume that $Log1$ and Log_2 are robust enough to have well-defined notions of logical connectives, trees,

---

[39] Delimited by a fixed angle.

and atomic formulas.[40] Then, based on the notion of atomic analogy presented in Chap. 8, we define the *(atomic) analogy between* $\mathscr{A}_1$ *and* $\mathscr{A}_2$ as the meta-statement $analog(\mathscr{A}_1, \mathscr{A}_2)$ (together with analogical replacements) constructed recursively from $\mathscr{A}_1$ and $\mathscr{A}_2$ in the following way: when there is a syntactic match in position and form between A_1 and A_2, we include it exactly as it appears in both; if not, we include a corresponding type of meta-variable[41] (which implicitly generate analogical replacements). Subsequently, we form explicitly $analog(\mathscr{A}_1, \mathscr{A}_2)$ with the suitable juxtaposition of symbols. So, we can recover A_1 and A_2 exactly from $analog(\mathscr{A}_1, \mathscr{A}_2)$ through the corresponding global analogical replacements generated by $\mathscr{A}_1$ and $\mathscr{A}_2$, respectively.[42]

10.12   (General) Conceptual Blending

In this section, we will define a wider notion of conceptual blending of (a kind of generalization of) $V-$diagram consisting of general (mathematical) concepts (see Chap. 7 for the original notion). Explicitly, let $\mathscr{C}_1, \mathscr{C}_2$, and $\mathscr{G}$ be (mathematical) concepts grounded in a common logic L'_{og}. Then, a $V-$diagram consists of two total metaphors $\kappa_1 : \mathscr{C}_1 \rightrightarrows_T \mathscr{C}_2$ and $\kappa : \mathscr{C}_2 \rightrightarrows_T \mathscr{C}_2$ (together with the three former concepts as well, see Fig. 10.1).

$$(10.1)$$

Definition 10.17 Let $\mathscr{V} = \langle \mathscr{G}, \mathscr{C}_1, \mathscr{C}_2, \kappa_1 : \mathscr{C}_1 \rightrightarrows_T \mathscr{C}_2, \kappa_2 : \mathscr{C}_2 \rightrightarrows_T \mathscr{C}_2 \rangle$ be a $V-$diagram over a common logic L'_{og}. A (mathematical) concept $\mathscr{B} = \langle L_{og}, \Sigma, A, \mathbb{M} \rangle$ is *a conceptual blending for V* (or *between* $\mathscr{C}_1$ *and* $\mathscr{C}_2$ along $\mathscr{G}$), if there exists two metaphors $\lambda_1 : \mathscr{C}_1 \rightrightarrows \mathscr{B}$ and $\lambda_2 : \mathscr{C}_2 \rightrightarrows \mathscr{B}$ satisfying the following conditions:

---

[40]For example, constants, variables, functional and relational symbols and terms, or equivalent notions. These notions can be defined, but are not completely necessary as a whole for the definition under discussion.

[41]For the general case, the collection of meta-variables is defined according to the level of sophistication of L_{og_1} and L_{og_2}, which is directly related to the level of expressiveness of Σ_1 and Σ_2.

[42]This notion can be generalized to n mathematical statements in a straightforward manner, generating $n-$aric replacements (for a concrete application, see, for example, Sect. 11.17, Chap. 11).

1. (Partial cross-space mappings) There exists an local bi-metaphor $\mu : \mathscr{C}_1 \rightrightarrows$ $\mathscr{C}_2$, whose conceptual scope generates a concept that is conceptually meta-isomorphic to $\mathscr{G}$.[43]
2. (Selective projection to the blend) λ_1 and λ_2 are (potentially) partial metaphors to the blend.
3. (Emerging structure) $\mathbb{M}$ contains at least one model not being in any of the corresponding semantic closures of the members of the $V-$diagram.
4. (Topology) $\lambda_1 \circ_{\rightrightarrows} \kappa_1 = \lambda_2 \circ_{\rightrightarrows} \kappa_2$.
5. (Unpacking) The constitutive concepts of $\mathscr{V}$ can be reconstructed from $\mathscr{B}$ through (conceptual) weakenings, particularizations (e.g., on symbolic units), and (generic) exemplifications.
6. (Integration) The formal union of the syntactic scopes of λ_1 and λ_2 plus its semantic closure is conceptually equal to $\mathscr{B}$.

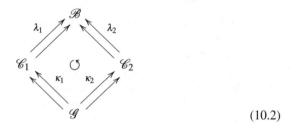

$$(10.2)$$

The former notion if a formal version of the minimal constitutive and governing principles of a genuine conceptual blend[44] expressed in terms of our cognitive meta-framework. We do not include any condition regarding the web principle, because in such a cognitive meta-framework, the conceptual manipulations that we can do with a blend preserving it as a (conceptual) unit, implicitly preserve the conceptual structure of the corresponding network giving origin to it.

10.13 (General) Conceptual Substratum and Conceptual Lining

For describing a more general formalization of the meta-notion of conceptual substratum (resp. conceptual lining), we will differentiate between a more formal and abstract conceptual substratum of a mathematical structure, and a more "pragmatic" one. For instance, due to heuristic considerations, we can say that the more pragmatic and useful conceptual substratum of the concept of polynomial rings over a field is a m.-s. configuration of the form $\ulcorner \sum_{i=1}^{n} a_i x^i, a_i \in k, n \in \mathbb{N} \lrcorner$.

---

[43] A bi-metaphor is simply a metaphor with "inverse" metaphor on both sides regarding the operation of *metaphorical composition* (denoted by $\circ_{\rightrightarrows}$), which is defined in the natural way.
[44] See, for example, [21].

The main reason is that we use mostly the former kind of expression for solving a lot of problems involving polynomials. On the other hand, it could happens that for other mathematical structures there are no such (useful) representations at hand, so one should use, for example, first-order descriptions for them. Subsequently, from this information one must abstract the corresponding conceptual substratum. In such a case, we can use an approach similar to the one developed in Sect. 9.4 Chap. 9.

Note that one could also define a conceptual substratum for polynomial rings (over a field) in terms of its grounding set-theoretical description involving infinite tuples with entries over the base field, among others. However, we prefer (in most cases) the former version of the conceptual substratum mainly due to the immediate cognitive practicality of such a m.-s. representation, which does not necessarily exclude the use of the more formal version, when the corresponding problem (e.g., conjecture) to solve requires such a m.-s. variant. Finally, it could be the case that both versions coincide for certain kinds of elementary notions, e.g., even numbers.

Definition 10.18 Let $\mathscr{S}$ be a mathematical structure. Then, *a pragmatic conceptual substratum of* $\mathscr{S}$ (denoted by $CS_P(\mathscr{S})$) consists of a finite collection of m.-s. configurations $\ulcorner \Delta_1, \cdots \Delta_k; \Delta'_1, \cdots, \Delta'_n \urcorner$ that codifies exactly the essential pragmatic features of the structure $\mathscr{S}$, where $\Delta_1, \cdots, \Delta_k$ represent the main m.-s. configurations of the substratum[45] and $\Delta'_1, \cdots, \Delta'_n$ represent the secondary m.-s. configurations, namely, the last ones represent additional information of the main sub-m.-s. configurations.[46] Additionally, *a purely formal conceptual substratum of* $\mathscr{S}$ (denoted by $CS_F(\mathscr{S})$) consists of a finite collection of m.-s. configurations $\ulcorner \Gamma_1, \cdots, \Gamma_m \urcorner$ that codifies exactly the theoretical features of the structure $\mathscr{S}$, and in general is based on an axiomatic representation of it, described within the corresponding underlying logic (e.g., first-order logic).

Finally, *a (general) conceptual substratum of* $\mathscr{S}$ (denoted by $CS(\mathscr{S})$) consists of (both) a pragmatic and a purely formal conceptual substratum.[47]

Dually, $\mathbb{D}$ is called the *conceptual lining* of $CS(\mathscr{S})$ (or any of its components $\ulcorner \Delta_1, \cdots \Delta_k, \Delta'_1, \cdots, \Delta'_n \urcorner$ or $\ulcorner \Gamma_1, \cdots, \Gamma_m \urcorner$).[48]

In general, at least one of the two types of conceptual substratum exists for any mathematical structure $\mathscr{S}$, otherwise $\mathscr{S}$ could not be called a "mathematical structure."

Both conceptual substrata (i.e., the pragmatic and the formal) satisfy implicitly the following (meta-heuristic) principles: a) The usage of logic symbols is avoided

---

[45] In other words, these are the central m.-s. configurations from an heuristic perspective.

[46] When it is clear from the context, we just use a colon (instead of a semicolon) to separate the main and the secondary m.-s. configurations.

[47] For the sake of simplicity, we usually call "conceptual substratum" a pragmatic or a purely formal substratum, depending on the specific cognitive generation to be done.

[48] In a many-sorted first-order setting, conceptual lining usually produces a new sort defined explicitly by a new notion that was conceptually covered. In further parts of this book, we do not mention this new sort unless it is completely necessary for more clarity on the presentation.

as much as possible. b) One can recover uniquely the original concept $\mathcal{M}$ from such m.-s. configurations.[49]

Note that the former notion also includes mathematical concepts given in a more spatial context like Euclidean geometry [34]. For instance, in a classic geometric sense and from a pragmatic point of view, the conceptual substratum of the (mathematical) concept of a triangle (in Euclidean geometry) is simply the two-dimensional drawing of a triangle, where such a representation is as generic as possible. This means that one prefers to draw a scalene triangle instead of a equilateral or isosceles one, due to the fact that scalene triangles represent the "majority" of triangles to be abstracted.[50]

In general, one can obtain several conceptual substrata of a concept if one represents it in different ways. In our formal AMI meta-framework, it would mean that we are considering, strictly speaking, different concepts, but from a (classic) semantic perspective equivalent, because they would have the same models.

So, if we express concepts in more than one syntactic system, we would obtain more than one conceptual substratum.

For instance, in the case of the set of the even natural numbers, we can have the following two conceptual substrata:

$$CS(2 \cdot \mathbb{N}) = \left\lfloor 2 \cdot n, n \in \mathbb{N} \right\rceil,$$

$$CS(2 \cdot \mathbb{N}_{Bin.Rep}) = \left\lfloor \sum_{i=1}^{m+1} a_i 2^i, m \in \mathbb{N}, a_i \in \{0, 1\} \right\rceil.$$

Here, the second substratum comes from the binary representation of the natural numbers. In fact, one can get a new additional substratum for any new even basis chosen in advance; i.e., there is a unique way for describing even natural numbers with the representation in base $\alpha = 2k$, for $k \in \mathbb{N}$.

Explicitly,

$$CS(2 \cdot \mathbb{N}_{Base2k}) = \left\lfloor 2b + \sum_{i=1}^{m+1} a_i \alpha^i, (b, a_i \in \mathbb{N}), (0 \le b \le \alpha/2), (0 \le a_i \le \alpha - 1) \right\rceil.$$

Thus, conceptual substratum (resp. conceptual lining) is highly sensitive to the language that we use for describing our (mathematical) notions.

Remark 10.2 In the formation of conceptual substrata and conceptual linings, our mind implicitly prioritizes the main m.-s. expression(s) and the secondary formal constraints regarding such m.-s. expression(s). So, we write, in a lot of cases,

---

[49] One explicit description of such m.-s. configurations was already developed in Sect. 9.4 within a (many-sorted) first-order logic setting.

[50] For a concrete example of how these notions are used, the reader can see the meta-generation done in Sect. 11.3, Chap. 11.

the main m.-s. expression(s) first and the subsequent clarifications afterwards. Furthermore, when we meta-generate mathematical notions and proofs using conceptual substrata and conceptual lining (see Chap. 11), we implicitly assume that the corresponding m.-s. configurations are described in the order that is most convenient for our specific purposes. Thus, at some stages of some cognitive meta-generations, a graphic cognitive operation of permutation of m.-s. configurations is assumed and applied without an explicit mention.

10.14 Conceptual Complement

This is the cognitive ability responsible for our pragmatic understanding of the formal "negation" of a (mathematical) concept or a fact. In other words, it allow us to construct an explicit (and "constructive") concept (resp. statement) formally equivalent to the negation of the original one, such that this new version has a more affirmative nature.[51]

Definition 10.19 Let $\mathscr{C} = \langle L_{og}, \Sigma, A, \mathbb{M} \rangle$ be a sub-concept of a concept

$$\mathscr{C}' = \langle L_{og}, \Sigma', A, \mathbb{M}' \rangle.$$

Then, a concept $\mathscr{C}^c_{\mathscr{C}'} = \langle L_{og}, \Sigma, A_1, \mathbb{M}_1 \rangle$ is called the *conceptual complement of* $\mathscr{C}$ regarding $\mathscr{C}'$, if $\mathscr{C}^c$ is a sub-concept of $\mathscr{C}'$, $\mathbb{M}' \cap \mathbb{M} = \emptyset$ and the formal union of $\mathbb{M}_1$ and $\mathbb{M}'$ coincides with $\mathbb{M}$.[52]

Moreover, in the case that $\mathscr{C}$ is a collection of mathematical formulas over a logic L_{og}, $\mathscr{C}^c$ consists of another collection of formulas obtained by doing a formal negation of the formulas of $\mathscr{C}$ and rewriting them such that the negations appear just at the atomic level; i.e., expressed in kind of generalized affirmative normal form.[53]

10.15 Counterfactual (Contradictory) Affirmation

It is a well-known fact in cognitive science that our minds use the generation of "imaginary" facts, emerging from a fantastic combination of ideas, for enlightening and sometimes for grounding "logical" and more "realistic" situations [10]. A particular materialization of this "fantastic" ability of the mind is given in the form of counterfactuals, which are affirmations based on the assumption of alternatives to a concrete past real scenario [20, 61]. In the context of mathematical reasoning,

---

[51] Remember the remarks presented in Sect. 3.3 Chap. 3.

[52] If needed, both can be seen as proper classes in a suitable formal grounding framework, e.g., NBG set theory [48, Ch.4].

[53] See the discussion in Sect. 9.4 Chap. 9 for the case of first-order logic.

the former ability is the cognitive basis of the proof method known as "reductio ad absurdum" (or proof by the sake of contradiction). In other words, one assumes the formal logic negation of the thesis, cognitively understood (in general) in terms of its conceptual complement, and one aims to deduce in a coherent way the (hypothetical) validity of a mathematical statement (let us say Ω) and its conceptual complement (formally $\neg\Omega$). The deductive efficacy of this method relies on the specific rules of the corresponding logic in consideration, e.g., in first-order logic and similar systems this is a consequence of the "explosion principle" [11].[54] For the purposes of the AMI program and based on the heuristic (metamathematical) principle consisting of the fact that conceptual complement as a primary tool for generating formal contradictions is virtually used everywhere in mathematical research, we deal with the former ability as an additional cognitive metamathematical ability called *counterfactual (contradictory) affirmation*.

Definition 10.20 Let $\Theta = (A_1 \wedge \cdots \wedge A_n \to T)$ be a mathematical conjecture involving the sentences $A_1, \cdots, A_n, T$ over a logic L_{og} described in a language Σ. Then, the *counterfactual (contradictory) affirmation of T within Θ* is simply the establishment of the (temporal) cognitive veracity of the conceptual complement of T in the conceptual context delimited by Θ (and implicitly by L_{og} and Σ).[55]

Acknowledgments The author sincerely thanks Jesús Benjosé, Juan D. Vélez, Carlos M. Parra, Margarita Toro, Juan H. Escalante, Carlos Mejía, Rodney Jaramillo, Jorge Mejía, Pedro Isaza, Diego Mejia, Sigifredo Herrón and Leandro Junes for all their commitment to teaching throughout the years.

References

1. Arcani-Hamed, N.: Space-Time is doomed. What replaces it? Messenger Lectures, University of Cornell (2010)
2. Barsalou, L.W.: Context-independent and context-dependent information in concepts. Memory & Cognition **10**(1), 82–93 (1982)
3. Bernal, A.N., Lopez, M., Sánchez, M.: Fundamental units of length and time. Foundations of Physics **32**(1), 77–108 (2002)
4. Boyer, C.B.: The history of the calculus and its conceptual development. Courier Corporation (2012)
5. Brenner, H.: Tight closure and vector bundles. In: G. Colomé-Nin (ed.) Three Lectures on Commutative Algebra, *University Lecture Series*, vol. 42, pp. 1–71. AMS (2008)
6. Brenner, H., Gomez-Ramirez, D.: On the connectedness of the spectrum of forcing algebras. Revista Colombiana de Matemáticas **48**(1), 1–19 (2014)
7. Brenner, H., Gomez-Ramirez, D.d.J.: Normality and related properties of forcing algebras. Communications in Algebra **44**(11), 4769–4793 (2016)

---

[54]This issue is a part of the cognitive substratum of one of the most outstanding formalizations of proof used today in classic logic and mathematics (see Sect. 3.3 Chap. 3).

[55]This establishment is temporal, typically because it happens within the generation of a specific mathematical proof.

8. Brenner, H., Monsky, P.: Tight closure does not commute with localization. Annals of Mathematics pp. 571–588 (2010)
9. Bruza, P.D., Kitto, K., Ramm, B.J., Sitbon, L.: A probabilistic framework for analysing the compositionality of conceptual combinations. Journal of Mathematical Psychology **67**, 26–38 (2015)
10. Byrne, R.M.: The rational imagination: How people create alternatives to reality. MIT press (2007)
11. Carnielli, W.A., Marcos, J.: Ex contradictione non sequitur quodlibet. Bulletin of Advanced Reasoning and Knowledge **1**, 89–109 (2001)
12. Clark, A., Chalmers, D.: The extended mind. analysis **58**(1), 7–19 (1998)
13. Cornell, G., Silverman, J.H., Stevens, G.e.: Modular forms and Fermat's last theorem. Springer-Verlag New York (2013)
14. Diaconescu, R.: Institution-Independent Model Theory. Studies in Universal Logic. Birkhäuser Basel (2008)
15. Dirac, P.: A new notation for quantum mechanics. Mathematical Proceedings of the Cambridge Philosophical Society **35**(3), 416–418 (1938)
16. Dubuc, E.J., de la Vega, C.S.: On the galois theory of grothendieck. arXiv preprint math/0009145 (2000)
17. Edwards, H.M.: Fermat's last theorem: a genetic introduction to algebraic number theory, *GTM*, vol. 50. Springer-Verlag New York (1996)
18. EGA, I., Grothendieck, A., Dieudonné, J.: Eléments de Géometrie Algébrique I Grundlehren d. math. Wiss. 166. Springer (1971)
19. Epstein, N.: A guide of closure operations in commutative algebra. In: Progress in Commutative Algebra 2: Closure, Finiteness and Factorization, pp. 1–37 (2012)
20. Epstude, K., Roese, N.J.: The functional theory of counterfactual thinking. Personality and Social Psychology Review **12**(2), 168–192 (2008)
21. Fauconnier, G., Turner, M.: Conceptual blending, form and meaning. Recherches en communication **19**(19), 57–86 (2003)
22. Fauconnier, G., Turner, M.: The Way We Think. Basic Books (2003)
23. Fauconnier, G., Turner, M.: The Way We Think. Basic Books (2003)
24. Garay, L.J.: Quantum gravity and minimum length. International Journal of Modern Physics A **10**(02), 145–165 (1995)
25. Gelbaum, B.R., Olmsted, J.M.: Theorems and counterexamples in mathematics. Springer-Verlag New York (2012)
26. Gibbs Jr, R.W.: The Cambridge handbook of metaphor and thought. Cambridge University Press (2008)
27. Goguen, J.: What is a concept? In: International Conference on Conceptual Structures, pp. 52–77. Springer, Berlin, Heidelberg (2005)
28. Graham, S.A., Namy, L.L., Gentner, D., Meagher, K.: The role of comparison in preschoolers' novel object categorization. Journal of Experimental Child Psychology **107**(3), 280–290 (2010)
29. Grattan-Guinness, I.: Companion encyclopedia of the history and philosophy of the mathematical sciences. Routledge (2002)
30. Gray, J.: The real and the complex: a history of analysis in the 19th century. Springer Graduate Mathematics. Springer International Publishing (2015)
31. Grayson, D.: Concept substitution: An instructional strategy for promoting conceptual change. Research in Science Education **24**(1), 102–111 (1994)
32. Grothendieck, A., Raynaud, M.: Revetements etales et groupe fundamental. Séminaire de Géométrie Algé-brique du Bois-Marie, 1960–1961, *Lecture Notes in Math.*, vol. 224. Springer-Verlag, Berlin (1971)
33. Hartshorne, R.: Algebraic Geometry. Springer-Verlag, New York (1977)
34. Hartshorne, R.: Geometry: Euclid and beyond. Springer Science & Business Media (2013)
35. Hochster, M., Huneke, C.: Tight closure. In: Commutative Algebra, Math. Sci. Research Inst. Publ. 15, Springer-Verlag, New York-Berlin-Heidelberg, 1989, 305–324

36. Holte, R.C.: A conceptual framework for concept identification. In: Machine Learning, pp. 99–102. Springer (1986)
37. Hoyos, C., Gentner, D., Bach, T., Meltzoff, A., Christie, S., Murphy, Z., Obee, A., San Juan, V., O'Driscoll, K., Ganea, P., et al.: The role of comparison in social cognition. In: Proceedings of the Annual Meeting of the Cognitive Science Society, vol. 36 (2014)
38. Janelidze, G.: Pure galois theory in categories. Journal of Algebra 132(2), 270–286 (1990)
39. Joyal, A., Tierney, M.: An extension of the Galois theory of Grothendieck, vol. 309. American Mathematical Soc. (1984)
40. del Junco, A., Rosenblatt, J.: Counterexamples in ergodic theory and number theory. Mathematische Annalen 245(3), 185–197 (1979)
41. Kurcz, I., Shugar, G., Danks, J.: A simulation approach to conceptual identification. Knowledge and Language (39), 49 (1986)
42. Lakoff, G., Núñez, R.: Where Mathematics Comes From: How the Embodied Mind Brings Mathematics into Being. Basic Books, New York (2000)
43. Lavie, N.: Selective attention and cognitive control: Dissociating attentional functions through different types of load. In Attention and performance XVIII, S. Monsell and J. Driver, eds. pp. 175–194 (2000)
44. Low, G., Cameron, L.: Researching and applying metaphor. Cambridge University Press (1999)
45. Mackie, M.A., Van Dam, N.T., Fan, J.: Cognitive control and attentional functions. Brain and cognition 82(3), 301–312 (2013)
46. Marker, D.: Model theory: an introduction, vol. 217. Springer Science & Business Media (2006)
47. Medin, D.L., Shoben, E.J.: Context and structure in conceptual combination. Cognitive Psychology 20(2), 158–190 (1988)
48. Mendelson, E.: Introduction to Mathematical Logic (Fifth Edition). Chapman & Hall/CRC (2010)
49. Morel, F., Voevodsky, V.: A 1-homotopy theory of schemes. Publications Mathématiques de l'Institut des Hautes Études Scientifiques 90(1), 45–143 (1999)
50. Ortony, A.: Metaphor and thought. Cambridge University Press (1993)
51. Osherson, D.N., Smith, E.E.: Gradedness and conceptual combination. Cognition 12(3), 299–318 (1982)
52. Padmanabhan, T.: Physical significance of Planck length. Annals of Physics 165(1), 38–58 (1985)
53. Paprotté, W., Dirven, R.: The ubiquity of metaphor: metaphor in language and thought, vol. 29. John Benjamins Publishing (1985)
54. Pierpont, J., et al.: Early history of galois' theory of equations. Bulletin of the American Mathematical Society 4(7), 332–340 (1898)
55. Polya, G.: Generalization, specialization, analogy. The American Mathematical Monthly 55(4), 241–243 (1948)
56. Pólya, G.: Mathematics and plausible reasoning: Induction and analogy in mathematics, vol. 1. Princeton University Press (1990)
57. Pólya, G.: Mathematics and plausible reasoning: Patterns of plausible inference, vol. 2. Princeton University Press (1990)
58. Posner, M.I., Snyder, C.R.: Attention and cognitive control. Cognitive psychology: Key readings 205 (2004)
59. Van der Put, M., Singer, M.F.: Galois theory of linear differential equations, *Grundlehren der mathematischen Wissenschaften*, vol. 328. Springer-Verlag Berlin Heidelberg (2012)
60. Ribenboim, P.: The new book of prime number records. Springer-Verlag New York (1996)
61. Roese, N.J.: Counterfactual thinking. Psychological bulletin 121(1), 133 (1997)
62. Rosenthal, A.: The history of calculus. The American Mathematical Monthly 58(2), 75–86 (1951)
63. Singh, S.: Fermat's enigma: The epic quest to solve the world's greatest mathematical problem. Anchor (2017)

64. Smith, E.E., Osherson, D.N.: Conceptual combination with prototype concepts. Cognitive science **8**(4), 337–361 (1984)
65. Steen, L.A., Seebach, J.A., Steen, L.A.: Counterexamples in topology, vol. 18. Springer-Verlag New York (1978)
66. Sturmfels, B.: Four counterexamples in combinatorial algebraic geometry. Journal of Algebra **230**(1), 282–294 (2000)
67. Wiles, A.: Modular elliptic curves and Fermat's last theorem. Annals of mathematics **141**(3), 443–551 (1995)
68. Wisniewski, E.J.: Construal and similarity in conceptual combination. Journal of Memory and Language **35**(3), 434–453 (1996)
69. Zeilberger, D.: The method of undetermined generalization and specialization: Illustrated with Fred Galvin's amazing proof of the Dinitz conjecture. The American mathematical monthly **103**(3), 233–239 (1996)

Part III
Towards a Universal Meta-Modeling of Mathematical Creation/Invention: Meta-Analysis of Several Classic and Modern Proofs and Concepts in Pure Mathematics

Chapter 11
Meta-Modeling of Classic and Modern Mathematical Proofs and Concepts

11.1 Introduction

Based on the initial taxonomy of cognitive mechanisms that we have developed so far, we will show how to meta-model the "creative" origin of a significant amount of proofs and concepts coming from several mathematical sub-disciplines that are, at first glance, not connected with each other, but, as we will show here, are objects of the same cognitive meta-principles.

It is fundamental to clarify at this point that by "meta-modeling cognitively" a notion/proof D, we mean to be able to reconstruct D from atomic conceptual "bricks" (e.g., elementary (mathematical) structures) using the former cognitively-inspired meta-framework. So, this reconstruction could be (but not necessarily must be) similar to the "historical origin" of D. For instance, the meta-generation of the basic notions of Fields and Galois Theory that we developed in Chap. 7 does not coincide in every aspect with the historic development of such notions [11, 40]. This is not surprising for at least two reasons.

Firstly, there are several manners in which our minds can generate the same concept/proof. This is an implicit working principle confirmed in the history of mathematics[1] through hundreds of examples where researchers found similar/identical concepts (resp. proofs) independently, working with different initial settings [22].

Secondly, the exponential growth of the amount of knowledge on the working mechanisms of the mind that we acquired over the last 40 years gives us a significant technical advantage on this endeavor in comparison with the corresponding original researchers (resp. historians and philosophers) in mathematics whose main work was often centered around solving specific conjecture (resp. on giving a historic

---

[1] And, in general, in the history of science.

© Springer Nature Switzerland AG 2020
D. A. J. Gómez Ramírez, *Artificial Mathematical Intelligence*,
https://doi.org/10.1007/978-3-030-50273-7_11

description of the origins of a concept/argument). Therefore, we have more technical tools for finding shorter and clearer ways to meta-generate such notions/arguments.

We focus here mainly on the cognitive generation of mathematical notions in order to make an initial contribution with the aim to fill the gap that exists in automated deduction regarding concept generation, which is one of the central issues of this outstanding research's discipline that should be enlightened more globally [3, 39, 43].

All the initial structures, proofs, and concepts cognitively meta-generated in this chapter can be found in [1, 10, 13, 17, 25, 28, 38] and [33].

One of the most fascinating and challenging aspects of the new cognitive foundations program is the semantic one. In other words, the origin and cognitive grounding of the specific mathematical structures or exemplifications that support ontologically more sophisticated conceptual constructions that our "mathematical" thinking produces.

In this chapter, we will use elementary initial mathematical structures, and on top of them, we will show how to generate cognitively a broad collection of more complex and fine-grained mathematical concepts (and proofs). Let us offer a more detailed description of these structures:

The cognitive and (neuro-)biological origin of the most common numerical systems and seminal notions used in mathematics (and further disciplines) like the natural, integer, rational, real, and complex numbers and the notion of set (and its fundamental operations) have been intensively studied in psychology, computer science, neuroscience, mathematical education, cognitive science, and philosophy for decades.[2] The work of R. Lakoff and G. Núñez presented in [27] is particularly interesting for us in relation with some of the goals of our new cognitive foundational program. Explicitly, [27] offers a quite valuable (initial) conceptual grounding of our classic numerical systems highly based on the mechanism of metaphorical thinking and additional relevant results in neuroscience, psychology, and linguistics, among others.[3]

From a more pragmatic and intuitive perspective, [27] is a relevant work regarding a cognitive generation of the standard numerical systems. On the other hand, from the point of view of the AMI program and our global taxonomy (described in Chap. 10), the usual formal set-theoretical constructions are additionally inspiring for finding specific cognitive meta-generations of each of these classic quantitative systems, and to show at the same time that the modern way of formally grounding such mathematical structures is mainly based on syntactic constructions at the end [17, 28].

---

[2]In particular, the sub-discipline of cognitive science known as numerical cognition was founded mainly with this aim.

[3]In particular, we share and agree with many theses and approaches of [27] strictly regarding the cognitive generation of a lot mathematical notions. However, regarding points of view involving general considerations about the (cognitive) reality of mathematics and related philosophical issues, we differ in some other aspects (see, for example, the initial sections of Chap. 4 for more explicit descriptions of our theses in this direction).

In Chap. 2 there is a concise and general description of most of the notions meta-generated in this chapter.[4]

Finally, from an algorithmic point of view, all the conceptual meta-generations presented in this chapter can be considered as a "pseudoprecode," on top of which one and after producing and analyzing a lot of them, one can subsequently generate global pseudocode simulating essential inner aspects of a UMAA.

11.2   The Classic Proof for Estimating the Cardinality of The Primes Numbers

The ancient proof of the formal "existence" of an infinity quantity of prime numbers is one of the simplest, but at the same time, ingenious (elementary) arguments that one can see in classic number theory [1]. Specifically, although the methods used there are of a high school level and do not require more that the most basic properties of the integers related with divisibility and factorization, the main "trick" of that proof still remains (at some degree) a "mystery" from the point of view of understanding the causing mechanisms giving rise to it.

Here, we aim to clarify this formal mystery in terms of all the cognitive meta-tools developed so far and with the additional help of a sensitive notation which will guide us in a more natural way to the desired result.

We will choose, in this special case, as the initial methodological mechanism the method of proof for the sake of contradiction. This election is not a mandatory one, since one can recover the core argument, if one decides to find a direct proof of the same statement in consideration.[5]

Under this assumption, we start to specify gradually what we want to achieve syntactically; i.e. a statement of the form $\Phi \wedge \neg\Phi$, where Φ is a sentence involving implicitly the fact that there are finitely many prime numbers. This (in principle, hypothetical) fact leads us to be able to find a generic (implicit) representation of the set of the prime numbers. Effectively, if we start to label the natural numbers as $p_1 = 2, p_2 = 3, p_3 = 5$; then we know that there exists a natural number n such that p_n would be the last prime number. So, the set of prime numbers could be syntactically described as $\{p_1, \cdots, p_n\}$, where we have an implicit label for a (generic) prime number as p_i, for a fixed $1 \leq i \leq n$. Let us choose a natural symbolic label (or conceptual lining) for the set of prime numbers according to the standard labels used for the natural ($\mathbb{N}$), real ($\mathbb{R}$), and complex ($\mathbb{C}$) numbers; i.e., let us choose the symbol $\mathbb{P}$ for denoting the primes, namely, $\mathbb{P} = \{p_1, \cdots, p_n\}$.

---

[4]Thus, for the non-specialist mathematician reader, it is highly recommended to possess a minimal knowledge of the content developed in Chap. 2 in order to grasp deeply into the cognitive construction of this chapter. It is also possible to read both chapter parallelly.

[5]One can find the most representative proofs in [42].

Let us start with a natural syntactic generalization process on the (hypothetical) conceptual substratum of $\mathbb{P}$; i.e., on the expression $\ulcorner p_i, 1 \leq i \leq n \urcorner$. This will generate the following conceptual substrata by increasing the m.-s. level:

$$\ulcorner p_i^a, a \in \mathbb{N}, a \geq 1 \urcorner,$$

$$\ulcorner p_i^{a_i}, i \in \{1, \cdots, n\}, a_i \in \mathbb{N}, a_1 \geq 1 \urcorner,$$

$$\ulcorner \prod_{i \to 1}^{n} p_i^{a_i}, n \in \mathbb{N}, n \geq 1, i \in \{1, \cdots, n\}, a_i \in \mathbb{N}, a_1 \geq 1 \urcorner.$$

By doing conceptual lining on the last conceptual substratum and by using the fundamental theorem of arithmetic we obtain the expression

$$\mathbb{N} \setminus \{1\} = \{ \prod_{i \to 1}^{n} p_i^{a_i}, n \in \mathbb{N}, n \geq 1, i \in \{1, \cdots, n\}, a_i \in \mathbb{N}, \sum_{i \to 1}^{n} a_1 \geq 1 \}.$$

Another way of doing the former syntactic generalization is the following. As seen before (Chap. 10), one can perform syntactic generalization on several operations like indexing, m.-s. product, exponentiation, and addition.

In our case, the most natural choice would be to use syntactic generalization on the product, since we are dealing with the basic "bricks" regarding the product operation; i.e. the prime numbers. $\mathbb{P}$ can be seen as $\mathbb{P}^{(1)}$, because its elements are exactly the numbers that can be written as the product of one prime number. Further, the straightforward syntactic generalization of this set would have the form

$$\mathbb{P}^{(2)} = \{q_1 q_2 : q_1, q_2 \in \mathbb{P}\},$$

namely, numbers that can be written exactly as the product of two primes.[6]

Continuing in this fashion, we obtain the following expression for a fixed number m of factors.

$$\mathbb{P}^{(m)} = \{ \prod_{j \to 1}^{m} q_j : q_j \in \mathbb{P} \}.$$

Now, (again) the fundamental theorem of arithmetic implies that any natural number $r > 1$ belongs to $\mathbb{N}^{(r)}$, for some $r \geq 1$. So, we can express this fact in the following direct m.-s. way:

---

[6]In the description of this set we temporarily change the conceptual substratum of the factors by expressions of the form $\ulcorner q_r, 1 \leq r \leq n, q \in \mathbb{P} \urcorner$, due to syntactic simplicity. Otherwise, we should use sub-indexes of depth two (i.e., expressions of the form p_{a_b}), which in the long run turns out to be an additional (and undesired) m.-s. complexity.

$$\mathbb{P}^{(\mathbb{N}_{\geq 1})} = \mathbb{P}^{(\bigcup_{m \in \mathbb{N}_{\geq 1}} \{m\})} := \bigcup_{m \in \mathbb{N}_{\geq 1}} \mathbb{P}^{(m)}$$

$$= \{\prod_{j \to 1}^{r} q_j : q_j \in \mathbb{P}, r \in \mathbb{N}_{\geq 1}\}.$$

Combining both approaches we can write

$$\mathbb{P}^{(\mathbb{N}_{\geq 1})} = \{\prod_{i \to 1}^{n} p_i^{a_i}, n \in \mathbb{N}, n \geq 1, i \in \{1, \cdots, n\}, a_i \in \mathbb{N}, \sum_{i \to 1}^{n} a_1 \geq 1\}.$$

So far, we have applied syntactic generalization at a maximum level within the (implicit) conceptual framework, where we are working in this problem; i.e., the natural numbers $\mathbb{N}$. Keeping in mind that we want to achieve a formal contradiction, we proceed to define the formal "negation" of the concept obtained so far, namely, the complement of $\mathbb{P}^{(\mathbb{N}_{\geq 1})}$ with respect of $\mathbb{N}$. An initial natural name or m.-s. label for this concept is $\neg\mathbb{P}^{(\mathbb{N}_{\geq 1})}$, in other words, the set of natural numbers that cannot be written genuinely as a finite product of prime numbers. This concept has as conceptual substratum the only natural number satisfying such property; i.e.,

$$CS(\neg\mathbb{P}^{(\mathbb{N}_{\geq 1})}) = \ulcorner 1 \urcorner.$$

We proceed now by considering the conceptual disjunction of the former two concepts, or eventually a special form of conceptual blending, that is,

$$\mathbb{P}^{(\mathbb{N}_{\geq 1})}\{\triangle\}\neg\mathbb{P}^{(\mathbb{N}_{\geq 1})}.$$

The idea behind this construction is that we firstly declare syntactically the "contradictory concept" (which would deliver us the formal contradiction) and we explore its conceptual substratum to figure out how it should explicitly look.

Because the second concept consists of just one element, the conceptual disjunction would be the empty set, since its elements have simultaneously the form $\prod_{i \to 1}^{n} p_i^{a_i}$ (where $n \in \mathbb{N}, n \geq 1, i \in \{1, \cdots, n\}, a_i \in \mathbb{N}$ and $\sum_{i \to 1}^{n} a_1 \geq 1$) on the one hand, and 1 on the other hand.

So, we should look for more sophisticated m.-s. conceptual blends.

Effectively,

$$CS(\mathbb{P}^{(\mathbb{N}_{\geq 1})}\{\triangle\}\neg\mathbb{P}^{(\mathbb{N}_{\geq 1})})$$

$$= \ulcorner \prod_{i \to 1}^{n} p_i^{a_i} \triangle 1, n \in \mathbb{N}, n \geq 1, i \in \{1, \cdots, n\}, a_i \in \mathbb{N}, \sum_{i \to 1}^{n} a_1 \geq 1 \urcorner.$$

We can consider the symbol $\triangle$ as a syntactic all-rounder that can be replaced, in principle, by the following natural symbols denoting the most basic possible blends, $+, *, -,$ and $exp(-, -)$.

The options involving the product and the exponential functions are discarded since they give rise to conceptual substrata generating again one of the original concepts. So, it remains $+$ and $-$. In this case, the addition has priority since the subtraction is, in fact, defined in terms of it. Therefore, we explore the option $+$, in other words, we do a *blending specification* by fixing syntactically $\triangle$ as $+$:

$$CS(\mathbb{P}^{(\mathbb{N}_{\geq 1})}\{+\}\neg\mathbb{P}^{(\mathbb{N}_{\geq 1})}) =$$

$$\ulcorner\prod_{i \to 1}^{n} p_i^{a_i} + 1, n \in \mathbb{N}, n \geq 1, i \in \{1, \cdots, n\}, a_i \in \mathbb{N}, \sum_{i \to 1}^{n} a_1 \geq 1 \urcorner.$$

Due to fact that this blending emerged from the concepts $\mathbb{P}^{(\mathbb{N}_{\geq 1})}$ and $\neg\mathbb{P}^{(\mathbb{N}_{\geq 1})}$, we proceed further by doing a conceptual comparison between $CS(\mathbb{P}^{(\mathbb{N}_{\geq 1})}\{+\}\neg\mathbb{P}^{(\mathbb{N}_{\geq 1})})$ and $\mathbb{P}^{(\mathbb{N}_{\geq 1})}$. The simplest way to do that is in terms of the containment relation; i.e., to verify that $CS(\mathbb{P}^{(\mathbb{N}_{\geq 1})}\{+\}\neg\mathbb{P}^{(\mathbb{N}_{\geq 1})}) \subseteq \mathbb{P}^{(\mathbb{N}_{\geq 1})}$. Effectively, a straightforward proof for this is to consider the core conceptual substratum of the blend (namely $\prod_{i \to 1}^{n} p_i^{a_i} + 1$) and to compare it again with the smallest element of $\mathbb{P}^{(\mathbb{N}_{\geq 1})}$, i.e. 1. We use again a syntactic all-rounder $\diamond$, which can be specified (at first glance) by the standard binary relations between natural numbers like $=, <, \leq, >, \geq$. Subsequently we obtain the expression:

$$\prod_{i \to 1}^{n} p_i^{a_i} + 1 \diamond 1,$$

which is equivalent to

$$\prod_{i \to 1}^{n} p_i^{a_i} + 1 - 1 \diamond 1 - 1,$$

that is

$$\prod_{i \to 1}^{n} p_i^{a_i} \diamond 0,$$

which suggests that we should do the m.-s. specification given by $\diamond \rightsquigarrow >$, because $a_1 \in \mathbb{N}$.

So, we conclude that

$$CS(\mathbb{P}^{(\mathbb{N}_{\geq 1})}\{+\}\neg\mathbb{P}^{(\mathbb{N}_{\geq 1})}) \subseteq \mathbb{P}^{(\mathbb{N}_{\geq 1})}.$$

Further, led by the contrafactual (contradictory) affirmation, we aim to prove exactly the conceptual complement of the last statement, that is,

$$\mathbb{P}^{(\mathbb{N}_{\geq 1})}\{+\}\neg\mathbb{P}^{(\mathbb{N}_{\geq 1})} \nsubseteq \mathbb{P}^{(\mathbb{N}_{\geq 1})}.$$

Explicitly, let us apply the simplest syntactic particularization on $\ulcorner \prod_{i\to 1}^{n} p_i^{a_i} + 0 \urcorner$, namely, $a_i \rightsquigarrow 1$. So, we should show that $\prod_{i\to 1}^{n} p_i^1 + 0 \notin \mathbb{P}_{\geq 1}$. To prove such a negative statement, it is syntactically more simple to use the method of proof by contradiction, since in such a case one has a concrete statement to use, namely, there exists a $b_i \in \mathbb{N}, i \in \{1, \cdots, n\}$ such that $\prod_{i\to 1}^{n} p_i^{b_i} = \prod_{i\to 1}^{n} p_i^1 + 1$, and $\sum_{i\to 1}^{n} b_i \geq 1$. This means that there exists $j \in \{1, \cdots, n\}$, with $b_j \geq 1$. Hence, $p_j \mid \prod_{i\to 1}^{n} p_i^1 + 1$. By applying the (outermost) syntactic simplification (left) on the left-hand side of the last conceptual comparison (i.e., erasing the additive symbol $+$ and the term at its left side (i.e., "1")), one obtains the conceptual comparison

$$p_j \mid \prod_{i\to 1}^{n} p_i^1,$$

which turns out to be true, almost by definition. Contrastingly, when we perform the corresponding (outermost) syntactic simplification (right) on the left-hand side of it (i.e., erasing the additive symbol $+$ and the term at its right side (i.e., "$\prod_{i\to 1}^{n} p_i^1$"), we get the conceptual comparison

$$p_j \mid 1,$$

which is true again due to the last two statements and the additivity of the divisibility relation. Now, this is a contradiction due to the fact that $1 \in \neg\mathbb{P}^{(\mathbb{N}_{\geq 1})}$. Therefore, $\mathbb{P}^{(\mathbb{N}_{\geq 1})}\{+\}\neg\mathbb{P}^{(\mathbb{N}_{\geq 1})} \nsubseteq \mathbb{P}^{(\mathbb{N}_{\geq 1})}$.

Finally, we can express the former (contradictory) facts in a summarized way as follows:

$$(\mathbb{P}^{(\mathbb{N}_{\geq 1})}\{+\}\neg\mathbb{P}^{(\mathbb{N}_{\geq 1})} \subseteq \mathbb{P}^{(\mathbb{N}_{\geq 1})}) \wedge (\mathbb{P}^{(\mathbb{N}_{\geq 1})}\{+\}\neg\mathbb{P}^{(\mathbb{N}_{\geq 1})} \nsubseteq \mathbb{P}^{(\mathbb{N}_{\geq 1})}),$$

in other words, we apply here a (conjunctive) conceptual combination of the two most complex (contradictory) syntactic concepts (i.e., statements).

Let us note that the suitable and gradual combination of the right cognitive mechanisms (as described before) gives rise naturally to the (famous) "out of the blue" element $\prod_{i\to 1}^{n} p_i + 1$, which is the key idea of the classic Euclid's proof.

11.3   Pitagoras' Theorem

In this section we offer a meta-description of one of the most famous theorems in classic geometry: the Pitagorean theorem.

First, let us consider a fixed (pragmatic) conceptual substratum $T_1 = ABC$ of the notion of right triangle (see Fig. 11.1), with sides of lengths a, b, and c, and inner angles α, β, and γ, where β corresponds to the right angle. So, from the specific figure we can add the quantitative information $a \geq b$ and $c \geq a$.

Let us take a conceptual duplication $T_2 = A_1 B_1 C_1$ of the former substratum T_1. Note that within the special context of geometrical reasoning, pragmatic substrata can be conceptually identified after doing simple rotations and translations. Now, let us consider a quite simple geometrical blend S_1 between T_1 and T_2, where we identify (through a suitable generic space) the side BC with the part of the side $A_1 B_1$ of length b and sharing B_1 (see Fig. 11.2)

Using the facts that $\beta = \pi/2$ and $\prec (ABB_1) + \pi/2 = \pi$, together with suitable conceptual replacements, we obtain $\prec (ABB_1) = \pi/2$.

Further, let us consider a third conceptual duplication $T_3 = A_2 B_2 C_2$ of T_1. Due to the fact that S_1 contains conceptually all the (geometrical) information of T_1, we can basically repeat the former blend, but now with input spaces S_1 and T_3 to form a second blend S_2 (see Fig. 11.3). In this case, we deduce similarly that $\prec (B_2 B_1 A_1) = \pi/2$.

We generate a fourth conceptual duplication $T_4 = A_3 B_3 C_3$ of T_1 and we construct a blend S_3 in a similar way as before between S_2 and T_3, but doing in this case a double identification. Explicitly, by doing elementary conceptual replacements we arrive at the fact that the length of the segment $B_2 B_3$ is exactly

Fig. 11.1 Pragmatic conceptual substratum of a right triangle

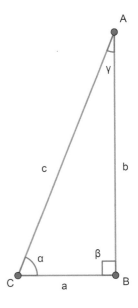

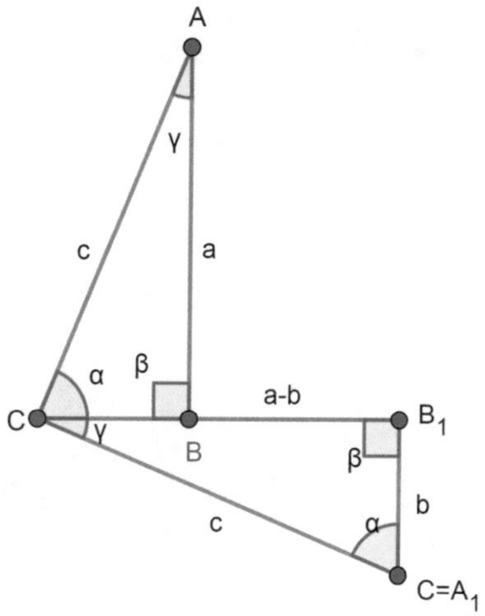

Fig. 11.2 (Geometric)
conceptual blending between
T_1 and T_2

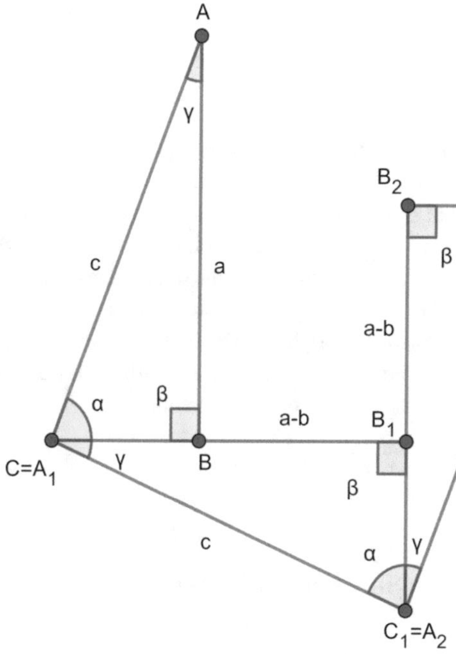

Fig. 11.3 (Geometric) conceptual blending between S_1 and T_2

$a - b$. On the other hand, one can prove that the lines containing the segments BB_1 and B_2B_3 (resp. BC_3 and B_1B_2) are parallel. Hence, we can identify naturally the segment C_3B_3 with the part of the segment AB that collapses A and C_3. Again, we deduce that $\prec (B_3B_2A_2) = \pi/2$.

Fig. 11.4 (Geometric) conceptual blending between S_2 and T_3

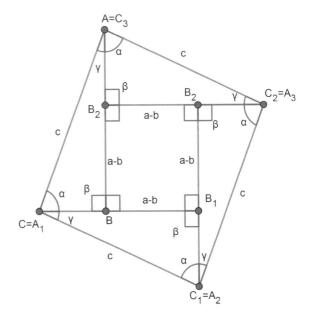

So. the last blend S_3 possesses an additional conceptual identification non-being available in the former ones (see Fig. 11.4).

Finally, due to the fact that $\alpha + \gamma = \pi/2$, we consider a square Q of length c and we proceed to identify conceptually Q and the geometrical figure consisting of the external sides of S_3 (both have the sides and the internal angles of the same lengths). So, we compare them conceptually through their surface; i.e., on the one hand, the surface of Q is c^2, and, on the other hand, the surface of S_3 is $4(ab/2)+(a-b)^2$. So, after identifying the former two quantities and subsequently after doing elementary conceptual replacements we arrive to the desired formula[7]

$$c^2 = a^2 + b^2.$$

11.4   Principle of Mathematical Induction

The principle of mathematical induction, in all its several forms, is one of the most used proof's strategies in mathematics. It is based on our central intuitions and perceptions of the kind of "familiar" mathematical structure known commonly as the natural numbers. Let us assume in this section the presentation given in

---

[7] Here, we left to the reader the cognitive meta-generation of the implicit facts used on the former steps, e.g. the deduction of the formula for the area of a right triangle, and the parallelism of the two pairs of lines described before.

[33, §1,Ch.3] in terms of the Peano's axioms. Let us denote by $\mathbb{N}_{pre}$ the concept described in [33, §1,Ch.3] without the principle of mathematical induction, and let $N(x)$ be the (predicate) statement "x is a (pre-)natural number." Then, from our seminal understanding of the natural numbers, we know that there is an initial natural number.[8] Let us denote such an initial (pre-)number by e. Further, we know that for any natural (pre-)number y, we can generate a "next" (or successor) (pre-)natural number y' in a unique way. And, finally the former two conditions should be necessary for generating any (pre-)natural number, in other words, any arbitrary (pre-)natural number can be generated from e by a finite application of the successor function. So, we can formalize the former information in a compact way by the formula

$$\Psi_N \cong (N(e) \wedge (\forall x)(N(x) \to N(x'))) \to (\forall y)(N(y))).$$

Let us denote by $\phi(w)$ a generic exemplification of a (first-order) formula described within the conceptual framework given by $\mathbb{N}_{pre}$. Let $\mathbb{N}_{pre}^+$ be the concept $\mathbb{N}_{pre}^+$ enriched with a predicate symbol ϕ^+ and an extra axiom codifying the new symbol as ϕ, i.e. $(\forall x)(\phi^+(x) \leftrightarrow \phi(x))$.

In addition, let us consider the (local) metaphor $\nu : \mathbb{N}_{pre} \rightrightarrows \mathbb{N}_{pre}^+$ given by

$$\nu := \lceil N \rightrightarrows \phi^+ \rceil.$$

So, the generic generalization at the meta-level[9] of the (conceptual) scope of ν corresponds with the concept $\mathbb{N}_{pre}$ enriched with the principle of mathematical induction as it is described in [33, §, Ch.3].[10]

11.5 Cartesian Product

From a mathematical point of view, one of the simplest ways of combining two sets is "putting them formally together" (e.g., set-theoretically). In other words, we create a new set "codifying" such a spacial grouping of sets. Specifically, from the perspective of ZFC set theory [33], one uses the axiom of pairing as the technical tool for giving a formal meaning to a m.-s. configuration of the form $\{a, b\}$, where a and b denote sets, the auxiliary symbols "{" and "}" are, from a cognitive point of view, useful symbols for describing from a graphic perspective that the corresponding new set is exactly formed by the two initial ones, and the comma ","

---

[8]In some mathematical disciplines it is fixed as one, and in others as zero.

[9]Namely, the generic generalization regarding ϕ, which corresponds to a higher formal level in comparison with the initial objects in consideration; i.e. (pre-)natural numbers.

[10]For the higher-order version of induction, one can proceed in a slightly similar fashion as before, using in addition suitable conceptual comparisons.

is the usual way to distinguish where the explicit m.-s. configuration of *a* ends and where the explicit m.-s. configuration of *b* begins. Furthermore, the way in which the axiom of pairing generates such a set is an implicit one; i.e., one should apply conceptual substratum to the obtained pairing set in order to get a more explicit expression of the form $\{a, b\}$.

Based on the spatial metaphor inspiring the former construction, the order of appearance of the sets is irrelevant, although the corresponding m.-s. configurations for denoting them are, strictly speaking, different (i.e., $\{a, b\}$ as m.-s. configuration is a different from $\{b, a\}$). However, in some mathematical contexts, for example, Cartesian geometry, the order of such binary combinations is highly important. So, one constructs a next m.-s. level with new auxiliary symbols "(" and ")," where the order of the sets effectively matters. Explicitly, this is done through the m.-s. declaration $(a, b) := \{\{a\}, \{a, b\}\}$ [33].

Finally, let *A* and *B* be sets, and let $a \in A, b \in B$ be generic elements of each of the sets, then one generates the Cartesian product between *A* and *B*, denoted by $A \times B$ as

$$CL(\{\{a\}, \{a, b\}\}, a \in A, b \in B).$$

On the other hand, one can straightforwardly verify that the Cartesian product is, at the same time, a formal conceptual blending of the corresponding input sets.

11.6 (Equivalence) Relations

Firstly, the notion of a set-theoretical relation between two sets *A* and *B* is generated from the former notion by the declaration of a conceptual comparison; i.e., a relation *R* between *A* and *B* is just another set satisfying $S \subseteq A \times B$. From a pragmatic point of view, relations are often described by means of m.-s. configurations involving generic exemplifications of the elements of the corresponding sets, e.g., $R_< \subseteq \mathbb{R} \times \mathbb{R}$, given by $(a, b) \in_<$ if and only if $a < b$, for all $a, b \in \mathbb{R}$. In other cases, relations are explicitly described pair by pair, for example, when the input sets are (mathematically) "finite."

Secondly, an outstanding form of relation with a broad use in several mathematical disciplines is the notion of equivalence relation; i.e. a relation *R* defined over ordered tuples of a set *X* (i.e., $R \subseteq X \times X$) satisfying the elementary conditions of reflexivity, symmetry, and transitivity, in other terms, for all $a, b, c \in R$, $(a, a) \in R, (a, b) \in R \to (b, a) \in R$ and $(a, b) \in R \wedge (b, a) \in R \to (a, c) \in R$, respectively.[11]

---

[11] The concrete generation of this super-concept of mathematical relation can be done by the interested reader using as a starting point the equality relation over the natural numbers and its most elementary properties.

Such a relation produces a uniquely defined partition over X, which is cognitively described, for instance, through conceptual identifications over a generic exemplification of the elements of X. In other words, first we fix a generic exemplification of an element of X, let us say x; further, we consider the conceptual identification of $\langle x_a \rangle_{a \in H_x}$ regarding the concept *being an element of* X, where $H_x = CL(y \in X, (x, y) \in R)$; finally, we describe single generic elements of the new partitioned set, sometimes denoted by X/R, by adding an auxiliary morpho-syntactic unit (e.g., $-$) written on top of the m.-s. configurations used for representing (generic) elements of X (e.g., $\bar{x}$).[12, 13] In the case that X possesses an additional structure, for example, being a group, a ring or an algebra, such an additional structure is inherited to the partitioned set in a similar fashion as before, e.g. $\bar{a} \bar{+} \bar{b} := \overline{a + b}$.

11.7 Mathematical Functions and Function Compositions

The two most important cognitive processes responsible for the origin of the notion of mathematical function are conceptual comparison and formal metaphorical reason. Effectively, let us assume that $\mathscr{W}$ and $\mathscr{V}$ are mathematical structures, e.g. mathematical concepts, sets, or mathematical variables. A natural way to create sophisticated conceptual comparisons between $\mathscr{W}$ and $\mathscr{V}$ is doing that in a local manner; i.e., to compare the defining elements of $\mathscr{W}$ (locally) with the defining elements of $\mathscr{V}$. In other words, one defines a "pointwise" generic or/and explicit description of the single comparisons involving all the elements of the initial mathematical structure $\mathscr{W}$ with the desired elements of the target mathematical structure $\mathscr{V}$. The classic morpho-symbolic units used for denoting this kind of directed comparison are a (main) symbol indicating the direction of the formal comparison (e.g., "$\rightarrow$"), a symbolic label for naming the function(al comparison) (e.g., f), and an auxiliary symbol for explaining schematically the function of new introduced label (e.g., colon "$:$"). The most typical atomic m.-s. configurations for describing the former conceptual interrelations at a global level are:

$$f : \mathscr{W} \rightarrow \mathscr{V}$$

and

$$\mathscr{W} \xrightarrow{f} \mathscr{V}.$$

---

[12]It is an elementary exercise to verify that the three former properties satisfied by R guarantee the coherence of the new m.-s. representations introduced for the partitioned set.

[13]The typical notation $\bar{x}$ can be enhanced by introducing explicitly the token used for denoting the corresponding relation, e.g. $\bar{x}^{[R]}$. However, this is usually not done due to simplicity on the notation and justified by the (pragmatic) fact that the relation is implicitly clear from the (conceptual) concept.

Contrastingly, the local version of the former m.-s. configurations are typically expressed as a generic functional equality requiring local generic labels for the elements of $\mathscr{W}$ and $\mathscr{V}$, which are often chosen having strong graphic similarities with the global symbols used for describing the corresponding mathematical structures (e.g., w and v[14]), and (sometimes) auxiliary symbols like parenthesis when the description is done at the meta-level:

$$v = f(w).$$

Moreover, if we want to use the set-theoretical formalization of the notion of mathematical function,[15] we can see in a more explicit way the cognitive nature of doing a formal "comparison" among the (generic) elements of the corresponding mathematical structures. For instance, by using tuples we can describe the former m.-s. configuration in the form of

$$(w, f(w)),$$

which suggests graphically a natural comparison between the involved elements at the simplest level; i.e., one simply puts both (generic) elements together using an additional formal mathematical structure (e.g., f as a special kind of mathematical relation). Moreover, such a comparison possesses the additional attribute of being "locally directed," namely, one has a fixed starting mathematical structure for doing the comparison and at the (local) level of elements one compares each of them with single elements of the second structure.

In most cases, the conceptual substratum of the corresponding function is located in the explicit description of this local m.-s. configurations (see Chap. 9).[16]

Our first conclusion at this point is that conceptual comparison plays a foundational role on the emergence of the notion of mathematical function. Let us explore in more detail the slightly different role that formal metaphorical reason plays in inspiring instantes of this fundamental mathematical concept.

Explicitly, in a lot of cases, we define functions between mathematical structures $\mathscr{A}$ and $\mathscr{B}$ in order to be able to "formalize" metaphorically inspired intuitions aiming to understand more deeply $\mathscr{B}$ in terms of $\mathscr{A}$. For example, suppose that we want to show that a mathematical structure $\mathscr{D}$ is finite, then a typical way of doing that is to find a natural number $m \in \mathbb{N}$ and a surjective function $g : \mathbb{N}_m =$

---

[14]This simple symbolic trick can have a lot of mnemotic advantages when one uses dozens of symbols throughout the generation of a specific mathematical proof, e.g. elementary and advanced exercises in modern algebraic geometry involving, for example, the formal gluing of affine schemes (see, for instance, [25]).

[15]For example, taking as a foundational framework ZFC set theory [34].

[16]In such cases the usage of m.-s. graphs represents considerable conceptual help for getting gradually inside by solving conjectures involving the corresponding representations (see Sect. 10.6 Chap. 10).

$\{1, \cdots, m\} \to \mathscr{D}$. One strong inspiration for proceeding in such a way is implicitly given by the quest of formalizing in more detail a metaphor of the form $\mathbb{N}_m \rightrightarrows \mathscr{D}$.

So, although doing metaphors between (mathematical) concepts is a more fundamental and primary operation of the mind, one can affirm that mathematical functions often materialize a lot of formal aspects that initial metaphors possess.

Another enlightening example is given by the concept $\mathscr{K}_{fg.alg}$ of finitely generated algebras over a field k (see [13]). It is very common in commutative algebra, and related areas, to try to understand (generic) specification of this concept implicitly in terms of metaphors of the form $\mathscr{P}_{ol}(n, k) \rightrightarrows \mathscr{K}_{fg.alg}$, where $\mathscr{P}_{ol}(n, k)$ denotes the concept of the ring of polynomials in n variables with coefficients in k. Furthermore, the concrete forms of these kinds of metaphors are described in terms of surjective functions $f : k[x_1, \cdots, x_n] \to T$, where T denotes a finitely generated k−algebra, or, in our terminology, the main sort of a generic exemplification of $\mathscr{K}_{fg.alg}$. With this conceptual trick, our mind reduces the algebraic understanding of T to the more intuitive comprehension of (formal quotients of) polynomial rings, which have more accessible forms for doing computations and symbolic manipulations (see, for example, [9]).

Hence, our second conclusion is that formal metaphorical reasoning serves, among many other things, as a "conceptual engine" for constructing mathematical functions in a wide number of situations. Nonetheless, the subsequent generation of the corresponding functions is essentially done in a lot of cases by means of the cognitive ability of conceptual comparison.

For the generation of the notion of function composition, let us consider a second function $g : V \to U$ described by

$$v = f(u).$$

If we do a simple conceptual replacement of u in the former equation based on the equality $u = f(w)$, we obtain the expression

$$u = f(g(w)).$$

Conceptual lining from the last expression generates a function from W (suggested by the generic symbol w) to U (suggested by generic symbol u). A natural name for the new function is suggested by the m.-s. sub-configuration "$f(g$." So, we choose a new symbol connecting "f" and "g" in the given order, e.g. "$\circ$" (as usual). Thus, we find an explicit description for such conceptual lining:

$$CL(u = f(g(w))) = f \circ g : W \to U.$$

In this way, we can reconstruct cognitively the notion of composition between functions.

An important m.-s. strengthening of the former concept is given by the additional conditions $W \subseteq V$ and $f = I_{nc} : W \hookrightarrow V$ is the set-theoretical inclusion. In this case, the composition $g \circ f$ is also known as the restriction of g to W and is also denoted by $g_{|W}$.

11.8   Topological Spaces

In this section, we will focus our attention in a seminal notion in topology, i.e. topological spaces starting with completely elementary notions.

11.8.1   Structures of Basic Sets

For the cognitive meta-generation of this concept, let us start by defining an even more elementary structure which seems to be more familiar from high school, namely, the notion of *basic sets* or *pre-sets*.

For a more precise description of this conceptual space, we use a many-sorted first-order (kind of) approach.

Definition 11.1 A *structure of basic sets* consists of the following data: A sort symbol V for the universe of basic sets;[17] a sub-sort symbol $U < V$, for denoting a global or "reference pre-set,"[18] a constant $\varnothing : V$, a binary relation symbol ("membership relationship") $\in: V \times V$; three binary operations $-\cup-, -\cap-, -\backslash- : V \times V \to V$, and an unary relation $(-)^c : V \to V$, satisfying the following conditions:

1. $(\forall a, b : V)(a \cup b : V)$
2. $(\forall a, b : V)(a \cap b : V)$
3. $(\forall a, b : V)(a \backslash b : V)$
4. $(\forall a : V)(a)^c : V)$
5. $(\forall a, b, c : V)(c \in a \cup b \leftrightarrow c \in a \vee c \in b)$
6. $(\forall a, b, c : V)(c \in a \cap b \leftrightarrow c \in a \wedge c \in b)$
7. $(\forall a, b, c : V)(c \in a \backslash b \leftrightarrow c \in a \wedge c \notin b)$
8. $(\forall a, d : V)(d \in (a)^c \leftrightarrow d : U \wedge d \notin a)$
9. $(\forall a : V)(a \notin \varnothing)$

Informally, the first four conditions simply state the well-definition of the corresponding meta-functions and the last four describe explicitly the internal relations defining them by means of the other (meta-)functions and (meta-)relations.

Since we are working in this case essentially in a many-sorted first-order setting, the condition requiring that U should be a sub-sort of V can be seen as a technical way of stating that U "belongs" to V (when one interprets U as a basic set on its own). Another way of doing this is adding a new constant $C_U : V$ and declaring explicitly through the binary relation $\in$ that the "elements" of U (i.e., the $u : V$ such that $u : U$) are exactly the "elements" of C_U (i.e., the $u : V$ such that $u \in C_U$). In other words, adding the axiom:

---

[17]One can think of V as an elementary version of the universal class in the NBG set theory [34].

[18]This sort is used mainly for defining complements of pre-sets.

$$(\forall u : V)(u : U \leftrightarrow u \in C_U).$$

For the sake of simplicity, we will adopt the shortest description, namely, $U < V$.

The former axiomatization can be seen as a more precise formalization of the informal descriptions of "sets" that are taught in high schools by means of Venn diagrams.

11.8.2 Substructures of Power Pseudo-Sets

In classic (ZFC) set theory there are two basic (syntactic) ways to compare sets, namely, through the membership relationship ($\in$) and through the containment relationship ($\subseteq$). A natural formal metaphor that can be done in this context is simply $\in \Rrightarrow \subseteq$. So, let us fix a set A. We know for the axiom of extensionality (see [34, Ch. 4] that A is completely determined by its elements; i.e., by all the sets x such that $x \in A$. In other words, "A is equal to the collection of all x such that $x \in A$." If we re-formulate the last statement by using the last metaphor ($\in \Rrightarrow \subseteq$), we will obtain the statement, "A is equal to the collection of all sets x such that $x \subseteq A$," which is not true. However, if we try to "repair" it (syntactically), we can replace the first occurrence of A with a formal (pseudo-)function of A, let us call it $F(A)$. Thus, we obtain through this *repaired metaphorical inference* "$F(A)$ is equal to the collection of all sets x such that $x \subseteq A$," which is exactly the statement giving origin to the (classic) notion of power set.[19]

We will give the formal definition emerging from the former construction at a general level:

Definition 11.2 A *structure of power pseudo-sets* (resp. *substructure of power pseudo-sets*) consists of the following data: A sort symbol V' for the universe of basic sets; two sub-sorts symbol $X, Y < V'$, for denoting the initial pseudo-set (X) and its corresponding power pseudo-set (Y), and two binary relations ("membership and containment relationships") $\in', \subseteq': V' \times V'$, satisfying the following conditions:

1. $(\forall a : V')(a : Y \leftrightarrow a \subseteq' X)(resp.(\forall a : V)(a : Y \rightarrow a \subseteq' X))$
2. $(\forall x, y : V')(x \subseteq' y \leftrightarrow (\forall z)(z \in' x \rightarrow z \in' y))$

The derived notion of substructure of power pseudo-sets emerges from the former one through the cognitive mechanism of conceptual comparison materialized by means of the (pseudo-)membership relationship $\subseteq'$.

---

[19]Our cognitively-inspired reconstruction is, of course, a modern way of generating such a concept, which might (not) coincide with the historical reconstruction, but which can offer new and more pragmatic insights going beyond pure mathematics.

11.8.3   Doing Local Metaphors

From the last two conceptual spaces, we can obtain a first version of pre-topological space by doing the simple local metaphors with bounded syntactic scope from the substructures of power pseudo-sets into the structures of basic sets: (1) $\Theta :=$ $\lceil U \rightrightarrows X, V \rightrightarrows Y \rfloor$ with syntactic scope the first axiom of the substructures of power pseudo-sets (i.e., $A1'$), the four first axioms defining structures of basic sets (i.e., $\Theta(A1)$, $\Theta(A2)$, $\Theta(A3)$ and $\Theta(A4)$), and $\Theta(U < V, -\cup -, -\cap -, -\setminus - :$ $V \times V \rightarrow V; (-)^c : V \rightarrow V, \varnothing \in V)$ (2) $\Xi = \lceil V \rightrightarrows V', \in \rightrightarrows \in' \rfloor$ with syntactic scope the second axiom of the notion of substructures of power pseudo-sets (i.e., $A2'$), $\Xi(X < V', Y < V', \in': V' \times V')$ and the last five axioms describing structures of basic sets (i.e., $\Xi(A5)$, $\Xi(A6)$, $\Xi(A7)$, $\Xi(A8)$, and $\Xi(A9)$).

By doing the conceptual conjunction of the syntactic scopes of the former two local metaphors, we obtain the following conceptual space, which will be denoted by $\mathscr{P}_{top}$:

Definition 11.3 A *pre-topological space* consists of the following data: A sort symbol V' for the universe and two sub-sort symbols fulfilling $X < Y$ and $X, Y < V'$; a constant $\varnothing : Y$, a binary relation symbol $\in': V' \times V'$; three binary operations $-\cup -, -\cap -, -\setminus - : Y \times Y \rightarrow Y$, and an unary relation $(-)^c : Y \rightarrow Y$, satisfying the following conditions:

1. $(\forall a, b : Y)(a \cup b : Y)$
2. $(\forall a, b : Y)(a \cap b : Y)$
3. $(\forall a, b : Y)(a \setminus b : Y)$
4. $(\forall a : Y)(a)^c : Y)$
5. $(\forall a : V')(a : Y \rightarrow a \subseteq' X)$
6. $(\forall a, b, c : V')(c \in' a \cup b \leftrightarrow c \in' a \vee c \in' b)$
7. $(\forall a, b, c : V')(c \in' a \cap b \leftrightarrow c \in' a \wedge c \in' b)$
8. $(\forall a, b, c : V')(c \in' a \setminus b \leftrightarrow c \in' a \wedge c \notin' b)$
9. $(\forall a, d : V')(d \in' (a)^c \leftrightarrow d : U \wedge d \notin' a)$
10. $(\forall a : V')(a \notin' \varnothing)$
11. $(\forall x, y : V')(x \subseteq' y \leftrightarrow (\forall z)(z \in' x \rightarrow z \in' y))$

In the former conceptual space, one can make a syntactical-semantic distinction between the first axioms and the last six based on the fact that the first group simply described "explicit" properties related to most of the functional and relational symbols and the second group described exactly "implicit" properties of the same symbols in terms of the "membership relation." So, depending on the conceptual context where we are cognitively situated,[20] the last six axioms can be avoided since they are implicitly presupposed.

---

[20]Depending on the global theory we use as a general foundational framework, e.g. ZFC set theory, lambda calculus, topos theory, category theory [7].

11.8.4 Doing a Syntactic Restriction to the Real Line

Let us now see the huge scope of single examples to shape and generate (in a lot of cases unconsciously) quite general notions. In our case, we will consider one of the most studied mathematical structures in analysis, namely, the real line, which, from a historical point of view, is an older concept than the one of topological space [35]. Effectively, in our terminology let us define the (formal) meta-exemplification $\mathbb{R}^1 = (\mathcal{L}'_{an}, \mathcal{L}'_o, \mathcal{A}_r, \mathcal{M}_r)$, where $\mathcal{L}'$ contains the standard symbols for describing the real line as an ordered field and the collection of all its subsets that can be written as open intervals ("open balls");[21] $\mathcal{L}'_o$ is a many-sorted first-order logic;[22] $\mathcal{A}_r$ are the (corresponding) axioms describing the real numbers as an ordered field, the collection of subsets formed entirely by open interval, and the natural definitions of complement and difference of (sub-)sets of the real numbers; and $\mathcal{M}_r$ contains mainly two structures, i.e., the collection of real numbers constructed (for example) in terms of Dedekind's cuts ($\mathbb{R}$), and the collection of the distinguished (open) subsets of $\mathbb{R}$ ($\mathcal{O}_r$).

Let us do the syntactic restriction (see Chap. 10) of the notion of pre-topological space to the real line, (i.e., $\mathscr{P}_{top_{\mathbb{R}^1}}$) with all the natural identifications of sorts et al., e.g. X, Y, and V' are identified with $\mathbb{R}$, $\mathcal{O}_r$, and V (the proper class containing all sets), respectively. So, one can easily see that $\mathscr{P}_{top_{\mathbb{R}^1}}$ contains all the symbols and axioms of $\mathscr{P}_{top}$ with the exception of $-\backslash-$ and $(-)^c$, and the corresponding axioms describing the properties of these operations, since, the corresponding functions are not total over open real subsets, and, therefore, they do not even fulfill the well-defined property described in such axioms.

In summary, $\mathscr{P}_{top_{\mathbb{R}^1}}$ is a conceptual space with the following data: A sort symbol V' for the universe and two sub-sort symbols fulfilling $X < Y$ and $X, Y < V'$; a constant $\varnothing : V'$, a binary relation symbol $\in' : V' \times V'$; two binary operations $-\cup-, -\cap- : Y \times Y \to Y$, satisfying the following conditions:

1. $(\forall a, b : Y)(a \cup b : Y)$
2. $(\forall a, b : Y)(a \cap b : Y)$
3. $(\forall a : V')(a : Y \to a \subseteq' X)$
4. $(\forall a, b, c : V')(c \in' a \cup b \leftrightarrow c \in' a \vee c \in' b)$
5. $(\forall a, b, c : V')(c \in' a \cap b \leftrightarrow c \in' a \wedge c \in' b)$
6. $(\forall a : V')(a \notin' \varnothing)$
7. $(\forall x, y : V')(x \subseteq' y \leftrightarrow (\forall z)(z \in' x \to z \in' y))$

---

[21] In this case it is not necessary to add the syntactic information describing the completeness of real numbers into the meta-exemplification because we implicitly add to it a concrete "materialization" of them in terms, for instance, of Dedekind's cuts [4].

[22] It is important to mention at this point that the methods we are creating/inventing in this work are not restricted to any kind of classic logic and are based on a more global and pragmatic understanding of the way in which the mind conducts actual mathematical research.

In the next step, we apply a syntactic generalization over the operators $\cup$ and $\cap$. First of all, let us consider usual mathematical expressions describing these operators acting over two sets, e.g. $A \cup B$ and $A \cap B$. Now, we can rewrite them in a more prototypical manner according to the principles developed in Chap. 9:

$$A_1 \cup A_2 = \bigcup_{i=1}^{2} A_i = \bigcup_{i \in I = \{1,2\}} A_i,$$

$$A_1 \cap A_2 = \bigcap_{i=1}^{2} A_i = \bigcap_{i \in I = \{1,2\}} A_i.$$

From these expressions we can apply a syntactic generalization over I; i.e., instead of considering to different morpho-syntactic expressions A and B with (m. s.) level one (see Chap. 10), we replace them with a more "unifying" generic expression A_i with (m. s.) level two[23] and replace the initial index set $\{1, 2\}$, with the generic (m. s.) expression I.

In this way, after doing all the syntactic-semantic meta-computations, one can obtain a new conceptual space $\mathcal{G}_{en}(\mathcal{P}_{top_{|\mathbb{R}^1}})$ characterized by the following data:

Sort symbols X, Y, and V' fulfilling $X < Y$ and $X, Y < V'$; a constant $\varnothing : V'$, a binary relation symbol $\in' : V' \times V'$; generic symbolic operators $\bigcup$ and $\bigcap$ satisfying the following conditions:[24]

1. $(\forall I : V')(\forall A : I \to Y)(\bigcup_{i \in I} A(i) : Y)$
2. $(\forall I : V')(\forall A : I \to Y)(\bigcap_{i \in I} A(i) : Y)$
3. $(\forall a : V')(a : Y \to a \subseteq' X)$
4. $(\forall I : V')(\forall A : I \to Y)(\forall x : V')((x \in \bigcup_{i \in I} A(i)) \leftrightarrow (\exists j : V')(j \in I \wedge x \in A(j)))$
5. $(\forall I : V')(\forall A : I \to Y)(\forall x : V')((x \in \bigcap_{i \in I} A(i)) \leftrightarrow (\forall j : V')(j \in I \to x \in A(j)))$
6. $(\forall a : V')(a \notin' \varnothing)$
7. $(\forall x, y : V')(x \subseteq' y \leftrightarrow (\forall z)(z \in' x \to z \in' y))$

If we apply again syntactic restriction to the real line with the natural identifications, (i.e., if we consider $\mathcal{G}_{en}(\mathcal{P}_{top_{|\mathbb{R}^1}})_{|\mathbb{R}^1}$) we obtain a conceptual space similar to the former one, but without the generalized intersection's operator and the axioms defining it. Now, let us apply a syntactic enrichment adjoining again the initial binary relation symbol $\cap$ together with the corresponding axioms describing it, exactly in the way that they appear before doing the syntactic generalization. So, we obtain a conceptual space characterized by the following data:

---

[23] In the (m. s.) expression A_i there are two (m. s.) unities A and i.

[24] In this case, we need to expand our logic ontology to a (many-sorted) higher-order one to be able to define arbitrary unions, intersections, and functions among (pseudo-)sets [33, Ch. 4].

Sort symbols X, Y, and V' fulfilling $X < Y$ and $X, Y < V'$; a constant $\varnothing : V'$, a binary relation symbol $\in'$: $V' \times V'$; a generic symbolic operator $\bigcup$ and a binary operator $\cap : Y \times Y \to Y$ satisfying the following conditions:

1. $(\forall I : V')(\forall A : I \to Y)(\bigcup_{i \in I} A(i) : Y)$
2. $(\forall a, b : Y)(a \cap b : Y)$
3. $(\forall a : V')(a : Y \to a \subseteq' X)$
4. $(\forall I : V')(\forall A : I \to Y)(\forall x : V')((x \in \bigcup_{i \in I} A(i)) \leftrightarrow (\exists j : V')(j \in I \wedge x \in A(j)))$
5. $(\forall a, b, c : V')(c \in' a \cap b \leftrightarrow c \in' a \wedge c \in' b)$
6. $(\forall a : V')(a \notin' \varnothing)$
7. $(\forall x, y : V')(x \subseteq' y \leftrightarrow (\forall z)(z \in' x \to z \in' y))$

This conceptual space is essentially equivalent to the notion of topological space. Effectively, the four explicit axioms characterizing the notion of a topological space are materialized in the condition $X < Y$ and the first three axioms of the former list. Moreover, the axioms $4, 5, 6$, and 7 are simply descriptions of the defining properties of the corresponding constant ($\varnothing$), the binary relation ($\subseteq$), the binary (local) operation ($\cap$), and the generic operator ($\bigcup$), which are conditions implicitly presupposed in the notion of topological space.[25]

Notice that our cognitively-inspired conceptual generation of topological spaces has an important similarity with the historical invention/creation of the concept, namely, in both cases the role of the real line (as a metric space) is central [35].

11.9   Base for a Topological Space

Let us consider as initial structures the real plane with the standard topology $\mathscr{R}^2 = (\mathbb{R}^2, T)$ and the usual collection of open sets $B = B(\mathbb{R}^2)$ consisting of all open balls in $\mathbb{R}^2$, which are of common usage in real analysis. A simple analysis of graphic (generic) conceptual substrata of elements of T and B offers the minimal formal relationships between generic exemplifications of them involving the fundamental operations of union and intersection and the membership and containment relations. Explicitly, let us describe such statements in a many-sorted first-order setting:

1. $B \subseteq T$.
2. For all $U \in T$ and for all $u \in U$, exists a $W \in B$ such that $u \in W$ and $W \subseteq U$.
3. For all $U, V \in T$ and for all $u \in U \cap V$, exists $W \in B$ such that $u \in W$ and $W \subseteq U \cap V$.

---

[25]Formally speaking, general topology is grounded in ZFC set theory; therefore, the notion of topological space involves implicitly more axioms, e.g., the ones involving the existence of power sets, arbitrary unions and intersections appearing in its standard definition [38].

Let us consider the mathematical structure consisting on the spaces $\mathscr{R}^2$ and $B(\mathbb{R}^2)$, together with the former axioms $A_i(\mathscr{R}^2, B)$ for $i = 1, 2, 3$; denoted by $\mathscr{B}(\mathscr{R}^2, B)$. Now, let consider the conceptual generalization

$$CG_{\mathscr{R}^2 \to TS_{[g]}, B \to B_{[g]}}(\mathscr{B}(\mathscr{R}^2, B)),$$

where $TS_{[g]}$ is a generic exemplification of a topological space and $B_{[g]}$ is a generic exemplification of the notion of set.

Thus, by applying a (global) generic generalization on the former structure, we obtain the standard notion of a basis for a topological space, namely,

$$\mathscr{B}_{asisTopSpace} = GG(CG_{\mathscr{R}^2 \to TS_{[g]}, B \to B_{[g]}}(\mathscr{B}(\mathscr{R}^2, B))).$$

11.10 Commutative Rings with Unity

The notion of commutative ring with unity can be essentially generated as the one of fields (see Chap. 7). Effectively, instead of blending the notions of (abelian) group and pointed abelian group (see Sect. 7.3 of Chap. 7), we can do essentially the same blending but with the difference that the second concept should be replaced by the one of *pointed commutative monoid*:

Definition 11.4 A pointed commutative monoid consists of a set C, a binary operation $*$, and a (distinguished) element $b \in C$ such that $(B \setminus \{b\}, *_{|C \setminus \{b\} \times C \setminus \{b\}})$ is an commutative monoid and, $b * c = c * b = b$ for all $c \in C$.

Afterwards, the resulting concept is blended with the one of distributive spaces exactly as in Sect. 7.3 of Chap. 7. So, we generate the notion of commutative ring with unity as the recursive blend of the notions of abelian group, pointed commutative monoid, and distributive space:

$$CommRingwithUnity = (AbGroup \vee_G PoinCommMonoid) \vee_{G_1} DistSpace.$$

Let us denote this concept by $\mathscr{R}_{ComUnit}$.

11.11 Isomorphisms (of Commutative Rings with Unit)

Let $\mathscr{S}_1$ and $\mathscr{S}_2$ be two exemplifications of a concept $\mathscr{C}$ (e.g., $\mathscr{R}_{CommUnit}$). A natural question is whether $\mathscr{S}_1$ and $\mathscr{S}_2$ can be conceptually identified as exemplifications of $\mathscr{C}$ (e.g., as commutative rings with unity). One of the main reasons for the emergence of this cognitive request is a *principle of cognitive simplicity*, in other words, our minds wish to use the minimal amount of (internal) memory (i.e., maximizing compression) and avoid redundancy in the most effective manner [8, 18].

In the case of commutative rings with unity, such a conceptual identification is obtained simply if we are able to "re-label" (generically) all the elements, operations, and outstanding constants of $\mathscr{S}_2$ exactly in terms of the corresponding structures of $\mathscr{S}_1$. In other words, we wish to express in a unique way any (generic) element/structure of $\mathscr{S}_2$ essentially as a (generic) element/structure of $\mathscr{S}_1$ in such a way that all the minimal conceptual relations given by the structures are coherently "traced" to $\mathscr{S}_2$. Specifically, let us denote such a re-labeling by $(-)'$, let s_1 and s_2 be generic exemplifications for the elements of $\mathscr{S}_1$ and $\mathscr{S}_2$, respectively, and let the languages of both concepts be written as $\langle S_1, 0, 1, +, * \rangle$ and $\langle S_2, 0_2, 1_2, +_2, *_2 \rangle$, respectively. Then, the re-labeling requires explicitly that $s_2 = s_1'$, $+_2 = +'$, $*_2 = *'$, $0_2 = 0'$, $1_2 = 1'$ and that $c' = a' +' b'$ (resp. $c' = a' *' b'$), as long as $c = a + b$ (resp. $c = a * b$); and $a' \neq b'$ as long as $a \neq b$, where a, b, and c are generic exemplifications of elements of $\mathscr{S}_1$. So, the former conditions are simply the cognitive substratum of the notion of isomorphism between commutative rings with unity. Effectively, let us change the symbol $(-)'$ by a function $f : S_1 \rightarrow S_2$, then the generic nature of s_1 and s_2, together with the condition $s_2 = s_1'$, guarantees the surjectivity of f; the injectivity is simply given by the condition $a \neg b \rightarrow a' \neg b'$, and the homomorphism conditions are given by the rest of constrains.

Note that the former re-labeling process is naturally extended to most of the algebraic structures appearing in abstract algebra, groups, Fields and Galois theory, and commutative algebra, among others.[26]

11.12 Sub-Rings and Ideals of Commutative Rings with Unity

Let us consider two versions of the notion of commutative ring with unity through conceptual replication

$$\mathscr{R}_{ComUnit1} = \langle \mathscr{L}_{MFOL}, \mathscr{L}_{an1} = \langle S, +_S, *_S, 0_S, 1_S \rangle, A_S, \mathbb{M}_1 \rangle,$$

and

$$\mathscr{R}_{ComUnit2} = \langle \mathscr{L}_{MFOL}, \mathscr{L}_{an2} = \langle R, +_R, *_R, 0_R, 1_R \rangle, A_R, \mathbb{M}_1 \rangle,$$

---

[26]For instance, in the case of $k-$vector spaces or $k-$algebras, re-labeling additionally requires a natural identification with the corresponding scalar operations; i.e. $a' = r' \bullet' b'$, as long as $a = r \bullet b$ (where r is a generic exemplification of elements of k).

where $\mathscr{L}_{MFOL}$ denotes a many-sorted first-order logic, and both S and R denote the main sorts denoting the carriers of the elements of commutative rings with unity.

Define the following conceptual comparison between these generic exemplifications as follows:

$$\Theta = \langle S \subseteq R, +_S = +_{R|S \times S}, 0_S = 0_R, 1_S = 1_R \rangle.$$

Then, the concept of *sub-ring of a commutative ring with unity* is obtained by means of the conceptual conjunction of the former conceptual spaces:

$$Sub\mathscr{R}_{ComUnit} = \langle\langle \mathscr{R}_{ComUnit1}, \mathscr{R}_{ComUnit2}, \Theta \rangle\rangle.$$

The choice of the former conceptual comparison is implicitly inspired by dozens of elementary examples in abstract algebra and real and complex analysis, e.g., $\mathbb{Z}$ considered as a sub-ring of $\mathbb{R}$ and $\mathbb{R}$ considered as a sub-ring of $\mathbb{C}$.

If we considered the ring of even numbers $2 \bullet \mathbb{Z}$ as a sub-ring without unity of $\mathbb{Z}$, then it is an elementary arithmetical fact to prove that if a is an even number and b an integer, then ab is again an even number. Thus, this suggests that the multiplication between even numbers fulfills a stronger condition than just the closure property; i.e. one can multiply even numbers by arbitrary integers and still obtain even numbers as a result. This elementary fact can be extrapolated in a suitable way with arbitrary commutative rings with unity to obtain the notion of ideal.

Effectively, let $2 \bullet \mathscr{X} \sqsubset \mathscr{X}$ be the exemplification consisting of $\mathbb{Z}$ considered as a commutative ring with unity and the subset $2 \bullet \mathbb{Z}$ also considered as a kind of substructure of $\mathbb{Z}$ with the corresponding restrictions of the binary operations $+_{\mathbb{Z}}$ and $*_{\mathbb{Z}}$.

The syntactic restriction of $Sub\mathscr{R}_{ComUnit}$ to $2 \bullet \mathscr{X} \sqsubset \mathscr{X}$ describes a concept similar to $Sub\mathscr{R}_{ComUnit}$ with the difference that it lacks the symbol 1_S, the corresponding axiom characterizing 1_S as a unit for $*_S$ and the comparison $1_S = 1_R$. Now, let us consider the axiom of $Sub\mathscr{R}_{ComUnit|2 \bullet \mathscr{X} \sqsubset \mathscr{X}}$ describing the closure property of $*_S$, i.e.,

$$B(x(S), *_S) := (\forall x : S)(\forall y : S)(x *_S y : S).$$

By taking a m.-s. generalization of B with respect to the variable x to the bigger sort R (considered as a super-sort of S) and simultaneously doing the corresponding generalization of the binary operation $*_S$ to $*_R$, we obtain

$$B(x(R), *_R) := (\forall x : R)(\forall y : S)(x *_R y : S).$$

Finally, extending this generalization to $Sub\mathscr{R}_{ComUnit|2 \bullet \mathscr{X} \sqsubset \mathscr{X}}$, we obtain exactly the concept of *ideals of a commutative ring with unity*, denoted by

$$Ideal\mathscr{R}_{ComUnit}.$$

11.13   Spectrum of Ideals and Prime Ideals, and Multiplicative Rings

Let us consider the notion of ideals of commutative rings with unity $Ideal\mathscr{R}_{ComUnit}$. Now, let us apply generic generalization on the morpho-syntactic part of the concept $\mathscr{R}_{ComUnit}$ seen as a sub-concept of $Ideal\mathscr{R}_{ComUnit}$. In other words, we fix (cognitively) a generic commutative ring with unity and the (global) binary operations and we let ideal's sort and the corresponding (local) operations free. By applying conceptual lining to the former conceptual m.-s. configurations, we obtain the notion consisting of the spectrum of ideals of the generic commutative ring with unity R_{gen}, denoted by $\mathrm{Spec}_{Ideal}(R_{gen})$. Further, we obtain the notion of the spectrum of ideal of a commutative ring with unity by applying conceptual lining to the former one; i.e.,

$$\mathrm{Spec}_{Ideals}(\mathscr{R}_{ComUnit}) = CL(\mathrm{Spec}_{Ideals}(R_{gen})).$$

For the meta-generation of the prime spectrum, let us start by considering the following exemplification:

$$MulPrim(\mathbb{Z}) = \langle L_{msfol}, L', A', \langle \mathbb{Z}, MulPrim_E, \in_E, P_{gen} \rangle \rangle$$

where L' is the standard language containing the classic binary operations, the sort of integers Z, the constants 0 and 1, the (local) belonging relation's symbol $\in'$, the special sort $MulPrim$, and a sort representing a (generic) set of multiples of a prime, $MulPrim$ is the collection of subsets of $\mathbb{Z}$ that consists of multiples of a fixed prime number (excluding ± 1), $\in'$ is the local binary relation between integers and elements of $MulPrim$, and P_{gen} represents a generic exemplification for multiples of primes. Furthermore, A' contains the standard ring axioms and the following axiom emerging classically from basic heuristics regarding the prime numbers and the re-interpretation of divisibility in term of the membership relation:

$$(\forall X : MulPrim)(1 \notin' X \wedge (\forall a, b : Z)(a * b \in' X \rightarrow a \in' X \vee b \in' X).$$

Let us consider the metaphor $\xi = MulPrim(\mathbb{Z}) \rightrightarrows \mathrm{Spec}(\mathscr{R}_{ComUnit})$, given by the natural identifications, sending P_{gen} to the sort of ideals I in $\mathscr{R}_{ComUnit}$, among others.

So, the (conceptual) scope of ξ generates the concept of the prime spectrum of a commutative ring with unity, because we add the translation of the former axiom to the condition of being ideal, which characterized exactly the standard notion of prime ideals in commutative algebra [13].

Another way of generating such a prime spectrum is obtained by using a kind of weakened generalization of the notion of prime numbers as follows:

Definition 11.5 Let $\mathscr{P} = \langle L_{msfol}, L = \langle Z, *, 1, \|, isprime, Prime \rangle, A_P \mathbb{M}_P \rangle$ be the concept consisting of a binary operation $*$ and the axioms A_P are given by the following specifications: 1 is a neutral element (i.e., for any $z \in Z$, $z * 1 = 1 * z = z$), upside-down divisibility relation $\parallel$ described by

$$e \parallel g := g|e \Leftrightarrow (\exists f \in Z)(g = e * f);$$

and an unary relation $isprime$ on Z defined as follows:

For all $p : Z$, $isprime(p)$ if and only if

$$p \neq 1 \wedge (\forall a, b \in Z) ((ab \parallel p) \rightarrow (a \parallel p \vee b \parallel p)).$$

Finally, the sorf $Prime$ is defined by

$$Prime = \{p : Z/isprime(p)\}.$$

Let us denote by $\mathrm{Spec}_{Ideals}(\mathscr{R}_{ComUnit})_{ext}$ the notion of $\mathrm{Spec}_{Ideals}(\mathscr{R}_{ComUnit})$ extended with sorts and basic axioms for the power set of the sort of rings, the product of ideals, a containment relation's symbol for ideals, and a (local) belonging relation between elements of the ring and subsets of it. Then, in [6] it was essentially showed the there exists a natural conceptual blending (see Chap. 7) between $\mathscr{P}$ and $\mathrm{Spec}_{Ideals}(\mathscr{R}_{ComUnit})_{ext}$ that produces an conceptual strengthening of the notion of the prime spectrum of a commutative ring with unity, together with the former additional sorts; i.e.,

$$\mathscr{P} \wedge_{\mathscr{G}_{en}} \mathrm{Spec}_{Ideals}(\mathscr{R}_{ComUnit}) = \langle \mathrm{Spec}(\mathscr{R}_{ComUnit})_{ext}, A_{CDR} \rangle,$$

where A_{CDR} is the axiom describing the condition that for any pair of ideals I and J, $I \subseteq J$ if and only J divides I as ideals (i.e., there exists an ideal H such that I equals the product of H and J). Commutative rings of unity satisfying this condition are called Containment Division Rings, or more classically, multiplicative rings [19].

Lastly, we obtain the concepts of a prime ideal (i.e., $PrimIdeal\mathscr{R}_{ComUnit}$), and the whole prime spectrum of a commutative ring with unity (i.e., $\mathrm{Spec}(\mathscr{R}_{ComUnit})$) by doing simple weakenings of the former conceptual blend.

11.13.1   Dedekind Domains

If we enrich the notion of commutative ring with unity with the Noetherian condition (i.e., every ideal can be generated by finitely many elements), then one can extend the former blending generating the prime spectrum and obtain the notion of Noetherian Containment-Division ring.[27] Using the fact that the notion of Dedekind domain is equivalent to the notion of Noetherian Containment-Division ring (see, for instance, [21] and [20]), one obtains the former concept similarly as a blend. Originally, the establishment of the former conceptual equivalence between Dedekind domains and multiplicative rings was a post-discovery motivated by the blending described in the former section (see also [6]).[28]

11.14   Local Rings

Let us consider the conceptual space $\mathscr{Z}_{27}$ given by the commutative ring with unity $\mathbb{Z}/27\mathbb{Z}$, together with its prime spectrum.[29] It is straightforward to see that the spectrum of ideals of this elementary structure which are not the whole ring, consists of exactly of three ideals; i.e. $(\overline{0})$, $(\overline{3})$, and $(\overline{9})$. $(\overline{3})$ possesses the property of being the only ideal containing all the other ideals that are not the whole ring; i.e. $(\overline{3})$ is the only maximal ideal. Let us denote by A_m the axiom describing the fact that $\mathbb{Z}/27\mathbb{Z}$ has exactly one (non-trivial) maximal ideal.

Finally, let us consider the axiom A_R given by the following conceptual generalization:

$$CG_{\mathscr{Z}_{27} \to \mathbb{R}_{CommUnit}}(A_m).$$

So, the notion of local ring is obtained as the strengthening with A_R of the concept of commutative rings with unity enriched with its spectrum of ideals.

11.15   Zariski Topology Over Prime Spectra

Consider the concepts $Ideal\mathscr{R}_{ComUnit}$ and $PrimIdeal\mathscr{R}_{ComUnit}$. Let us do a natural conceptual blend between these concepts where the generic space is the notion $\mathscr{R}_{ComUnit}$ and the corresponding conceptual morphisms are the natural embeddings viewing $\mathscr{R}_{ComUnit}$ as a sub-concept of each of the input concepts. So, the resulting blend is the composite notion of ideal and prime ideal over the same commutative ring with unity. Let us consider a conceptual comparison enriched with the following data: first, we add a new (super-)sort P_R for the power set of the

---

[27]Here, one can codify the Noetherian condition in a first-order many-sorted setting by adding enough sorts "simulating" the additional sets needed, e.g. (finite) power set of the ring.

[28]As an implicit exercise, the more curious reader can cognitively generate the Noetherian condition for a commutative ring with unity.

[29]We can also assume that the specification of the axioms is given in a many-sorted first-order framework, for example.

(ring) sort R, two binary relations $\in'$, $\notin'$: $R \times P_R$ denoting the (local) belonging and not belonging relations,[30] the axioms describing the fact that $\in'$ and $\notin'$ are complementary relations, the comparison $I \not\subseteq P$ (where I is the sort denoting an ideal and P is the sort denoting a prime ideal), two constants I_c, P_c : P_R used as additional formal "representatives" of I and P, respectively; together with the corresponding axioms defining the former specifications. Among them, the axiom specifying the non-containment of I and P can be described as follows:

$$(\forall x : R)(x \in' I_c \rightarrow x \notin' P_c).$$

Let us denote the former concept by

$$(Ideal\mathscr{R}_{ComUnit} \wedge_{\mathscr{R}_{ComUnit}} PrimIdeal\mathscr{R}_{ComUnit})_{[I \not\subseteq P]}.$$

Consider the generic exemplification of the sub-concept ideal of a commutative ring with unity:

$$GE_{(Ideal\mathscr{R}_{ComUnit} \wedge_{\mathscr{R}_{ComUnit}} PrimIdeal\mathscr{R}_{ComUnit})_{[I \not\subseteq P]}}(Ideal\mathscr{R}_{ComUnit}).$$

If we consider the abstract conceptual lining of the former generic concept, we obtain a new sort (let us call it) U_I containing (generically) all the (abstract) prime ideals which do not contain I_{gen}. Furthermore, let us take the morpho-syntactic part of the generic sub-concept $GE(Ideal\mathscr{R}_{ComUnit})$ within the last one. In other words, we omit the abstract generic model sort I_{gen} together with all the related functional/relational sorts depending on it. Let us consider again the conceptual lining of the resulting generic concept. In this case, we obtain a new sort $ZarTop_R$ containing all the generic sets of the form U_I, for all ideals I of R_{gen}. Lastly, by considering the generic generalization of the notion $\mathscr{R}_{ComUnit}$ considered as a sub-concept of the (morpho-syntactic part of) the last generated concept, we obtain the desired notion of the Zariski topology of (the prime spectrum of) a commutative ring with unity, denoted here by $\mathscr{Z}_{ar}\mathscr{T}_{op}(\mathscr{R}_{CommUnit})$.

11.16 Multiplicative Sets and Localizations of a Commutative Ring (with Unity)

The idea of constructing a new commutative ring starting with a fixed one R by "inverting" some elements of it, can be seen as (a type of) generalization of the

---

[30]From a cognitive perspective, our minds understand the negation of a binary relation as an additional binary relation since, pragmatically speaking, formal negation plays the role of an artificial "dual tool" that leads our consciousness to search for an "equivalent" formal expression in terms of similar atomic structures, e.g. an additional binary relation.

classic construction of the rational numbers starting from the integers [5]. In other words, one wishes to create a mathematical structure with the same basic algebraic properties of the integers (e.g., commutativity, associativity, existence of additive inverses), but where there exist additionally multiplicative inverses for any non-zero integer. Furthermore, an elementary generalization over exponents of the m.-s. configurations appearing in the fundamental theorem of arithmetic [1] produces intermediate rings of fractions of the form

$$\{\prod_{i=1}^{m} p_1^{a_1} \cdots p_m^{a_m} : p_i \in \mathbb{N} \wedge p_r \in \mathbb{Z} \wedge r \in \{i_1, \cdots, i_s\}\},$$

where $\{p_{i_1}, \cdots, p_{i_s}\}$ is a fixed set of prime numbers that generates a subset T of the integers being multiplicatively closed, i.e.,

$$(\forall a : T)(\forall b : T)(a * b : T);$$

and containing the multiplicative unity 1; i.e., $1 \in T$.

This can be naturally generalized through a formal metaphor from the integers (considered as an exemplification and with main sort Z to describe the algebraic axioms of the integers) enriched with the former subset and axioms, to the notion of commutative rings (with unity) with an additional sub-sort $S < R$ and a basic belonging relation:

$$\kappa = \lceil Z \rightrightarrows R, T \rightrightarrows S, \subseteq \rightrightarrows \subseteq \rfloor$$

to produce the notion of a *multiplicative set* of a commutative ring (with unity) R, which is simply a multiplicatively closed subset of R (containing the multiplicative unity of R).

Further, exactly as in the case of the rationals generated from the integer numbers as "formal quotients," we firstly look for generating the abstract "fractions" as equivalent classes (see Sect. 11.6) of tuples (r, s) in $R \times S$ (see Sect. 11.5). The way of defining the equivalent classes is metaphorically inspired by the rational numbers. In fact, remember that the classic way of defining such a (equivalence) relation is $(r_1, s_1) \approx (r_2, s_2)$ if and only if $s_2 r_1 = s_1 r_2$, for $r_1, r_2 \in \mathbb{Z}$ and $s_1, s_2 \in \mathbb{Z} \setminus \{0\}$. Let $P_r = \{A_1, \cdots, A_m, T\}$ be a (standard) proof for the transitivity property of $\approx$.[31]

By gradually considering the (validity of the) image of P_r under κ, we see that one can take valid deductive steps until the fact

$$r_2(s_1 r_3 - s_3 r_1) = 0,$$

---

[31] Here, we understand the notion of proof in the classic first-order logic setting as a sequence of formulas gradually constructed using only the corresponding permitted inferential rules [33].

where some of the starting hypotheses are $(r_1, s_1) \approx (r_2, s_2)$ and $(r_2, s_2) \approx (r_3, s_3)$, with $r_1, r_2, r_3 \in R$ and $s_1, s_2, s_3 \in S$.

On the other hand, one can see by using elementary (non-domain) commutative rings with unity that, from the former equation, it cannot be deduced in general that $s_1 r_3 - s_3 r_1 = 0$. So, one uses such an equation in order to re-define the desired (equivalence) relation $\approx$, which is, strictly speaking, a m.-s. generalization of the classic one, where one simply adds a level to the m.-s. graph of the initial term $s_1 r_3 - s_3 r_1$ by multiplying by an additional element of the multiplicative system, e.g. $r = r_2$ (Sect. 10.6 Chap. 10). One generalizes morpho-syntactically P_r with the updated definition of $\approx$ to verify that such a relation is, in fact, transitive. In this way, one can reconstruct from a cognitive metamathematical perspective the notion of the localization of a commutative ring R with unity with respect to a multiplicative system S.

11.17 The (Meta-)Notion of Category

In this section, we present a basic cognitive reconstruction of the concept of category [30]. This fundamental conceptual space is very special since it can be seen, from a classic point of view, in two ways: firstly, from a historical perspective it emerges straightforwardly as a metamathematical construction, since it was born through the identification of formal syntactic commonalities among several mathematical theories [12]. On the other hand, when it is used as a foundational framework for mathematics [16], or without the aim of obtaining any kind of metamathematical result, daily research with categories possesses essentially the same kind of cognitive phenomenology as the research in other mathematical fields like number theory or algebraic geometry, etc. Even more, due to the wide explanatory scope of our interdisciplinary methodological approach, going beyond of classic "metamathematics," we are able to use our cognitively-based framework to (meta-)analyze this seminal definition.

First of all, due to the fact that ZFC set theory serves (virtually) as a foundational framework for modern mathematics, a lot of classic mathematical fields like analysis, topology, algebraic topology and geometry, and abstract algebra, were (in several aspects) re-written using a set-theoretical syntactic and semantic approach. In fact, currently, most of the standard texts dealing with the former disciplines adopt (implicitly) such a foundational approach (see, for instance, [13, 25, 29, 36, 37] and [1]).

Let us consider the conceptual substratum of the most fundamental way of comparing sets in ZFC, namely, the membership relationship's symbol

$$\ulcorner \in \urcorner.$$

From this fundamental way of comparing the conceptual atoms of ZFC, (i.e., sets) one constructs a more sophisticated similar alternative summarized by the morpho-syntactic unit

$$\to,$$

Namely, one generates an alternative and more metaphorically inspired way of comparing sets by associating two sets A and B, with a third set f, which compares the former ones in a "functional" way (denoted suitably with the symbols $\to$ and the auxiliary symbol :). So, one does all the corresponding formal constructions within ZFC in order to be able to generate a simple (working) meaning for an expression of the form

$$f : A \to B,$$

which implicitly means that this new functional comparison between A and B is described through a new set f codifying it.

The same kind of functional expressions were able to find concrete formal "materializations" within several independent and interdependent mathematical disciplines, for example, the ones described before. For instance, in different conceptual contexts, they were called (monoid, groups, ring) homomorphisms, continuous or smooth functions (see, for example, [12]).

In this case, a quite natural analogical phenomena occurred. Effectively, during this modern set-theoretical characterizations of several mathematical disciplines, one could find the following kind of statements:[32]

"Let G and H be groups and let $\phi : G \to H$ be an homomorphism."
(Group Theory).
"Let V and W be $k-$vector spaces and let $\psi : V \to W$ be an $k-$linear map."
(Linear Algebra).
"Let X and Y be topological spaces and let $g : X \to Y$ be a continuous function."
(Topology).

Let us consider (the generalized notion of (atomic)) analogy between the former (mathematical) statements (see Sect. 10.11, Chap. 10). Remember that the cognitive mechanism of formal analogy essentially preserves the (syntactic) commonalities among several conceptual environments and (abstractly) re-labels the sub-conceptual units that are different in each case, but that play at least a similar positional role within the corresponding conceptual environment.

In this way, let us describe explicitly the result of doing a formal analogy of the former statements considered as conceptual environments; we will use the syntactic units A_1, A_2, $\mathfrak{C}$, f and *formal arrow* for the re-labeling process.

The resulting statement written again in natural language is:

"Let A_1 and A_2 be $\mathfrak{C}$ and let $f : A_1 \to A_2$ be a *formal arrow*."

Now, the last statement implicitly starts to suggest initial fundamental conceptual data about what $\mathfrak{C}$ should be and the way in which its "sub-conceptual unities" (or "elements") should be compared. In fact, if we look for the (mathematical) syntactic substratum among the formal commonalities found in the former statements, we immediately see that the answer takes the form of

---

[32]Here, we use paraphrasing for describing such statements.

$$\ulcorner \quad \rightarrow \quad \urcorner.$$

Thus, a perceptual shifting occurs from "$\in$" to "$\rightarrow$" for the further understanding of what $\mathfrak{C}$ could be. In addition, this implies that the next step would be to identify the minimal collection of concepts and properties surrounding the syntactic unity "$\rightarrow$" and that are common to all the conceptual surroundings as a whole (i.e., the underlining theories). In this way, one finds exactly as showed before, the minimal axioms defining a category:

"Let A_1, A_2, and A_3 be $\mathfrak{C}$ and let $f : A_1 \rightarrow A_2$ and $h : A_2 \rightarrow A_3$ be *formal arrows*, then $h \circ f : A_1 \rightarrow A_3$ is a *formal arrow*."

"Let A_1, A_2, A_3, and A_4 be $\mathfrak{C}$ and let $f : A_1 \rightarrow A_2$ and $h : A_2 \rightarrow A_3$ and $k : A_3 \rightarrow A_4$ be *formal arrows*, then $(k \circ h) \circ f = k \circ (h \circ f)$."

"Let A_1 be $\mathfrak{C}$, then there exists a *formal arrow* $Id : A_1 \rightarrow A_1$ such that for all A_2 (being $\mathfrak{C}$) and for all $f : A_1 \rightarrow A_2$ and $j : A_2 \rightarrow A_1$ *formal arrows*, it holds that $Id \circ f = f$ and $j \circ Id = j$."[33]

Further axioms involving formal arrows and described in terms of the membership relation are not included mainly because they are specific and cannot be found among a large collection of conceptual environments. For instance, the fact that for two objects A_1 and A_2 there is always a "product" $A \times B$, cannot be integrated into this notion because there are specific conceptual environments where this is not the case (e.g., the category of fields).

However, if there are large sub-collections of conceptual environments (e.g., specific mathematical theories) fulfilling larger families of similar axioms, then an identical analogical meta-procedure as before will generate sub-notions like (co-)complete or abelian categories [41].

It is worth noting that our priority is to find a cognitively-inspired reconstruction of the concept of category which shows in more detailed the most prominent cognitive abilities involved and how they are combined in the most efficient way. This has a bigger priority than the aim of describing the exact way in which this notion was invented by S. Eilenberg and S. Maclane [12]. Nonetheless, in this special case both reconstructions are similar from a cognitive perspective.

One of the most fundamental exemplifications of the notion of category is $\mathscr{S}\!ets$, which consists of sets (in the context of ZFC) as objects, functions between them as formal arrows, and composition of functions as the operation $\circ$. This concept emerges by means of the formal metaphor *sets are objects of a category*, given by the former description.

---

[33] We write here (using paraphrasing) the seminal way in which the axioms of a category emerge from the conceptual commonalities of the corresponding axioms lying in each of the initial conceptual surroundings. Therefore, we write "being $\mathfrak{C}$," which is a schematic way of abstracting the commonality between the corresponding underlying statements "is a group," "is a vector space," or "is a topological space."

11.18 Functors Between Categories

Similar to the former construction, the notion of functor between categories emerges by means of an analogical procedure between several situations starting from a mathematical concept (e.g., a field, a (pointed) topological space, a differential manifold), one constructs a second mathematical concept by using generic procedures that are invariant from the specific concept chosen and only depend on more general principles of the corresponding conceptual environments (e.g., the corresponding constructions of the Galois group, the fundamental group, the tangent and cotangent bundles (or the DeRham cohomology groups), respectively).[34]

Thus, after doing a similar cognitive procedure as in the case of the notion of categories, we formally abstract an axiomatic description of the form:

"For any A being $\mathfrak{C}_1$ (i.e., belonging to $\mathfrak{C}$), $F(A)$ is $\mathfrak{C}_2$."

"If A_1 and A_2 are $\mathfrak{C}_1$ and $h : A_1 \rightarrow A_2$ is a *formal arrow* (in $\mathfrak{C}_1$), then $F(h) : F(A_1) \rightarrow F(A_2)$ is a *formal arrow* (in $\mathfrak{C}_2$)."

"For any A being $\mathfrak{C}_1$, $F(Id_{A_1}) = Id_{F(A_1)}$."

"If A_1, A_2, and A_3 are $\mathfrak{C}_1$ and $h_1 : A_1 \rightarrow A_2$ and $h_2 : A_2 \rightarrow A_3$ are *formal arrow* (in $\mathfrak{C}_1$), then $F(h_2 \circ h_1) = F(h_2) \circ F(h_1)$."[35]

Again, the resulting numbers of axioms is so small due to the considerably high number of conceptual environments that such a notion wants to cover. By reducing the quantity of such conceptual environments, one can generate similarly sub-notions like faithful and flat functors [30, 31].

11.19 Polynomial Rings, (Finitely) Generated Algebras Over a Field and Quotients of Commutative Rings with Unity

From a physical perspective, we wish to understand complex phenomena in terms of quite simple, irreducible and atomic principles and structures, e.g., quantum of action (Plank's constant) [23].[36] In a similar way, our minds implicitly (and sometimes unconsciously) take a *quantized meta-approach* to (mathematical) creation/invention.

Effectively, let R be a commutative ring (with unity). Let us imagine that we want to understand algebraic properties of R from a formally "quantizable" bottom-up perspective, explicitly using both binary operations of R. In other words, we would

---

[34]This reconstruction corresponds essentially with the historical way in which such a notion emerged [12].

[35]This is the explicit description of covariant functors; the dual version for contravariant functors is almost identical and based on "dual" instances appearing in several conceptual environments. For the sake of our exposition we will focus exclusively on the covariant construction.

[36]See also the introductory discussion done in Chap. 5.

like to start with an *m.-s. quantum of element of* R and subsequently construct more sophisticated m.-s. configurations based on such an initial m.-s. "quantum" unit.

So, we explicitly declare a new m.-s. symbol, let us say "x," to be an (algebraic) m.-s. quantum for R; i.e., x is a m.-s. unit (in general of level 1 (see Sect. 10.6 Chap. 10)) in-decomposable and morpho-syntactically independent regarding its binary operations of R.[37]

Having as starting point x, one constructs gradually all the corresponding m.-s. configurations (involving it) with the help of (a fixed collection of) generic exemplifications $\{a_0, a_1 \cdots, a_n\}$ of R. Thus, exactly as in the case of the m.-s. graphs for the ring of polynomials with integer coefficients (see Sect. 10.6 Chap. 10), one gets a generic expression of the form[38]

$$\sum_{i=0}^{n} a_i x^i.$$

By adding the generic condition $a_0, \cdots, a_n \in R$ to this m.-s. configuration and taking conceptual lining, we obtain (cognitively) the initial set needed for generating the notion of ring of polynomials in one variable with coefficients in R, i.e., $R[x]$. The corresponding extended binary operations are generated in the natural way based on the m.-s. independence of x and in the former generic m.-s. representation.[39]

Another way of generating this concept is starting with the two (input) notions of commutative ring with unity and ring of polynomials in one variable with integer coefficients and doing a formal conceptual blending between them having as a generic space the concept of (the commutative ring with unity of) the integer numbers, together with the natural conceptual morphisms.

Further, we can consider a finite number of m.-s. quantum units $\{x_i\}_{i \in \{1, \cdots, k\}}$, and then we generate as before a generic expression of the form

$$\sum_{\underline{j}=(j_1, \cdots, j_k)}^{\text{finite}} a_{\underline{j}} \prod_{r=1}^{k} x_r^{j_r},$$

similarly, we generate the notion of the ring of polynomial in finitely many variables over R, denoted by $\mathscr{P}ol_{FinVar}$.

---

[37]These conditions mean that neither x nor any m.-s. configuration involving it and generic exemplifications of R can be re-expressed non-trivially in terms of other m.-s. configurations involving x, any elements of R or/and generic exemplifications of elements of R. Trivial re-expressions in this context are the ones consisting of replacements involving only explicit elements of R.

[38]One starts gradually with expressions of the form $x, a_0, a_1 x^1, x^2, a_2 x^2 + x$ until one develops a generic representation valid for any polynomial.

[39]This meta-verification can be left as an exercise to the interested reader.

Finally, let us consider an ideal I in $R[x_1, \cdots, x_k]$, we will generate cognitive-metamathematically the quotient ring $R[x_1, \cdots, x_k]/I$. In fact, we will do the corresponding construction for a general commutative rings with unity R and an ideal $I \subseteq R$.

In particular, we want to extend the qualitative formal aspects of the classic operation of division among the ring of integers and the intuitions about the elementary construction of the integers modulo n to arbitrary commutative rings of unity. For example, the fact that the smallest non-trivial commutative ring with unity $\mathbb{Z}/2\mathbb{Z}$ (i.e., $1 \neq 0$) can be identified with the elementary partition of $\mathbb{Z}$ as the even and the odd numbers based on their residue modulo 2, is implicitly supported by the fact that we conceptually identify all the integers with residue 0 and all the integers with residue 1 modulo 2.

In general, let us do conceptual identifications of the collection of all elements of I, $\langle a \rangle_{a \in I}$ regarding I. Moreover, we want to extend such identification to other subsets of R. So, let us consider a generic exemplification r of an element of R. A natural conceptual blend between the mathematical structures r and I is given by the addition operation; i.e., one wants to construct a structure of the form $a + I$. Thus, a straightforward manner of doing this is to take a generic exemplification i of I, generating the generic m.-s. configuration $a + i$, adding the condition $i \in I$ and taking conceptual lining of the resulting juxtaposition $\ulcorner a + i, i \in I \urcorner$. Further, we do again a conceptual identification of the collection of all elements of $a + I$ regarding $a+I$. Lastly, we apply conceptual lining to the m.-s. expression $\ulcorner a+I, a \in R \urcorner$ for obtaining the central set behind this quotient construction; i.e., R/I. The corresponding binary operations are extended through the simple metaphor

$$\Delta = \ulcorner a = b+c \rightrightarrows a+I = (b+I) +_I (c+I), a = b * c \rightrightarrows a+I = (b+I) *_I (c+I) \urcorner,$$

where $+_I$ and $*_I$ are the new binary operations to be defined in the quotient structure and a, b, and c are generic exemplifications of R.

By contrast, it is an elementary meta-construction to generate the notion of commutative algebra over a field (denoted by $\mathscr{C}omm Alg_{Field}$) from the concepts of commutative rings with unity, field, and the exemplification of the real plane $\mathbb{R}^2$, considered as a ring enriched with the standard scalar (component-wise) multiplication by real numbers.

So, the notion of finitely generated algebra over a field can be generated as the super-concept of $\mathscr{R}_{CommUnit}$ whose exemplifications can be conceptually identified with the formal quotient of an exemplification of a polynomial ring in several variables ($\mathscr{P}ol_{FinVar}$) through an exemplification of an ideal of it ($Ideal(\mathscr{P}ol_{FinVar})$) regarding $\mathscr{C}omm Alg_{Field}$. Let us denote this concept by $FG\mathscr{C}omm Alg_{Field}$.

11.20   Algebraic Sets

We can obtain the classic concept of algebraic sets starting with (the analytical version of) one of the most omnipresent structures in mathematics, whose origin goes back even before Euclid's Elements, i.e. the circle

$$C = \{(x_1, x_2) \in \mathbb{R}^2 : P_1 = x_1^2 + x_2^2 - 1 = 0\}.$$

Let us denote the former mathematical structure as $\mathscr{C}_{ir}(P_1)$, where P_1 is the exemplification $x_1^2 + x_2^2 - 1$ of the ring of polynomial in two variables with real coefficients. Firstly, let us apply an exemplified generalization to $\mathscr{C}_{ir}(P_1)$ regarding the exemplification P_1. We obtain the concept $\mathscr{C}_{ir2} = \mathscr{C}_{ir}(g_1(x_1, x_2))$, where $g_1(x_1, x_2)$ is a generic exemplification of $\mathbb{R}[x_1, x_2]$. Secondly, let us write the former concept as $\mathscr{C}_{ir2}(\mathbb{R})$, where $\mathbb{R}$ is considered as an exemplification of the notion of mathematical field. Again, we apply an instantiated generalization of the former concept to obtain the notion $\mathscr{C}_{ir3} = \mathscr{C}_{ir2}(K)$, where K denotes a generic exemplification of the notion of field. Further, based on the meta-generation of the polynomial rings in finitely many variables, let us write the last concept as $\mathscr{C}_{ir3}(2)$, where 2 denotes a concrete exemplification of $\mathbb{N}$ used an index quantifying the number of variables of the corresponding polynomial ring. So, by doing the conceptual generalization of $\mathscr{C}_{ir3}(2)(\mathbb{N})$ (considered as depending on $\mathbb{N}$ due to the appearance of the exemplification 2) consisting of replacing (the exemplification) 2 by a generic exemplification n of $\mathbb{N}$, i.e. $CG_{2 \to n}(\mathscr{C}_{ir3}(2)(\mathbb{N}))$, we obtain a generic exemplification of the notion of algebraic set defined by a single polynomial equation. Let us denote this concept temporally by $\mathscr{C}_{ir4}(g_\alpha(x_1, \cdots, x_n))$. Now, let us use a finite chain of partial duplications of the polynomial m.-s. structure $g_\alpha(x_1, \cdots, x_n)$ combined conceptually in a conjunctive way, namely, $g_1(x_1, \cdots, x_n) \wedge \cdots \wedge g_m(x_1, \cdots, x_n)$.[40] Thus, by doing a conceptual replacement of the last expression into $g_\alpha(x_1, \cdots, x_n)$, we obtain a generic exemplification of the concept of algebraic set. Finally, by doing a generic generalization of the former structure we generate the classic notion of algebraic set denoted here by $\mathscr{A}_{lg}\mathscr{S}_{et}$.

11.21   Ideals of Polynomials Associated to Algebraic Sets

Let $R = k[x_1, \cdots, x_n]$ be the ring of polynomials in $n \in \mathbb{N}$ variables with coefficients in a field k. Let us consider one of the most elementary and well-known

---

[40]In this case, the partial duplication leaves all the symbolic units involving the variables intact, i.e. $x, 1, \cdots, n$.

ideals of R; i.e. the ideal of polynomials in R with zero as independent term,[41] namely,

$$I = \{f \in k[x_1 \cdots, x_n] : f(0, \cdots, 0) = 0\}.$$

In other words, if we consider the single point $(0, \cdots, 0)$ as the algebraic set defined by the polynomials $\{x_1, \cdots, x_n\}$ (denoted by V_0), then we can write I in a slightly different way:

$$I = \{f \in k[x_1 \cdots, x_n] : f(a_1, \cdots, a_n) = 0 \text{ for all } (a_1, \cdots, a_n) \in V_0\}.$$

Let us denote this mathematical structure by $I(V_0)$, where V_0 is considered as an exemplification of $\mathscr{A}_{lg}\mathscr{S}_{et}$. By doing a morpho-syntactic generalization of $I(V_0)$, replacing V_0 by a generic exemplification V of an algebraic set, we obtain a (generic) mathematical structure denoted by $I(V)$. Finally, we apply generic generalization to the former structure to obtain the notion of ideal of vanishing polynomials associated to an algebraic set, denoted by $\mathscr{I}(\mathscr{A}_{lg}\mathscr{S}_{et})$.

11.22 Rings of Coordinates of Algebraic Sets

Based on the former algebraic set consisting of the $n-$dimensional origin $(0, \cdots, 0) \in k^n$, we take as a starting structure the corresponding evaluation function (see Sect. 11.7) $F : k[x_1 \cdots, x_n] \to k$ generated by such a single point; i.e., for all $p \in k[x_1, \cdots, x_n]$, $F(p) = p(0, \cdots, 0)$. F has the special property that all the polynomials that are sent to zero by it, are exactly the polynomials belonging to the ideal I defined in the former section, in other words, polynomials with zero independent term. Now, let us consider the quotient ring (see Sect. 11.19) $k[I] = k[x_1, \cdots, x_n]/I$. Thus, we can see a clear conceptual identification between $k[I]$ and k (i.e., algebraically speaking, both structures are isomorphic as $k-$algebras). In this case, this means that two polynomial functions in $k[I]$ are considered "equal" if and only if their values at the origin coincide. Another way of expressing this fact is saying that $k[I]$ is the ring consisting of all the possible functions that are given by polynomials and can be distinguished essentially through their value at the origin (in fact, for any element in k exactly one of such a function can be constructed). So, doing basically the same cognitive metamathematical construction as in the former section, but having as a starting point the structure $k[I]$, we can generate the concept of ring of coordinates of an algebraic set V, being denoted here by $\mathscr{R}_{in}\mathscr{C}_{oor}(\mathscr{A}_{al}\mathscr{S}_{et})$.

---

[41]This structure can be generated directly from (the structure) R by combining a generic exemplification g of elements of R and combining it conceptually with the condition $g(0, \cdots, 0) = 0$, which is again a product of the conceptual identification of two more elementary structures.

11.23 Pre-Sheaves

First of all, let $\mathcal{T}$ be the concept of topological space and let $GE(\mathcal{T})$ be the generic exemplification of $\mathcal{T}$. Now, let us define the metaphor $GE(\mathcal{T})$ *is a category*, given by

$$\lceil \text{ open sets } \rightrightarrows \text{ objects, } \subseteq \rightrightarrows \rightarrow_1, \nsubseteq \rightrightarrows \rightarrow_2 \rfloor,$$

where $\rightarrow_1$ denotes a generic symbol for inclusion maps; i.e., $\rightarrow_1$ can be applied to the parameters (e.g., sets) A and B as $A \rightarrow_1 B$ if and only if $A \subseteq B$ and in such a case $\rightarrow_1$ would denote the inclusion map. In addition, $\rightarrow_2$ denotes a generic symbol for the empty map; i.e., $\rightarrow_2$ can be applied to the parameters A and B if and only if $A \nsubseteq B$ and in this case $A \nsubseteq B$ denotes the empty map between them.

The conceptual scope of this metaphor generates the concept $\mathcal{T}_{Cat}$ of *topological spaces as categories*.

Second, let $\mathcal{C}_{ontFunc}(Cat_1, Cat_2)$ be the concept of contravariant functor from the category Cat_1 to Cat_2. Then, the notion of *pre-sheaves with values on the category* Cat_2 is the conceptual particularization

$$\mathcal{C}_{ontFunc}(\mathcal{T}_{Cat}, Cat_2).$$

Even more specifically, if $\mathcal{R}$ denotes the category of commutative rings with unity, then the notion of *pre-sheaves of commutative rings with unity* is the conceptual particularization

$$\mathcal{C}_{ontFunc}(\mathcal{T}_{Cat}, \mathcal{R}).$$

11.24 Sheaves with Values on the Category of Sets

First of all, let us consider the real line $\mathbb{R}^1$ and let us fix an open set $U \subseteq \mathbb{R}$, or, in other words, let us consider a generic exemplification of the notion of open subset of $\mathbb{R}$. Now, suppose that we want to compare U and $\mathbb{R}$ in a functional way; i.e., let us consider a generic exemplification of the concept of function from U to $\mathbb{R}$:

$$f : U \rightarrow \mathbb{R}.$$

Assume that we would like to understand the former abstract exemplification from a more global perspective; i.e., we are interested in formalizing the concept having the former m.-s. configuration as conceptual substratum. Then, the next natural step is to find the conceptual lining of such a representation within the foundational framework, for example, of ZFC set theory:

$$CL(f : U \rightarrow \mathbb{R}) = \mathbb{R}^U = Hom(U, \mathbb{R}).$$

Now, let us consider a second generic exemplification V of the notion of open subset of $\mathbb{R}$. Suppose that we can compare such exemplifications through the containment relation; i.e. it holds

$$V \subseteq U;$$

or, in other words, let us write such a conceptual comparison in a functional way:

$$i : V \rightarrow U.$$

The m.-s. configurations $f : U \rightarrow \mathbb{R}$ and $i : V \rightarrow U$ suggest, implicitly through the m.-s. commonality given by "U," the formation of the corresponding composition of functions:

$$f \circ i : V \rightarrow \mathbb{R}.$$

After applying a m.-s. simplification to $f \circ i$ towards a generic expression (e.g., h) and then applying conceptual lining, the former expression generates the notion of

$$Hom(V, \mathbb{R}).$$

In the context of real analysis, the former composition simply describes the restriction of functions to open subsets of the domain. Therefore, an additional notation is normally employed for this kind of construction:

$$f \circ i = f_{|V}.$$

Globally, the former kind of functional is usually denoted by a m.-s. configuration using the domains U and V involved, e.g.

$$\rho_{U,V} : Hom(U, \mathbb{R}) \rightarrow Hom(V, \mathbb{R}).$$

Let us denote by $Hom(\mathbb{R}^1, \mathbb{R})$ the (global) conceptual lining of the former structures of $Hom(-, \mathbb{R})$, varying over all the open subset of $\mathbb{R}$ and including all the corresponding functionals (denoted generically by ρ_-).

Inspired by the mechanisms generating the notion of topological space as a category, we define the metaphor $Hom(\mathbb{R}^1, \mathbb{R})$ *is a category*, given by

$$\lceil Hom(-, \mathbb{R}) \rightrightarrows \text{objects}, \subseteq \rightrightarrows \rightarrow_1, \not\subseteq \rightrightarrows \rightarrow_2 \rceil,$$

with similar interpretations for the symbols $\rightarrow_1$ and $\rightarrow_2$ as in the previous section.

Let us consider the following exemplification of the concept $\mathscr{C}_{ont\,Func}(\mathscr{T}_{Cat}, Cat_2)$:

$$\mathscr{C}_{ont\,Func}(\mathbb{R}^1, Hom(\mathbb{R}^1, \mathbb{R}));$$

and let us define the canonical exemplification of this (double) exemplification given by the functor sending each open subset $U \subseteq \mathbb{R}$ to $Hom(U, \mathbb{R})$. This exemplification can be named *the pre-sheaf of real functions from open real subsets*, denoted by

$$Hom^{(PreSh)}(\mathbb{R}^1, \mathbb{R}).$$

Let us consider two generic exemplifications U_1 and U_2 of open real subsets, and two generic exemplifications $f_1 : U \to \mathbb{R}$ and $f_2 : U_2 \to \mathbb{R}$ of $Hom(U_1, \mathbb{R})$ and $Hom(U_2, \mathbb{R})$, respectively.

We will explore the existence of canonical conceptual blends for the generic expressions $f_1 : U \to \mathbb{R}$ and $f_2 : U_2 \to \mathbb{R}$ within the conceptual context given by the former pre-sheaf.

Firstly, a (typical) conceptual combination $U \cap U_2$ between U and U_2 suggests the generation of a conceptual comparison between the m.-s. configurations $f_{1|U_1 \cap U_2}$ and $f_{2|U_1 \cap U_2}$.

Let us suppose additionally that one can compare the former expressions by means of the equality relation; i.e.

$$f_{1|U_1 \cap U_2} = f_{2|U_1 \cap U_2}.$$

So, the former equation exhorts to think about the generic space (within the setting of conceptual blending, Chap. 7) as an expression of the form

$$g : U \cap U_2 \to \mathbb{R},$$

fulfilling $g = f_{1|U_1 \cap U_2} = f_{2|U_1 \cap U_2}$.

Seminal principles governing conceptual blending like topology, unpacking, and integration (see [15]) imply that a (minimal) blend for this conceptual V–diagram should be given by a m.-s. configuration of the form $h : U \cup U_2 \to \mathbb{R}$ (unpacking), and fulfilling the conditions $h_{|U_1} = f_1$ and $h_{|U_2} = f_2$. Such a function exists in $Hom^{(PreSh)}(\mathbb{R}^1, \mathbb{R})$, and is defined as usual point-wise over $U_1 \cup U_2$ and coherently over $U_1 \cap U_2$ due to the condition $f_{1|U_1 \cap U_2} = f_{2|U_1 \cap U_2}$. Such a blend is clearly unique if we impose the former requirements.

Let us denote by $A_{\{1,2\}}$ the axiom, written for example in a many-sorted first-order framework, describing the existence of such a unique blending under the former conditions, and where $\{1, 2\}$ denotes the index set for the m.-s. expressions involving the corresponding open real subsets $\{U_1, U_2\}$.

Let $A_{\{1,2,3\}}$ be the axiom describing the existence of an identical kind of blend, but for three open real subsets U_1, U_2, and U_3; i.e. if the pair-wise restrictions of functions defined on each U_i coincide, then there exists a unique global function defined over the common union and coinciding with each of the original ones on each U_i. So, one can again directly verify that $A_{\{1,2,3\}}$ is a true sentence over $Hom^{(PreSh)}(\mathbb{R}^1, \mathbb{R})$. Let us replace $\{1, 2, 3\}$ by a generic index set and let us denote by A_I the corresponding m.-s. generalization of $A_{\{1,2,3\}}$ on I. Then, it is an elementary exercise to prove that A_I is also a true sentence over $Hom^{(PreSh)}(\mathbb{R}^1, \mathbb{R})$. Moreover, A_I describes the existence of a unique kind of global conceptual blending for arbitrary families of m.-s. configurations of the form $f_i : U_i \to \mathbb{R}$, bounded by the conditions $f_{i|U_i \cap U_j} = f_{j|U_i \cap U_j}$ for all $i, j \in I$. Let us denote by $Hom^{(Cat+Blen)}(\mathbb{R}^1, \mathbb{R})$ the mathematical structure $Hom^{(PreSh)}(\mathbb{R}^1, \mathbb{R})$ plus the axiom A_I.

Finally, with the aim of generalizing the former construction to arbitrary pre-sheaves with values in the category $\mathscr{S}_{ets}$, we describe the axiom A_I in the form:

$$A_I = A_I(\mathbb{R}^1, Hom^{(Cat)}(\mathbb{R}^1, \mathbb{R}), Hom^{(PreSh)}(\mathbb{R}^1, \mathbb{R})).$$

In this way, we emphasize the dependence of A_I on a topological space, a category, and a contravariant functor between them.

Now, let us consider the m.-s. generalization of A_I to arbitrary topological spaces, the category of sets and contravariant functors between them:

$$A' := Mor\,Syn(A_I) = A_I(\mathscr{T}_{Cat}, \mathscr{S}_{ets}, \mathscr{C}_{ont\,Func}(\mathscr{T}_{Cat}, \mathscr{S}_{ets})).$$

In this way, we obtain the notion of *sheaves over the category* $\mathscr{S}_{ets}$ (see, for instance, [25, Ch. 2]) as the conceptual strengthening of $\mathscr{C}_{cont\,Funt}(\mathscr{T}_{Cat}, \mathscr{S}_{ets})$) adding the axiom A'; i.e.,

$$\mathscr{S}_{heaves}(\mathscr{S}_{ets}) = \langle \mathscr{C}_{ont\,Func}(\mathscr{T}_{Cat}, \mathscr{S}_{ets}), A' \rangle.$$

Due to the fact that A' describes both the existence and the uniqueness of the (global) blended function, it includes exactly both the locality and the gluing standard conditions for being a sheaf [25, Ch. 2].

Remark 11.1 Note that one can also reconstruct the seminal concept of sheaves with values on the category of sets more in accordance with the standard manner in which it gradually emerged from specific interconnections between mathematical structures coming originally from algebraic and differential topology, and algebraic geometry, among others [32]. In other words, one can proceed as during the cognitive generation of the (meta-)notion of category (see Sect. 11.17), by abstracting a formal analogy between particular statements describing the unique existence of (global) functional blends for (complex/differential) manifolds, vector bundles, rings of coordinates of (classic) varieties, and spaces of (continuous/smooth) functions [24, 26, 32, 44, 45].

We choose the conceptual generation presented in this section due to their simplicity and to illustrate the conceptual power that simple exemplifications like the pre-sheaf of real functions over open real subsets possess to generate foundational notions as the one of sheaves with values on the category of sets.

Remark 11.2 (Notions with a Higher Technical Sophistication)

In the following sections, we will continue with the cognitive meta-generation of well-known mathematical notions with a higher level of sophistication and abstraction.[42]

11.25   Stalk of a Pre-Sheaf at a Point

We begin with the elementary structure of the real line $\mathbb{R}^1$ with the standard topology T. Let $\mathscr{C}_{ons\,Func}(T, \mathbb{R}^1)$ be the pre-sheaf of constant functions defined over open sets. A notion closely related with $\mathscr{C}_{cons\,Func}(T, \mathbb{R}^1)$ is the (pre-)sheaf of locally constant real functions, denoted here by $\mathscr{L}_{oc}\mathscr{C}_{onst\,Func}(\mathbb{R}^1)$, which is defined for an open subset $U \subseteq \mathbb{R}$ as follows:

$$\mathscr{L}_{oc}\mathscr{C}_{onst\,Func}(U, \mathbb{R}) := \{f : U \to \mathbb{R} : f \text{ satisfies the condition } *\},$$

where the condition $(*)$ states that for all $u \in U$, exists an open set $V \in T$ and a constant function $g_u \in \mathscr{C}_{ons\,Func}(V, \mathbb{R})$ such that for all $v \in V$, $f(v) = g_u(v)$.

Suppose that we are interested in the behavior of the former functions at a (fixed) exemplification of a real point $x \in \mathbb{R}$. We will identify two locally constant functions $f_1 : U_1 \to \mathbb{R}$ and $f_2 : U_2 \to \mathbb{R}$, if they have the same value at x; i.e. $f_1(x) = f_2(x)$. In addition, conceptually speaking, we do not focus on the domains of definition of these functions in further detail, the only relevant condition that we have to preserve about their domains is that both should contain the point x. So, the mathematical formalization of the former idea of restricting the focus of attention to a single point is naturally obtained by considering the (generic) conceptual combination of all locally constant functions defined over open neighborhoods of x given by the set-theoretical union, i.e.,

$$W := \bigcup_{U \subseteq \mathbb{R},\, p \in U} \mathscr{L}_{oc}\mathscr{C}_{onst\,Func}(U, \mathbb{R}),$$

and doing the conceptual identification through the generation of an equivalent relation $\approx$ on W defined by the following condition:

for all $f_1 : U_1 \to \mathbb{R}$, $f_2 : U_2 \to \mathbb{R} \in W$, $f_1 \approx f_2$ if $f_1(x) = f_2(x)$.

Due to the specific nature of the space of functions considered here, we see that the former condition holds if and only if there exists a open V contained in both domains (i.e., $V \subseteq U_1 \cap U_2$) such that $f_{1|V} = f_{2|V}$. The former condition suggests that we can extend the study of the behavior of (not necessarily locally constant) functions not only restricted just to x but also allowing neighbor points near from x.

---

[42]The non-specialist reader is strongly encouraged to consult the references given at the beginning of this chapter and to internalize them before even reading the present meta-generations.

Effectively, we can now formalize the fact that two real-valued functions behave the same near the point x if there exists an open neighbor V in the intersection of their domains where both functions coincide. Let us denote by

$$(\mathscr{L}_{oc}\mathscr{C}_{onstFunc}(\mathbb{R}^1))_x$$

the former conceptual construction with the updated condition describing the functional identification in terms of a common open set.

Let us consider the conceptual generalization

$$CG_{(\mathscr{L}_{oc}\mathscr{C}_{constFunc}(\mathbb{R}^1),x)\to((P_{reSheaf})_{[g]},\alpha)}((\mathscr{L}_{oc}\mathscr{C}_{onstFunc}(\mathbb{R}^1))_x),$$

where $((P_{reSheafBasis})_{[g]}, \alpha)$ is a generic exemplification of the notion of pre-sheaf of sets together with a distinguished (generic) point α.

By considering the generic generalization of the former conceptual space, we obtain the notion of stalk of a pre-sheaf at a point.

11.26 Sheaf Associated to a Pre-Sheaf (Described Over a Basis)

Let us start with the elementary structure of the real line $\mathbb{R}^1$ with the standard topology T and with the typical basis Θ consisting of open intervals. Let $\mathscr{C}_{onsFunc}(\Theta, \mathbb{R}^1)$ be the pre-sheaf of constant functions defined over open intervals. Let us consider here again the sheaf of locally constant real functions, whose definition can be slightly re-written for an open subset $U \subseteq \mathbb{R}$ as follows:

$$\mathscr{L}_{oc}\mathscr{C}_{onstFunc}(U, \mathbb{R}) := \{f : U \to \mathbb{R} : f \text{ satisfies the condition } *\},$$

where the condition $(*)$ states that for all $u \in U$, exists an basic open $V \in \Theta$ and a constant function $g_u \in \mathscr{C}_{onsFunc}(V, \mathbb{R})$ such that for all $v \in V$, $f(v) = g_u(v)$.

Let us note that there is a natural conceptual identification between the field of the real numbers $\mathbb{R}$ and the stalk of $\mathscr{C}_{consFunc}(\Theta, \mathbb{R}^1)$ at any fixed point $a \in \mathbb{R}$.

If we want to generalize the former (paired) construction

$$\left\langle \mathscr{C}_{consFunc}(\Theta, \mathbb{R}^1), \mathscr{L}_{oc}\mathscr{C}_{onstFunc}(\mathbb{R}^1) \right\rangle$$

to arbitrary pre-sheaves, we should be able to replace the common codomain of all the constant functions (i.e., $\mathbb{R}$) with the help of former conceptual identification between $\mathbb{R}$ and $(\mathscr{C}_{onstFunc}(\Theta, \mathbb{R}^1))_a$, for any $a \in \mathbb{R}$. Explicitly, for any open set $W \subseteq \mathbb{R}$, and any point $w \in W$, the value $f(w)$ should be defined mainly in terms of structures depending on the pre-sheaf of rings (defined over a (fixed) basis for the grounding topological space). So, the most natural structure based on those assumptions is the stalk of such a pre-sheaf on w. Therefore, the general codomain should contain authentic copies of each of the stalks on all the points defined on the corresponding open set.[43] Thus, we form a "generic conceptual combination" of all these (localized) spaces materialized in the form of the disjoint union of all of them. For instance, in our initial conceptual space, let us replace $\mathbb{R}$ as common codomain of the functions in $\mathscr{L}_{oc}\mathscr{C}_{onstFunc}(W)$ by

$$\coprod_{w \in W} (\mathscr{C}_{onstFunc})_w.$$

Let us denote the updated structure by

$$\mathscr{L}_{oc}\mathscr{C}_{onstFunc}(\mathscr{C}_{constFunc}(\Theta, \mathbb{R}^1)).$$

Now, let us consider the conceptual generalization

$$CG_{\mathscr{C}_{constFunc}(\Theta, \mathbb{R}^1) \to (P_{reSheafBasis})_{[g]}}(\mathscr{L}_{oc}\mathscr{C}_{onstFunc}(\mathscr{C}_{constFunc}(\Theta, \mathbb{R}^1))),$$

where $(P_{reSheafBasis})_{[g]}$ is again a generic exemplification of the notion of pre-sheaf of rings defined over a basis.

The last conceptual space defines a generic sheaf given locally (on a basis) by sections of the generic pre-sheaf of rings. Therefore, by applying a generic generalization to the former space, we obtain exactly the notion of sheaf associated to a pre-sheaf described over a basis. Finally, note that exactly the same kind of method can be applied to the generation of the notion of sheaf associated to a pre-sheaf without the restriction of being defined only over open sets belonging to a basis of the (grounding) topological space.[44]

11.27   (Locally) Ringed Spaces

Note that one can do essentially the same cognitive generation done to obtain $\mathscr{S}_{heaves}(\mathscr{S}_{ets})$, but the category $\mathscr{S}_{ets}$ by the category $\mathscr{R}_{ings}$.[45]

---

[43]Further details of this cognitive generation can be done with the machinery developed in the former chapters and are left as an exercise for the reader.

[44]After reading the whole chapter, the more curious reader can do the enlightening exercise of finding cognitive meta-generations of the structures non-explicitly meta-generated assuming only the basic notions and properties about sets and the natural numbers, guided by the classic formal constructions of the subsequent mathematical entities (as shown in [17], for example).

[45]This concept is basically obtained from $\mathscr{R}_{CommUnit}$ through an elementary conceptual weakening eliminating the commutativity condition and the existence of the multiplicative unity.

In this case, the corresponding "restrictions" are given by morphisms of rings. The new notion $\mathscr{S}_{heaves}(\mathscr{R}_{ings})$ is exactly the notion of ringed space used widely in modern mathematics. So, we will denote this concept by $\mathscr{R}_{inged}\mathscr{S}_{paces}$.

Finally, be strengthening the former notion with the condition that any stalk (Sect. 11.25) is a local ring (Sect. 11.14), we obtain the notion of locally ringed space, that we will denote by $\mathscr{L}_{ol}\mathscr{R}_{inged}\mathscr{S}_{paces}$.

11.28 Algebraic Sets as Ringed Spaces

In order to present a general (cognitive) generation of the (classic) approach in algebraic geometry of considering algebraic sets as ringed spaces, we will use additional (elementary) notions and facts, whose generation is left to the reader as an exercise and that can be obtained in a similar manner to the former constructions presented in this chapter.

Firstly, if k is an algebraic closed field and $V \subseteq k^n$ is an algebraic set, then, due to Hilbert's Nullstellensatz [25, Ch.1], there exists a natural conceptual identification between V and (the points of) the spectrum of maximal ideals of $K[V]$, denoted by $\mathrm{MaxSpec}(k[V])$, considered with the natural topology derived from the Zariski topology of $\mathrm{Spec}(k[V])$. So, through the former identification, V acquires the classic topology T_V given by closed sets of the form $W = V(J) \subseteq V$, where J is an ideal of $k[V]$ (which corresponds to an ideal J' of $k[x_1, \cdots, x_n]$ containing $I(V)$). Secondly, T_V has a (distinguished) basis consisting of the special open sets $D(f) = V \setminus V(\{f\})$, for some $f \in k[x_1, \cdots, x_n]$. Thirdly, there is an additional conceptual identification between (the topological space) $D(f) \subseteq V$ and $\mathrm{MaxSpec}(k[V]_{\bar{f}})$, where $k[V]_{\bar{f}}$ denotes the localization of $k[V]$ of the element $\bar{f}$ given by the class of f in $k[V]$.

Now, let us consider the metaphor *algebraic sets considered as topological spaces enriched with their ring of coordinates and considered with the distinguished basis are pre-sheaves of rings*, given by

$$\kappa = \left[V \rightrightarrows k[V], D(f) \rightrightarrows k[V]_{\bar{f}} \right],$$

sending the restriction maps between basic open sets into the natural maps given between the corresponding localizations. So, the conceptual (scope) of κ describes a pre-sheaf of rings whose associated sheaf of rings over V (see Sect. 11.26) generates the notion of algebraic sets viewed as a ringed spaces. Let us denote this conceptual space as $\mathscr{R}_{inged}\mathscr{S}_{paces}(\mathscr{A}_{lg}\mathscr{S}_{et})$.

11.29 Affine Schemes

Let us extend the construction in the former section to arbitrary commutative rings with unity enriched with their prime spectrum (implicitly considered with the Zariski topology). In fact, let us consider the following partial metaphor from $\mathcal{R}inged\,\mathcal{S}paces\,(\mathcal{A}lg\,\mathcal{S}et)$ to $\mathcal{Z}ar\,\mathcal{T}op\,(\mathcal{R}CommUnit)$:[46]

$$\chi = \lceil k[V] \rightrightarrows \mathrm{Spec}(R), k[V] \rightrightarrows R \rceil .$$

To understand the meaning of the syntactic scope of χ regarding the ringed space structure of $(R, \mathrm{Spec}(R))$, let us focus on the "conceptual image" of t(he pre-sheaf of rings defined over the distinguished basis of $\mathrm{MaxSpec}(k[V])$. Explicitly, one can generate such a conceptual image by considering an extended form of the local metaphor κ used in the former section, namely,

$$\kappa' = \lceil \mathrm{Spec}(R) \rightrightarrows R, D(f) \rightrightarrows R_f \rceil$$

which is based on the conceptual identification between $D(f)$ and $\mathrm{Spec}(R_f)$. So, the sheaf of rings associated with the former pre-sheaf over the distinguished basis (Sect. 11.26) essentially generates the seminal notion of affine scheme[47] denoted here by $\mathcal{A}ff\,\mathcal{S}chemes$ (for domain-specific references, please see [14] and [2]).

11.30 Schemes

Let us consider as initial mathematical structure the real projective plane $P^2(\mathbb{R})$ viewed as an (algebraic) variety. We know that there is a covering of $P^2(\mathbb{R})$ by 3 open sets V_1, V_2, and V_3 such that the restriction[48] of $P^2(\mathbb{R})$ to each V_i ($i = \{1, 2, 3\}$) is conceptually identifiable (i.e., isomorphic) with the real affine plane $A^2(\mathbb{R})$.

Now, let us denote by

$$\mathcal{A}_3' := \mathcal{A}(P^2(\mathbb{R}), A^2(\mathbb{R})^{[1]}, A^2(\mathbb{R})^{[2]}, A^2(\mathbb{R})^{[3]})$$

---

[46] Let us keep in mind the conceptual identification between $k[V]$ and $\mathrm{Spec}(k[V])$. Moreover, we assume that both conceptual spaces included the notion of localization and that this metaphor simply identifies both concepts invariably.

[47] The formal notion of affine schemes as presented in our cognitive framework, is obtained after applying some elementary generic exemplifications (generalizations) and weakenings, which are left as a verification to the more curious reader.

[48] The notion of restriction in this context can be generated by means of suitable combinations of conceptual comparisons given through containment and local metaphors.

the formal axiom stating the former fact; i.e., there exists an open covering of $P^2(\mathbb{R})$ with 3 open sets V_1, V_2, and V_3 such that the restrictions of $P^2(\mathbb{R})$ are isomorphic to the corresponding (conceptual duplication of the notion of) affine real line, namely, $P^2(\mathbb{R})_{|V_i} \simeq A^2(\mathbb{R})^{[i]}$, for $i = \{1, 2, 3\}$.

First, let us consider the following conceptual generalization

$$Gen\mathscr{A}' := CG_{P^2(\mathbb{R}) \to GE(\mathscr{L}_{oc}\mathscr{R}_{inged}\mathscr{S}_{paces})}(\mathscr{A}'),$$

where $GE(\mathscr{L}_{oc}\mathscr{R}_{inged}\mathscr{S}_{paces})$ denotes a generic exemplification of a locally ringed space.

Second, let us apply the metaphor $\eta = \lceil A^2(\mathbb{R}) \rightrightarrows \mathrm{Spec}(\mathbb{R}[x_1, x_2]) \rfloor$ from the concept $\mathscr{L}_{oc}\mathscr{R}_{inged}\mathscr{S}_{paces}$ enriched with the axiom $Gen\mathscr{A}'$ to $\mathscr{L}_{oc}\mathscr{R}_{inged}\mathscr{S}_{paces}$. The conceptual scope of η regarding to $Gen\mathscr{A}'$ has the form

$Gen\mathscr{A}'' =$

$Gen\mathscr{A}'(\mathscr{L}_{oc}\mathscr{R}_{inged}\mathscr{S}_{paces}, \mathrm{Spec}(\mathbb{R}[x_1, x_2])^{[1]}, \mathrm{Spec}(\mathbb{R}[x_1, x_2])^{[2]}, \mathrm{Spec}(\mathbb{R}[x_1, x_2])^{[3]}),$

where $\mathrm{Spec}(\mathbb{R}[x_1, x_2])^{[i]}$ are conceptual duplications of $\mathrm{Spec}(\mathbb{R}[x_1, x_2])$, for $i = \{1, 2, 3\}$.

Third, let us generalize further the former conceptual duplications:

$$Gen\mathscr{A}''' = CG_{\mathrm{Spec}(\mathbb{R}[x_1, x_2])^{[i]} \to GE(\mathscr{A}_{ff}\mathscr{S}_{chemes})^{[i]} \text{ for } i = \{1,2,3\}}(Gen\mathscr{A}''),$$

where $GE(\mathscr{A}_{ff}\mathscr{S}_{chemes})^{[i]}$ denotes generic exemplifications of the notion of affine scheme, for $i = I_3 = \{1, 2, 3\}$.[49]

Fourth, let us rewrite in $Gen\mathscr{A}''' = Gen\mathscr{A}'''_{I_3}$ the occurrences of all the generic exemplifications $GE(\mathscr{A}_{ff}\mathscr{S}_{chemes})^{[i]}$, by adding a sub-formula of the form $\cdots (\exists x_i \in AffSchemes) \cdots$ in the suitable position, where $AffSchemes$ denotes a new sort describing affine schemes, and by replacing all the occurrences of $GE(\mathscr{A}_{ff}\mathscr{S}_{chemes})^{[i]}$ in $Gen\mathscr{A}'''_{I_3}$ by x_i, for all $i \in I_3$.[50] Fifth, let us generalize the former axiom for an arbitrary index set; i.e.

$$\mathscr{A}'''_I := CG_{I_3 \to I}(\mathscr{A}'''_{I_3}),$$

where I denotes a (generic exemplification of the notion of (index)) set.

Finally, the fundamental notion of scheme is obtained by strengthening the composed concept of locally ringed space and affine scheme with $\mathscr{A}'''_I$.

---

[49]We can assume that the axiomatization of the former axioms are given in a many-sorted first-order framework with enough sorts to reproduce all the intrinsic properties. We can also assume that I_3 is considered as an additional indexing sort.

[50]It is a straightforward calculation to verify that both statements are equivalent; i.e. semantically both describe the same condition.

An additional way of meta-generating this notion is by using a suitable form of generic conceptual blending (Sect. 10.10, Chap. 10) involving the notions affine schemes and isomorphism between locally ringed spaces.

Acknowledgments The author deeply thank Maria Batjacob, Ingrid Gonzalez, Johanna Guenther, Dorthe Berssenbuegge, Sindy Pinzón, Ana Carolina Sanchez, Adriana Jimenez, Jheny Quintero, July Jaramillo, and Juliana Fischer.

References

1. Apostol, T.M.: Introduction to analytic number theory. Springer Science & Business Media (2013)
2. Balaji, T.V.: An introduction to families, deformations and moduli. Universitätsverlag Göttingen (2010)
3. Beeson, M.J.: The mechanization of mathematics. In: Alan Turing: Life and legacy of a great thinker, pp. 77–134. Springer (2004)
4. Bloch, E.D.: The real numbers and real analysis. Springer-Verlag New York (2011)
5. Bloch, E.D.: The real numbers and real analysis. Springer Science & Business Media (2011)
6. Bou, F., Corneli, J., Gomez-Ramirez, D., Maclean, E., Peace, A., Schorlemmer, M., Smaill, A.: The role of blending in mathematical invention. Proceedings of the Sixth International Conference on Computational Creativity (ICCC). S. Colton et. al., eds. Park City, Utah, June 29-July 2, 2015. Publisher: Brigham Young University, Provo, Utah. pp. 55–62 (2015)
7. Cantini, A., Casari, E., Minari, P.: Logic and Foundations of Mathematics: Selected Contributed Papers of the Tenth International Congress of Logic, Methodology and Philosophy of Science, Florence, August 1995, vol. 280. Springer Netherlands (2013)
8. Chater, N., Vitányi, P.: Simplicity: A unifying principle in cognitive science? Trends in cognitive sciences **7**(1), 19–22 (2003)
9. Cox, D.A., Little, J., O'shea, D.: Using algebraic geometry, vol. 185. Springer Science & Business Media (2006)
10. Coxeter, H.S.M.: The real projective plane, vol. 1. Springer Science & Business Media (1992)
11. Edwards, H.M.: Galois theory., *GTM*, vol. 101. Springer-Verlag New York (1984)
12. Eilenberg, S., MacLane, S.: General theory of natural equivalences. Transactions of the American Mathematical Society **58**, 231–294 (1945)
13. Eisenbud, D.: Commutative Algebra with a View Toward Algebraic Geometry. GTM. Springer-Verlag (1995)
14. Eisenbud, D., Harris, J.: The geometry of schemes, vol. 197. Springer Science & Business Media (2006)
15. Fauconnier, G., Turner, M.: Conceptual blending, form and meaning. Recherches en communication **19**(19), 57–86 (2003)
16. Feferman, S.: Categorical foundations and foundations of category theory. In: Logic, foundations of mathematics, and computability theory, pp. 149–169. Springer (1977)
17. Feferman, S.: The number systems: foundations of algebra and analysis, vol. 333. American Mathematical Soc. (2003)
18. Feldman, J.: The simplicity principle in perception and cognition. Wiley Interdisciplinary Reviews: Cognitive Science **7**(5), 330–340 (2016)
19. Gilmer, R.W., Mott, J.L.: Multiplication rings as rings in which ideals with prime radical are primary. Transactions of the American Mathematical Society **114**(1), 40–52 (1965)

20. Gomez-Ramirez, D., Fulla M. Rivera, I., Velez, J., Gallego, E.: Category-based co-generation of seminal concepts and results in algebra and number theory: Containment division and Goldbach rings. JP Journal of Algebra, Number Theory and Applications **40**(5), to appear (2018)

21. Gomez-Ramirez, D., Smaill, A.: Formal conceptual blending in the (co-)invention of (pure) mathematics. In: R. Confalonieri, A. Pease, M. Schorlemmer, T. Besold, O. Kutz, E. Maclean, M. Kaliakatsos-Papakostas (eds.) Concept Invention: Foundations, Implementation, Social Aspects and Applications, pp. 221–239. Springer International Publishing, Cham (2018)

22. Grattan-Guinness, I.: Companion encyclopedia of the history and philosophy of the mathematical sciences. Routledge (2002)

23. Griffiths, D.J., Schroeter, D.F.: Introduction to quantum mechanics. Cambridge University Press (2018)

24. Grothendieck, A.: Sur quelques points d'algèbre homologique. Tohoku Mathematical Journal, Second Series **9**(2), 119–183 (1957)

25. Hartshorne, R.: Algebraic Geometry. Springer-Verlag, New York (1977)

26. Hirzebruch, F., Borel, A., Schwarzenberger, R.: Topological methods in algebraic geometry, vol. 175. Springer Berlin-Heidelberg-New York (1966)

27. Lakoff, G., Núñez, R.E.: Where mathematics comes from: How the embodied mind brings mathematics into being. AMC **10**, 12 (2000)

28. Landau, E.: Grundlagen Der Analysis: With Complete German-English Vocabulary, vol. 141. American Mathematical Soc. (1965)

29. Lang, S.: Algebra (revised third edition). Graduate Texts in Mathematics 211, Springer-Verlag, New York (2002)

30. Mac Lane, S.: Categories for the working mathematician, *GMT*, vol. 5. Springer Science+Business Media, New York (2013)

31. MacLane, S., Moerdijk, I.: Sheaves in geometry and logic: A first introduction to topos theory. Springer Science & Business Media (2012)

32. MacLane, S., Moerdijk, I.: Sheaves in geometry and logic: A first introduction to topos theory. Springer Science & Business Media (2012)

33. Mendelson, E.: Introduction to Mathematical Logic (Fifth Edition). Chapman & Hall/CRC (2010)

34. Mendelson, E.: Introduction to Mathematical Logic. Fifth Edition. Chapman & Hall/CRC, Boca Raton (2010)

35. Moore, G.H.: The emergence of open sets, closed sets, and limit points in analysis and topology. Historia Mathematica **35**(3), 220–241 (2008)

36. Munkres, J.: Topology. Second Edition. Prentice Hall, Inc (2000)

37. Munkres, J.R.: Elements of algebraic topology, vol. 2. Addison-Wesley Menlo Park (1984)

38. Munkres, J.R.: Topology. Prentice Hall (2000)

39. Nonnengart, A., Weidenbach, C.: Computing small clause normal forms. Handbook of automated reasoning **1**(335–367), 3 (2001)

40. Pierpont, J., et al.: Early history of Galois' theory of equations. Bulletin of the American Mathematical Society **4**(7), 332–340 (1898)

41. Popescu, N.: Abelian categories with applications to rings and modules, vol. 3. Academic Press London (1973)

42. Ribenboim, P.: The new book of prime number records. Springer-Verlag New York (1996)

43. Robinson, A.J., Voronkov, A.: Handbook of automated reasoning, vol. 1. Elsevier (2001)

44. Serre, J.P.: Faisceaux algébriques cohérents. Annals of Mathematics pp. 197–278 (1955)

45. Tennison, B.R.: Sheaf theory, vol. 20. Cambridge University Press (1975)

Chapter 12
The Most Outstanding (Future) Challenges Towards Global AMI and Its Plausible Extensions

12.1 On the New Cognitive Foundations for Mathematics Program

Apart from all the challenges established by the new cognitive program (see Chap. 3), one seminal aspect that should be tackled at a higher grounding level is the following: a lot of fundamental and enlightening aspects of the way in which a mathematician discovers/creates a proof lies deeply behind the cognitive representations that (s)he has about the corresponding mathematical structures in consideration, e.g., fixed generic representations of mathematical concepts and conceptual substrata of concrete exemplifications involving both the hypothesis and the thesis of the corresponding conjecture. The current foundational frameworks that we possess in mathematics (e.g., ZFC set theory) do not require (in general) that the mathematician describes symbolically m.-s. representations of these enlightening structures, but instead it would be "enough" if (s)he writes down the purely "formal version" of the deductive chain of facts leading to the solution, independently whether it genuinely shows the seminal "tricks" of the solution.

This disseminated and singular descriptive heuristic of current mathematical practice has the huge limitation that it tends to "hide" exactly those fundamental and causal aspects of the way in which the (genuine) creative ideas of a proof cognitively emerges. Therefore, refinements (resp. completely new versions) of the m.-s. representations used to describe final (public) versions of mathematical proofs should be developed in such a way that they can take into account simultaneously also the descriptions of the cognitive causal ideas originating and subsequently shaping the corresponding proofs.

This is particularly important for exemplifications. In fact, as we saw in Sect. 10.3, Chaps. 10 and 11, from the point of view of generating fruitful mathematical concepts and arguments, elementary and tiny exemplifications are cognitively and metamathematically as important and powerful as sophisticated mathematical

© Springer Nature Switzerland AG 2020
D. A. J. Gómez Ramírez, *Artificial Mathematical Intelligence*,
https://doi.org/10.1007/978-3-030-50273-7_12

structures since, in a lot of cases, the former turn out to be the causal origins of the latter.

A second central challenge has to do with the generation and subsequent coherent integration of deductive (logic) frameworks where several forms of exemplifications play a more central role in the meta-analysis of how our minds solve mathematical problems. In fact, (first- and higher-) order logic and proof theory are foundational frameworks that explicitly do not stress the pragmatic importance of exemplifications within their most fundamental notions. On the other hand, (modern) model theory involves exemplifications at a more central basis, however, due to the purely formal (logic) origin of this subject, it simply does not take into consideration explicitly the cognitive dimension that exemplifications possess within (metamathematical) deductive processes. So, the creation of a new theory of exemplifications within cognitive metamathematics with strong and refined influences of model theory turns out to be highly desired and useful. Explicitly, along almost all the cognitive constructions of mathematical structures in Chap. 11, we saw how foundational the role of exemplifications was, and how phenomenologically natural the construction of more complex structures was possible starting with elementary instances of them.

12.2   On a Final Global Taxonomy of the Fundamental Cognitive Metamathematical Mechanisms

If we perform a careful counting process of all the cognitive mechanisms initially described in Chap. 10, differentiating along the way between different forms of the same main global mechanism (e.g., between generic exemplification and exemplification), then we find that our starting taxonomy of metamathematical cognitive abilities consists of 23 different mechanisms.

Contrastingly, based on the considerably large amount of evidence that cognitive psychology and, in general, cognitive science has provided regarding the classification of the (fundamental) cognitive abilities used by the mind, we claim that the former list is basically final; i.e. the most fundamental abilities are all already included there. Nonetheless, this also takes into account the possibility that a few new (secondary and more domain-specific) abilities can be discovered during the meta-analysis of several additional mathematical structures and proofs beyond the ones presented in Chap. 11.

It could also be the case that such new abilities could be grounded as already described in Chap. 10, so the former taxonomy remains foundational. For example, it is an open problem to determine if conceptual blending can be strictly reduced to metaphorical or analogical reasoning on a global basis. In any case, our initial taxonomy could also be slightly enhanced more for pragmatic and computational reasons, than for an ontological need. In other words, if we find out that the addition of a secondary cognitive mechanism would facilitate the description of (global) algorithmic principles, then the former taxonomy will be updated accordingly.

An additional important aspect concerns the identification and subsequent refinement of those classic techniques coming, for example, from automated deduction, which simulate at a better degree syntactically and semantically the actual way in which mathematics is cognitively done . For instance, as we saw in Chap. 7, the well-known classic technique of Skolemization and the cognitive mechanism of (functional) conceptual substratum seem to have structural relations for first-order frameworks. And, similarly, one can develop more cognitive-inspired and equally powerful versions of the sequent calculus with the inspiration of initial formalizations of the conceptual substratum ability.

12.2.1 Precise Characterization of (Local) Conceptual Substrata

The cognitive aspect which is most sensible to change in our AMI program when we move from one specific mathematical sub-discipline to an another is not so much related to the general constitution of the taxonomy described in Chap. 10, but, on the other hand, it involves more centrally the concrete form of the m.-s. configurations describing the (local) conceptual substrata of the corresponding mathematical structures, together with the associated m.-s. directed graphs of their pragmatic (and sometimes formal) parts. Illustrative examples are elementary number theory and elementary geometry, where the substrata of the first sub-discipline are (very often) formed by polynomial configurations involving factorizations and additions of (graphically) disjoint numerical symbolic atoms. Meanwhile, for the second one, the substrata are given by more sophisticated graphic combinations of geometrical symbolic atoms which are, in general, not pictorially disjointed.

So, a suitable cognitive heuristic describing such (pragmatic) conceptual substrata will be found, in general, after a robust meta-analysis of dozens of domain-specific proofs, concepts, and heuristics.

12.3 On the Computational Aspects of Artificial Mathematical Intelligence

A lot of quantitative and qualitative aspects of (theoretical) computer science, in particular of computational complexity theory, are implicitly grounded in the classic understanding that we have about the natural numbers and the notion of (mathematical) infinity. Classic issues like the halting problem and Gödel incompleteness theorems, and their consequences regarding the limitations the computer programs could have in the conceptual solution of mathematical problems (at arbitrary levels of abstraction), are closely related to the most common and well-accepted assumptions that we have about the natural numbers and the (cognitive) process of counting "without end."

Assume that we give a closer and new glance to the former (classic) issues from the point of view of the physical numbers (Chap. 5) and subsequently to the goal of generating concrete forms of software able to support initial aspects of the artificial mathematical intelligence program as a whole. Then, we would be able to differentiate among the specific constraints of the computational dimension of AMI program more clearly between the effective (e.g., procedural) and merely mental ones emerging from the purely meta-physical (quantitative) aspects of the natural numbers and (several forms of) the incompleteness results based on those.[1] Moreover, such a renewed form of *a new physical foundations for the computational substratum of the AMI program* could have important consequences on the way in which the most theoretical sub-disciplines of computer science are currently conceived.

A second and more pragmatic challenge involves exploring the improvement and subsequent integration of existing software which model particular cognitive abilities (see, for instance, [15, 17, 22, 24]), together with (initial) sound implementations of each of the remaining cognitive abilities developed in Chap. 10. Additionally, such an integrated software should be enhanced in such a way that it can take into account the diversity and a minimal amount of semantic content of the wide range of m.-s. configurations that could appear in several mathematical disciplines, e.g., geometry, number theory, analysis, and algebra (see Chap. 9).

In addition, at this point it would be worth identifying among the most outstanding mathematical software currently used the most compatible with the foundational principles of the AMI program, and subsequently to explore the plausibility of enhancing initial versions of the AMI software with inspiration coming from them.

Presently, there is a relatively large amount of quite valuable and useful mathematical software covering local aspects of the mathematical practice in several sub-disciplines, some of them running with open source licenses and others restricted for commercial purposes (see, for instance, "The Guide to Available Mathematical Software,"[2] [3, 5, 16, 21, 23, 26], and the relevant references there).

Furthermore, the concrete task of exhaustively looking for relevant methodological and axiomatic commonalities between the corresponding software and the core principles of AMI program and cognitive metamathematics could be comparable in terms of spatio-temporal resources with the generation of an initial software for supporting the AMI main goal from scratch.

Moreover, one of the compatibility challenges in this regard is related to the cognitive plausibility of such existing software, since a huge amount either possesses a sophisticated representation language, not always close to mathematical

---

[1]For instance, arguments involving natural numbers strictly "bigger" than ω, which possess essentially a mental nature which is not physically grounded.

[2]https://gams.nist.gov.

practice, or focuses more on formal verification and less on cognitively-based human-style generation of arguments and concepts.

The third challenge, and one of the most fundamental ones, involves the identification of the algorithmic pillars of the AMI program on a global scale. In fact, due to the fact that the "formal laboratory" of cognitive metamathematics possesses (humanly discovered/created) mathematical concepts, (counter-)exemplifications and proofs as main "abstract substances" of experimentation, the (careful) cognitive meta-analysis of hundreds, or even thousands, of such mathematical entities seems to be necessary for getting the algorithmic quintessence of the AMI program. Specifically, it is highly desirable to start this meta-process with the most out-standing mathematical structures and results of fixed mathematical disciplines, for gradually materializing such computative (local) maxims. These could be subse-quently enhanced through the further meta-analysis of secondary formal samples until a starting (local) implementation of AMI can be done, initially restricted to the particular mathematical sub-discipline in consideration.

Examples of some of these seminal mathematical disciplines are real and complex analysis [11, 20]; (abstract and commutative) algebra [6, 7]; differential and algebraic geometry [9, 10, 14]; (algebraic) topology [18, 19]; graph theory [4]; and (algebraic and analytic) number theory [1, 13], among many others.

The former collection of meta-tasks can be potentially done not only by professional mathematicians (and related researchers) but also by undergraduate and graduate university students with a strong background in mathematics.

In this regard, a fourth challenge emerges with a more social nature, namely, the establishment of a coherent global multidisciplinary community of researchers worldwide around the central purposes of the AMI program. In fact, all the out-standing and multifaceted work done during the last eight decades in (automated and interactive) theorem proving, (logical methods in) computer science, computational logic, computational creativity, artificial intelligence, and mathematical software development, together with the unified results presented here, are more than enough evidence of the real plausibility and importance of artificial mathematical intelligence for the near future. So, we hope that (gradually) any graduate and amateur mathematician, together with any researcher using conceptual mathematics on a normal basis, could reserve even a small part of his/her intellectual curiosity to topics related with the AMI program.

Just imagine for a moment that you are located spatio-temporally 100 years in the past and you see two scientists. The first one is completely involved in developing more effective methods for doing faster arithmetic, algebraic, (analytic) geometric, and trigonometric calculations of any kind by hand, e.g., computing large additions, subtractions, multiplications, divisions, estimating products of polynomials with a large degrees, highly precise logarithms and trigonometrical functions and Fourier transforms, among others. The second one is working on the construction of a mechanical device able to solve all the former problems as fast as possible.

What do you think the probability of success if for each of them on a long-term perspective? Although, it could be the case that regarding the common purpose in mind (i.e., solving a specific kind of mathematical problem) the first scientist would come up with a larger amount of "short-term success evidence" materialized in the form of some (arithmetic, algebraic, algebraic) new tricks for solving new, larger formal computations a little faster. Nonetheless, in the long run, the second scientist would be able to invent a device with a higher level of efficacy; i.e. a solving engine which is able to solve virtually any (elementary) problem of the former kind.[3]

Presently, we are located spatio-temporally more or less at the next epoch of the former situation regarding the generation of an artificial mathematical solving agent with a global nature and being able to solve conceptually and at a cognitive basis arbitrarily abstract mathematical (and math-related) problems.

12.4 A New and More Human Way of Doing Mathematical Research

Let us imagine that the visions described in the preface and in the introductory chapter related with the fulfillment of the AMI program are materialized in the near future. What would it mean for the current way in which mathematics is done in practice?

The consequences are better than one can imagine. In fact, there are millions of quite important phenomena in nature at the micro-, macro- and mostly at the mecro-level that should be studied using more formal and mathematically grounded models. Challenges that involve thousands of variables and parameters aiming to describe complex systems and sophisticated behaviors related with issues in the industry, financial markets, sociology, politics, economics, ecology, city planning, sustainability, human relationships, among many others.

Now, with a robust version of UMAA (Universal Mathematical Artificial Agent), a lot of researchers would be able to generate and to use more mature, precise, and complex mathematical frameworks for modeling the former kind of phenomena, because they would know that the conceptual solution of the subsequent mathematical problem could be (in most of the cases) solved effectively by the UMAA.

In this way, we could accelerate top research involving vital needs of our societies in a faster and exponentially more accurate way. Even for pure mathematicians, this will represent a huge advantage because they could invest more time in modeling mathematically real natural phenomena having visible effects in our understanding of the universe, and create new abstract theories on top of that.[4]

So, the AMI program goes toward the enhancement of our mathematical understanding of the universe in a more human way. This is based on the belief that mathematics is an invaluable tool for improving the quality of our lives collectively as well as individually. And so, they are far more than a goal on their own.

---

[3]In fact, this is basically the case with the creation and subsequent improvement of several kinds of scientific calculators and mathematical software throughout the last decades.

[4]Like in the case of mathematical physics, for example.

Just think for a moment of the time, some decades ago, when we did not even have calculators for doing large (basic) arithmetic operations. At that time, some people spent most of their lives doing such calculations. Now, they are done in seconds and we can focus our (scientific) attention on qualitatively higher academic and pragmatic issues. So, with the proposal of this new multidisciplinary AMI program, we are preparing conceptually the next step "upstairs."

12.5 Plausible Extensions of the AMI Program

The fact that abstract mathematics has such a wide and omnipresent influence in the sciences, together with our current need for understanding in a more accurate way the highly sophisticated complex systems and natural phenomena that surround us in our daily lives, lead us to extend our initial AMI vision more pragmatically. Explicitly, if we focus our attention in scientific disciplines highly supported by mathematical structures,[5] then it is clear that the real fulfillment of the AMI vision would have a strong qualitative impact on any of them, which could enhance exponentially the precision in our understanding about particular objects of study. So, we will describe hereunder some of the most important cases, that, in fact, can be (and sometimes should be) carried out in parallel with the achievement of the AMI (original) program.

12.5.1 Artificial Physical/Chemical/Biological Intelligence

Since the very beginning, mathematics and physics have grown in constant support and inspiration of each other. In global terms, one can say that mathematics represents the grounding syntactic (and very often semantic) linguistic framework from which physics describes formally its models and theories. On the other hand, physical phenomena offers an invaluable source of inspiration for the creation of new mathematical structures and theories.[6] So, a coherent and, in a certain sense, more restrictive variant of the AMI vision in terms of conceptual generation in (theoretical) physics is quite plausible as well, i.e., *Artificial Physical Intelligence*. Explicitly, if we restrict ourselves to those sub-areas of modern physics that are based on more deterministic paradigms and whose specific objects of study possess formalizable (pragmatic) conceptual substrata, then a slight modification of the AMI program would basically be enough for its fulfillment.

---

[5]For instance, computer science, physics, financial mathematics, engineering, statistics, (mathematical) psychology, (mathematical) biology and chemistry, and (mathematical) sociology, among others.

[6]For further support in this direction, see Sect. 3.1, Chap. 3.

Finally, due to the fact that physics grounded (in a micro perspective) a considerable amount of processes studied in chemistry and biology, and in the rapid growth of (relatively new) scientific disciplines like computer chemistry and mathematical biology, extensions of our AMI vision to some sub-disciplines of these two fundamental natural sciences are similarly feasible, namely, *Artificial Chemical Intelligence* and *Artificial Biological Intelligence*.

Something even more motivating than the realizations of the former visions as a way of enhancing our global understanding of the cosmos, is to reflect on the positive applications that robust materializations could have in our (pragmatic) understanding of nature (in general), and of our bodies and minds (in particular), for example, in areas like medicine or ecology.

12.5.2   Artificial Financial Intelligence

Other closely related academic disciplines where the AMI vision can be extended coherently are finances (e.g., financial mathematics) and economics. Effectively, the last decades have seen a considerable large spectrum of applications of AI techniques to financial and economical phenomena, which, to a large extent, have developed sophisticated techniques for solving (artificially) problems involving the stochastic aspects of financial inquiries using, for example, methodologies involving fuzzy logic and neuronal networks (see, for instance, [2, 8, 25]).

On the other hand, the mathematical substratum of the deterministic part of contemporary financial mathematics involves simple (mathematical) structures belonging to elementary abstract algebra and real analysis, and, very often, modeled with logical frameworks implicitly supported on a first-order setting [12].

So, a financial version of the AMI program, namely, *Artificial Financial Intelligence* should involve not only a precise characterization of the (pragmatic) conceptual substrata of its atomic objects of study,[7] but also a suitable methodological integration between the classic AI techniques used before and the (updated) methodologies coming from the AMI (financial) program.

> A very large part of space-time must be investigated, if reliable results are to be obtained
>
> Alan Turing

Acknowledgments The author sincerely thank Alberto Vélez, Jairo Gómez, Eustelly Zapata R., and Orlando Múnera.

---

[7]For instance, supplies, flows, interests, marks, amortizations and bones (of different types); conjugated laws, among many others.

References

1. Apostol, T.M.: Introduction to analytic number theory. Springer Science & Business Media (2013)
2. Bahrammirzaee, A.: A comparative survey of artificial intelligence applications in finance: artificial neural networks, expert system and hybrid intelligent systems. Neural Computing and Applications **19**(8), 1165–1195 (2010)
3. Biere, A., Heule, M., van Maaren, H.: Handbook of satisfiability, vol. 185. IOS press (2009)
4. Bondy, J.A., Murty, U.S.R.: Graph theory with applications, vol. 290. Macmillan London (1976)
5. Char, B.W., Geddes, K.O., Gonnet, G.H., Leong, B.L., Monagan, M.B., Watt, S.: Maple V library reference manual. Springer Science & Business Media (2013)
6. Eisenbud, D.: Commutative Algebra with a View Toward Algebraic Geometry. GTM. Springer-Verlag (1995)
7. Fraleigh, J.B.: A first course in abstract algebra. Pearson Education India (2003)
8. Goonatilake, S., Treleaven, P.C.: Intelligent systems for finance and business. John Wiley & Sons, Inc. (1995)
9. Grothendieck, A., Dieudonné, J.: Eléments de Géométrie Algébrique I. Springer (1971)
10. Hartshorne, R.: Algebraic Geometry. Springer-Verlag, New York (1977)
11. Hewitt, E., Stromberg, K.: Real and abstract analysis: a modern treatment of the theory of functions of a real variable. Springer-Verlag (2013)
12. Janssen, J., Manca, R., Volpe, E.: Mathematical finance: deterministic and stochastic models. John Wiley & Sons (2013)
13. Janusz, G.J.: Algebraic number fields, vol. 7. American Mathematical Soc. (1996)
14. Kobayashi, S., Nomizu, K.: Foundations of differential geometry, vol. 2. Interscience publishers New York (1969)
15. Martinez, M., Abdel-Fattah, A., Krumnack, U., Gómez-Ramírez, D., Smail, A., Besold, T., Pease, A., Schmidt, M., Guhe, M., Kühnberger, K.U.: Theory blending: Extended algorithmic aspects and examples. Annals of Mathematics and Artificial Intelligence pp. 1–25 (2016)
16. MatLab, M.: The language of technical computing. The MathWorks, Inc. http://www.mathworks.com (2012)
17. Mossakowski, T., Maeder, C., Codescu, M.: Hets user guide (version 0,99) (2014). URL www.informatik.uni-bremen.de/agbkb/forschung/formal_methods/CoFI/hets/UserGuide.pdf
18. Munkres, J.: Topology. Second Edition. Prentice Hall, Inc (2000)
19. Munkres, J.R.: Elements of algebraic topology, vol. 2. Addison-Wesley Menlo Park (1984)
20. Palka, B.P.: An introduction to complex function theory. Springer Science & Business (1991)
21. Robinson, A.J., Voronkov, A.: Handbook of automated reasoning, vol. 1. Elsevier (2001)
22. Schwering, A., Krumnack, U., Kuehnberger, K.U., Gust, H.: Syntactic principles of heuristic driven theory projection. Cognitive Systems Research **10**(3), 251–269 (2009)
23. Stein, W., Joyner, D.: Sage: System for algebra and geometry experimentation. ACM SIGSAM Bulletin **39**(2), 61–64 (2005)
24. Boy de la Tour, T., Peltier, N.: Computational Approaches to Analogical Reasoning: Current Trends, chap. Analogy in Automated Deduction: A Survey, pp. 103–130. Springer-Verlag, Berlin, Heidelberg (2014)
25. Trippi, R.R., Turban, E.: Neural networks in finance and investing: Using artificial intelligence to improve real world performance. McGraw-Hill, Inc. (1992)
26. Wolfram, S.: The Mathematica Book, Wolfram Media, 2003. Received: November **2** (2015)

Printed in the United States
by Baker & Taylor Publisher Services